THE TECHNICAL DRAWING WORKBOOK

JAMES J. LUCKOW

DAWSON COLLEGE

ADDISON-WESLEY PUBLISHING COMPANY

Reading, Massachusetts • Menlo Park, California • New York
Don Mills, Ontario • Wokingham, England • Amsterdam • Bonn
Sydney • Singapore • Tokyo • Madrid • San Juan • Milan • Paris

Figure 2-12 © 1991 JAL Training

ISBN 0-201-62330-7
1 2 3 4 5 6 7 8 9 10-BAH-97969594

To my mother and father,
who taught me all the
important lessons in life.

PREFACE

The purpose of *The Technical Drawing Workbook* is to introduce you to the basic principles of technical drawing. Upon completion, you should be able to make a detail drawing that can be used by a craftsperson to make an object. This book introduces you to the principles of projection and basic dimensioning as used on the drawing board or on CAD (computer-aided drafting). It is designed for self-study or may be used in conjunction with class lectures. No prerequisite knowledge is required to complete this workbook—it can be used as a stand-alone text and workbook. It may also be used as a workbook supplement to a more comprehensive textbook on technical drawing.

How well we communicate through our drawings depends on how well we understand one another. Remember, technical drawing (or drafting) is a language, a graphic language meant to communicate ideas. Countries and even areas of a country often differ in their names of objects (for example, saying petrol for gasoline). So that everyone speaks the same "language" in drafting, certain *standards* have been established around the world. Three common standards have been used and explained throughout this book: international standards published by the *International Organization for Standardization (ISO)*, American standards published by the *American National Standards Institute (ANSI)*, and Canadian standards published by the *Canadian Standards Association (CSA)*. The Appendix at the back of the book has more information about these standards. The important point to remember is that you must follow the standards and directives of your employer or instructor, should they differ from this book.

Woven throughout each chapter are *activities* and *exercises* that will assist you in mastering the understanding and skills you need as you move through the book. The activities are intended as informal practice, whereas the exercises can be considered to be more formal assignments.

To make instrument drawings, some users of this book may use T-squares, some may use a parallel straightedge (guided by a wire attached to the drawing board), and others may use drafting arms or track-type drafting machines. The basic principles taught in this book apply to all this equipment. This book assumes that you will be using a parallel straightedge.

The text refers to both the inch and metric systems. Each size is given in inches and is then followed by an appropriate metric equivalent in brackets, for example, 5" (125 mm). Although 1" is closer to 25.4 mm, for convenience we are using 25 mm for 1" and the closest whole millimeter for fractions of an inch on the illustrations in this book. The numbers are not meant as a direct conversion. Rather, you are expected to refer to either all the inch sizes or all the millimeter ones.

As you work through the book, you will notice some repetition of hints about using drafting instruments. Such duplication is intentional to make the chapters more self-contained and to allow you or your instructor to change the order of the chapters without missing important introductory notes.

TO THE INSTRUCTOR

The Technical Drawing Workbook may be completed in a one-semester college course of 12 to 16 weeks. Each chapter should take approximately three to five hours of class work (plus homework) to complete, depending on the number of exercises that you assign. Some chapters are longer than others, but the shorter chapters usually have more complex concepts to learn and therefore may take as much time to cover as the longer ones. Chapter 4, for example, is quite long, but you do not necessarily have to teach each construction, because the steps are very detailed and may be followed by the student independently.

Instructors who wish to change the order of certain chapters to suit their pedagogical plans should feel free to do so, as each chapter is self-contained. Chapters 8 and 9 should be studied sequentially, however, because they represent the first and second parts of the same concept: auxiliary views. The same is true for Chapters 10 and 11, which cover different aspects of section view drawings.

Each page of a chapter usually introduces only one concept, so that, as students work through a chapter, they are introduced to new concepts in small "bite-size chunks."

To provide good, clear, reproducible drawings, this text features CAD, rather than hand-drawn, illustrations. By referring to the CAD drawings, students are offered an ideal of how their final drawing should look.

Many technical drawing textbooks introduce the many line types in one of the first chapters. This book refrains from this practice because many students feel obligated at the beginning of a course to memorize the various line types without understanding their purpose. This book favors the pedagogical approach of introducing material only as it is required for the concept being taught. For this reason, a complete list of line types appears only in the Appendix, for reference as the instructor or student sees fit. Similarly, each piece of equipment is discussed as it is needed, and a summary listing of equipment uses appears in the Appendix.

Because of the nature of a workbook in which many of the activities and exercises are done on the pages supplied, the workbook will necessarily have a limited number of exercises. This should not be seen as a disadvantage, since the students in this workbook are provided with the opportunity of working on a problem where it is visually laid out for them. Additional problems will be made available from the author to teachers who adopt this book.

The book is organized so that students need only a pencil to complete Chapters 1 and 2. Thus they can attend at least a few classes to learn what specific equipment to buy, the quality to look for, and the price to expect to pay. (Instructors may show the drafting instruments during the first class and discuss their purchase.) The early emphasis on freehand sketching also allows the student to concentrate on learning a few technical drawing skills such as lettering, as well as isometric and orthographic view principles, without concern for manipulating drawing instruments.

By Chapter 3, it is assumed that all students have purchased their equipment. In this chapter, students are introduced to the architect's, metric, and engineer's scales so that drawing using drafting instruments can begin. Chapter 4 introduces students to geometric constructions and offers them the first opportunity to practice with most of their drafting instruments. The chapter contains detailed steps for each construction that many students will be able to follow on their own. While some students work independently, you will have more time to work with students who are experiencing difficulties. Some instructors find that the helix and conical helix are somewhat advanced concepts for a beginning-level course. For this reason, they are presented at the end of the chapter, after the main body of construction exercises.

Chapter 5 covers isometric views using drafting instruments. In this chapter, students gain additional practice with drafting equipment and drawing techniques through the vehicle of copying familiar objects (isometric views that were introduced in Chapter 1). Title blocks are introduced here. Instrument drawings are first introduced through isometric projection rather than through orthographic projection so that students can concentrate on using instruments while copying isometric views. (To permit students to focus specifically on using the drawing instruments, the topic of centering isometric drawings is not covered.) Orthographic instrument drawings are introduced in Chapter 6, after students have acquired some experience with their instruments and can more easily concentrate on the more difficult concepts of orthographic projection.

In some of the exercises for Chapter 5, the objects are drawn isometrically so that the front surface is not always the one with the "characteristic shape." When students use these isometric views to draw orthographic ones in Chapter 6, they will be required to use their spatial skills to mentally rotate some of these objects to get the characteristic shape as the front view. In Chapter 6, the student is shown how to make orthographic views using drafting instruments. This chapter provides opportunities to develop skills in orthographic projection, to further practice using drafting instruments, and to apply some of the geometric constructions introduced in Chapter 4. Students should refer to Chapter 2 for any questions on the theory of orthographic projection.

Chapter 7 covers the principles of dimensioning so that sizes can be added to the orthographic views. In the exercises, no dimensions are provided. Instead, students are asked to scale the drawings to determine sizes. In this way, they are required to make decisions concerning the placement of the dimensions, rather than simply copying their location from an isometric drawing. If the class size is small enough and time permits, students should have the first exercises corrected before proceeding to the next one. In this way, they will be prevented from repeating the same mistakes on subsequent exercises.

Chapters 8 and 9 introduce auxiliary views so that students can enhance their knowledge of orthographic projection by learning how to draw the true shape of a sloping surface. Chapters 10 and 11 introduce various section views that permit students to draw objects with internal features. The chapters on sections may be taught before the chapters on auxiliary views (Chapters 8 and 9).

Finally, in Chapter 12, students are shown how to combine all the skills learned in the course to make detail drawings similar to the ones used by industry. Some isometric views are given without sizes marked on them. As in Chapter 7, students have to measure the isometric view and make the orthographic views to a suitable scale. An optional exercise suggested in Exercise 12–3 is to have your students check one another's work. This exercise provides the students with additional practice in drafting skills from the perspective of a checker.

ACKNOWLEDGMENTS

Many people provided me with valuable assistance in putting this book together. I would like to thank my wife, Anne, for her word processing, proofreading, and pedagogical consultations. I would also like to thank my daughter, Debra, for her word processing and CAD drawings, and my son, David, for his technical advice on using the computer to prepare this book. Professor Henry Rzepczyk kindly contributed some of the exercises in Chapter 4. Many thanks to the very professional team at Addison-Wesley: Denise Descoteaux, Editor; Melissa Honig, Assistant Editor; Amy Willcutt, Senior Production Coordinator; Peter Blaiwas, Cover Design Supervisor; and Joyce Grandy, Copyeditor.

I would also like to thank the following reviewers, whose comments have been invaluable while revising this text:

Bill Bilbrey, Durham Technical Community College
Ralph D. Denton, Tidewater Community College
Dr. Peter R. Frise, Carleton University
John Frostad, Green River Community College
H. Thomas Gillespie, Portland Community College

Bill Havice, Fort Hays State University
Carol Hill, Hocking College
Kevin D. Humphreys, Newport News Shipbuilding
Marcy Miller-Seller, Ivy Tech College
Joseph E. Moore, Norfolk State University
Robert L. Queen, Columbus State Community College
Michael A. Satterfield, Ivy Tech College

The next edition of this book will be modified according to comments received. Feedback, suggestions, and constructive criticism are most welcome. Please send your comments to

Dr. James Luckow, Ed.D.
Dawson College
Mechanical Engineering Technology Department
350 Selby Street
Montreal, Quebec
Canada
H3Z 1W7
(514) 931-8731 Ext. 5009

The author and publishers have tried to ensure that only original material has been used in this book. However, some examples have been submitted by fellow teachers. If any previously copyrighted work has been inadvertently included, it will be acknowledged or omitted in future editions.

CONTENTS

CHAPTER 3: SCALES 69

CHAPTER 4: GEOMETRIC CONSTRUCTIONS 87

CHAPTER 5: ISOMETRIC VIEWS USING DRAFTING INSTRUMENTS 137

CHAPTER 12: DETAIL DRAWINGS 311

APPENDIX 321

LETTERING AND SKETCHING

1

OBJECTIVES

When you have finished this chapter, you should be able to

1. Identify the equipment necessary for making technical drawings,
2. Form letters according to acceptable drafting standards,
3. Sketch isometric views.

EQUIPMENT

To complete this chapter, all you need is a pencil. Any type of pencil may be used for now. Specific pencil types are discussed in this chapter.

WHAT IS TECHNICAL DRAWING?

Technical drawing, often referred to as drafting (sometimes spelled draughting), is a visual language that communicates ideas through the use of drawings. Drafting may be thought of as a universal language because you can usually interpret a drawing even if you do not speak the same language as the person who made the drawing.

If you have ever tried to tell someone over the telephone how to make an object, you would have soon discovered that the process would be much simpler, and less error prone, if you could transmit a drawing with dimensions (or sizes). This would allow the person to see from a picture what the final product should look like.

Everything that is manufactured or built started as an idea, then became a sketch or a drawing. The drawing may have gone through various stages of development, from a freehand sketch, to a design drawing done with instruments, and finally to detailed and assembly drawings used by the craftsperson in the shop. Patent drawings may have been used to register a new concept or idea.

Technical drawing is used in all areas of technology, each with its own set of subdivisions. Electronic drafting, for example, may be divided into schematic drawings, printed circuit board (PCB) design, and chassis layout. Mechanical drafting has such divisions as piping drafting, plant layout, jig and fixture drawing, and hydraulic schematics.

Drafting skill is required by many trades and professions. Technical drawings may take the form of sketches drawn by a machinist, instrument drawings made by a mechanical drafter, or design drawings made by an engineer. The job description of a drafter (also draftsman, draftswoman, or draftsperson) can vary greatly in salary as well as responsibility, depending on the company, level of education, and years of experience.

All areas of drafting depend on certain basic concepts of view placement, line quality, dimensioning, lettering, neatness, and use of drafting instruments. These introductory ideas as well as basic mechanical drafting concepts are introduced in this book.

Below is a list of drafting equipment that you will need to complete the exercises in this book. In the same way that a surgeon uses precision surgical instruments to perform operations, a drafter uses drafting instruments to make accurate drawings.

NOTE:

1. When adjusting your drafting table, be careful not to hurt yourself or a person sitting near you by letting the drawing board drop too quickly.

2. Before starting to work through this book, familiarize yourself with the contents of the Appendix at the end of this book. You will then know what information is available for future reference.

STUDENT EQUIPMENT LIST

Personal preferences concerning specific pieces of equipment may vary from person to person. The following list should be modified accordingly:

a. Compass—6" (150 mm) bow compass

b. Triangle (also called a set square)—8" (200 mm) × 45°

c. Triangle—10" (250 mm) × 30°–60°

d. Scale—architect's (inch)—Scale markings are 3/32, 3/16, 1/8, 1/4, 1, 1/2, 3/8, 3/4, 1 1/2, 3, 16.

e. Scale, metric—Typical scale markings include 1:1, 1:100, 1:5, 1:20.

f. Scale, engineer's (inch)—Typical scale markings include 10, 20, 30.

g. Lead holders or technical pencils, 2 required—Choose either the "clutch type" or the "ratchet type." With the clutch-type lead holder, such as Turquoise 10, the two holders should have different colored tops so that you can differentiate between them. With this type of pencil, you will also need a *pencil sharpener* and a supply of leads, such as H and 3H. If you choose the ratchet-type technical pencil, buy 0.3 mm and 0.7 mm technical pencils with H lead. This type of pencil does not have to be sharpened.

Throughout this book, we will refer exclusively to the technical pencil lead thicknesses of 0.3 mm and 0.7 mm. For those using the clutch type lead holder, where a 0.3 mm lead is specified, use your 3H lead to make thin lines. Where a 0.7 mm lead is specified, use your H lead to make thick lines. In general, the thick lines should be at least twice as thick as the thin lines.

h. Protractor

i. Drafting tape or masking tape, 1/2" (13 mm) or 1" (25 mm) wide

j. Circle template—inch or metric

k. Erasing shield

l. French curve

m. Sandpaper pad—You will use this to sharpen your compass lead (not your pencil). Keep it in a plastic bag since it will get dirty!

n. Eraser

o. Mailing tube—1 1/2" (38 mm) diameter and 25" (635 mm) long, to hold your drafting paper

p. Drafting paper—C size (17" × 22") or A2 (420 mm × 594 mm), 10 sheets required

q. Dividers

r. Ames lettering guide

s. Adjustable triangle

t. Drafting brush

You will also need a drafting table with a T-square, parallel straightedge, or drafting arm. Each piece of equipment is discussed in the chapter in which it is used.

LEADS

Drafting is usually done with pencils. Drafting may also be done with ink or on a CAD (computer-aided drafting) system. Pencil types vary, from the familiar wooden pencil to mechanical pencils available with different mechanisms. Mechanical pencils range from the "clutch type," which uses a fairly thick lead, to the "ratchet type," which uses fairly thin lead, such as 0.3 mm, 0.5 mm, or 0.7 mm.

The leads used in mechanical pencils are available in various degrees of hardness. They are categorized as follows:

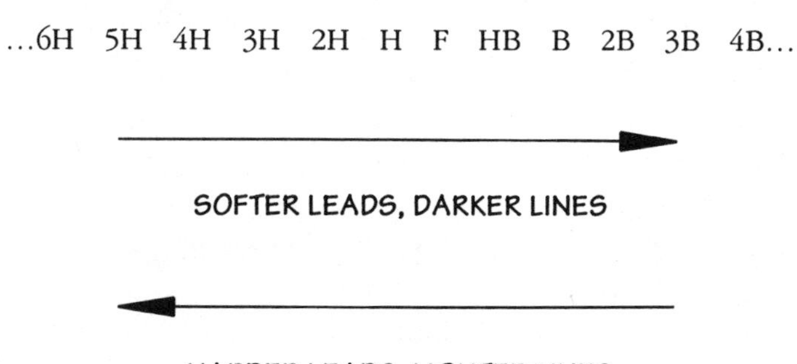

The arrows drawn above illustrate that as we move to the right on the above scale, the leads get softer and produce a darker line. As we move to the left, the leads get harder and produce a lighter line. For example, an HB lead is softer and darker than a 2H lead, and a 4H lead is harder and lighter than a 2H lead. With a clutch-type lead holder, you can do most drafting with an H lead for dark lines and lettering, and a 3H lead for light lines. With a ratchet-type technical pencil, you would use a thin lead (0.3 mm) holder and a thick lead (0.7 mm) holder, both with H leads. You may find it helpful to use a harder lead in the 0.3 mm technical pencil than in the 0.7 mm technical pencil. The various thicknesses (or grades) of lead are used to draw contrasting line types. These line types are illustrated on page 322 in the Appendix.

LETTERING

Lettering is very important in drafting. It demonstrates the skill of the drafter, but more importantly, neat lettering is necessary for a clear, easy-to-read drawing. Drawings with poor lettering quality may lead to an incorrect interpretation of the drawing, resulting in errors when the part is made in the workshop.

In drafting, whenever you form any numbers, words, or letters, always use *guidelines* to maintain consistent letter height.

RULES ABOUT LETTERING

a. When lettering, always use as guidelines very *light* lines drawn with a straightedge.

b. Space guidelines 1/8" (3 mm) apart for whole numbers and letters.

c. For fractions, space guidelines 1/4" (6 mm) apart.

d. Use your 0.3 mm lead to make light guidelines—do not erase them.

e. Letters and numbers should be *dark*; use your 0.7 mm (H) lead.

f. Use horizontal fraction lines, as shown in Fig. 1–1.

$3\frac{1}{2}$ $4\frac{1}{4}$ $5\frac{7}{8}$ $6\frac{1}{4}$

FIGURE 1–1

If your lettering is poor, then 1 1/16 can be read as either 1 and 1/16 or 11/16.

g. Ensure that numbers above and below the fraction line do not touch the line.

$\frac{1}{2}$ INCORRECT $\frac{1}{2}$ CORRECT

h. Ensure that numbers above and below the fraction line are vertically aligned.

$\frac{1}{8}$ INCORRECT $\frac{1}{8}$ CORRECT

i. Do *not* erase guidelines.

j. To keep your work area clean, avoid using a ballpoint pen when taking notes, because it may leave ink blotches on the drawing surface of your drafting table. These marks could transfer to your drawing.

k. Form letters *freehand*. Do not use a straightedge to form each letter, although you may use a vertical straightedge as a guide for vertical strokes. Lettering templates are not recommended, because they tend to slow you down. However, some drafters may use a lettering template for large lettering (over 3/8" or 10 mm, such as appears in the drawing title).

l. Do your lettering as the *last* step of your drawing, because you will be using a soft lead, which can lead to smudges.

m. Do make the strokes of letters *all* vertically aligned or *all* sloping to the right.

n. If you want a certain word to stand out, such as the title of a drawing, use 1/4" (6 mm) letters.

o. Remember to make letters *dark*.

p. Use single continuous strokes, not short brushing strokes, to form letters.

q. If you have an Ames lettering guide, you may use it to help draw parallel lines (guidelines) between which you will place your letters. Follow the instructions accompanying the lettering guide. If you do not have one, then simply use the lines printed in this book for practice.

r. Leave an extra line space between lines of lettering, as shown in Fig. 1–2. The vertical space between lines may be between half the height of the letters and the full height of the letters.

s. Make decimal points bold and clear. It is important that they be clearly visible.

NOTE: Of necessity, the guidelines reproduced in this book use dark lines. Remember that you should draw them *lightly*. One practice is to use blue lead to draw guidelines since this color will not reproduce on a copy of the drawing.

LETTERING IS A VERY IMPORTANT PART OF THE DRAFTING
PROCESS. NEATLY FORMED LETTERS CONVEY THE PROPER
MESSAGE AND AVOID THE POSSIBILITY OF ERRORS.
REMEMBER TO SKIP LINES WHEN LETTERING AND TO USE
ONLY UPPER CASE LETTERS. GUIDELINES MUST BE USED
FOR ALL LETTERS. THEY MUST BE DRAWN VERY LIGHTLY.

FIGURE 1–2

EXAMPLES OF VERTICAL CAPITAL LETTERS AND NUMBERS

The first three lines below show the style of letters and numbers you must use on all drawings and assignments. This style of lettering is called single-stroke Gothic. Uppercase letters are normally used in drafting, except for metric abbreviations such as mm or cm. Civil engineering and architectural drawings may use lowercase letters.

1. A B C D E F G H I J K L M N
2. O P Q R S T U V W X Y Z
3. 1 2 3 4 5 6 7 8 9 0

The next line shows letters and numbers formed only by straight lines.

4. A E F H I K L M N T V W X Y Z

This line shows letters and numbers with straight lines and curves.

5. B D G J P Q R U 2 5

This final line shows letters and numbers that use only curved lines.

6. C O S 3 6 8 9 0

Practice your lettering by copying some of the rules of lettering from pages 4 and 5 onto the space below, using only the uppercase letters shown above and on the previous page. Make your guidelines as light as possible. Those printed on this sheet are darker than normally used so that they can be reproduced for you to use. Normally, you would not want the guidelines to be reproduced (although in practice they are often seen as light lines).

Because of limitations of the word processor, fractions in this manual may be written with a sloping line (1/8, for example). Remember that this is not good practice, and you should always draw a horizontal fraction line, as explained on page 4.

If you want additional practice, use page 8.

REMEMBER: Use only the uppercase letters shown on pages 5 and 6.

The following guidelines allow you to form letters 1/4" high (6 mm), as used on a title block. Larger letters are more difficult to form. A template may be used for very large letters over 3/8" (10 mm) high.

If you think that you need additional practice with lettering, then use this page.

SKETCHING

Before making a detailed drawing with instruments, you may want to do a freehand sketch. This sketch will give you an idea of what your final drawing will look like and will help you to plan your work on good drafting paper. With experience, you may be able to skip this step, but sketching is still a valuable skill to help you to communicate ideas quickly and easily. It is also useful when you are in a machine shop or on a construction site and want to sketch some object so that you can later draw it properly, whether using instruments or a CAD system.

NOTES ABOUT SKETCHING

a. When sketching, first use 0.3 mm lead to lightly draw thin lines. Darken these lines with 0.7 mm lead only when you have a good overall layout of how the final sketch should look.

b. When sketching, the paper is not taped down, so you can rotate the paper to get a comfortable position in which to draw.

c. When sketching circles, it is sometimes difficult to make them round—they may look oval, or egg-shaped. As an aid to sketching circles, lightly sketch a square with each side the same size as the diameter of the circle and then round the corners.

d. It is helpful to use graph paper when sketching, or you may put the graph paper under blank paper to act as a guide.

e. Draw *lightly* at first with your 0.3 mm technical pencil, then finish with your 0.7 mm technical pencil. Place the pencil point at the start of where you want to draw the line. Look at the endpoint where you want to finish the line, and then draw a single straight line or use a brushing stroke. An example of sketched lines is shown on page 160. Practicing sketching techniques will improve the appearance of your sketches.

EXERCISE 1-1

To practice the techniques of sketching, copy each of the shapes below by sketching on the grid. Make your sketch the same size as the given drawing. Do *not* use any kind of straightedge. All you need for this exercise is a pencil.

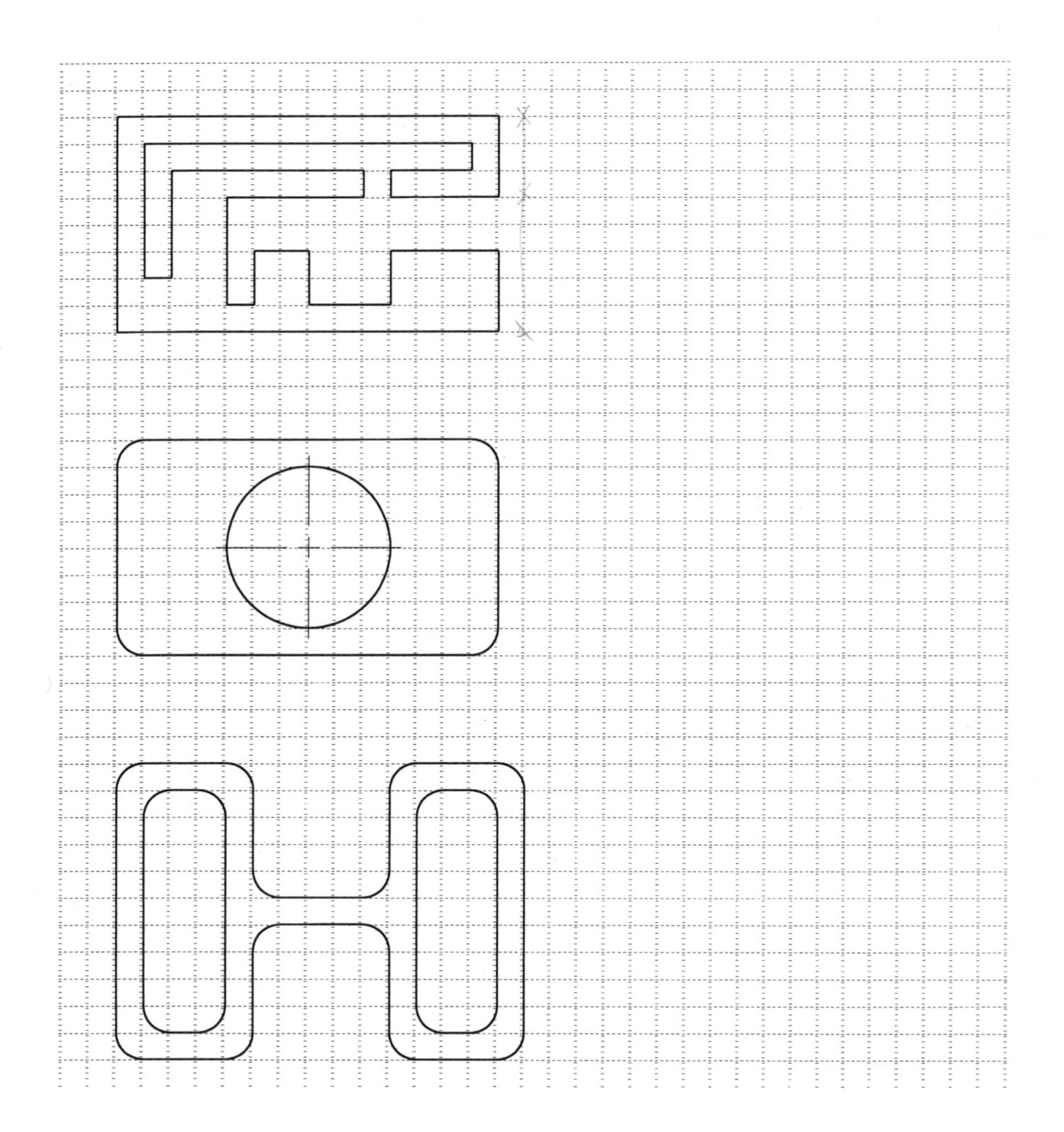

DRAWN BY DATE

EXERCISE 1–2

To further practice your sketching skills, copy the shapes below by sketching in the blank space to the right. Note that this exercise is more challenging than the one on the previous page, since here you do not have a grid on which to draw. All that you need to complete this page is a pencil. Make freehand sketches without using a straightedge. If you have graph paper, you can put it under the blank space to help you sketch.

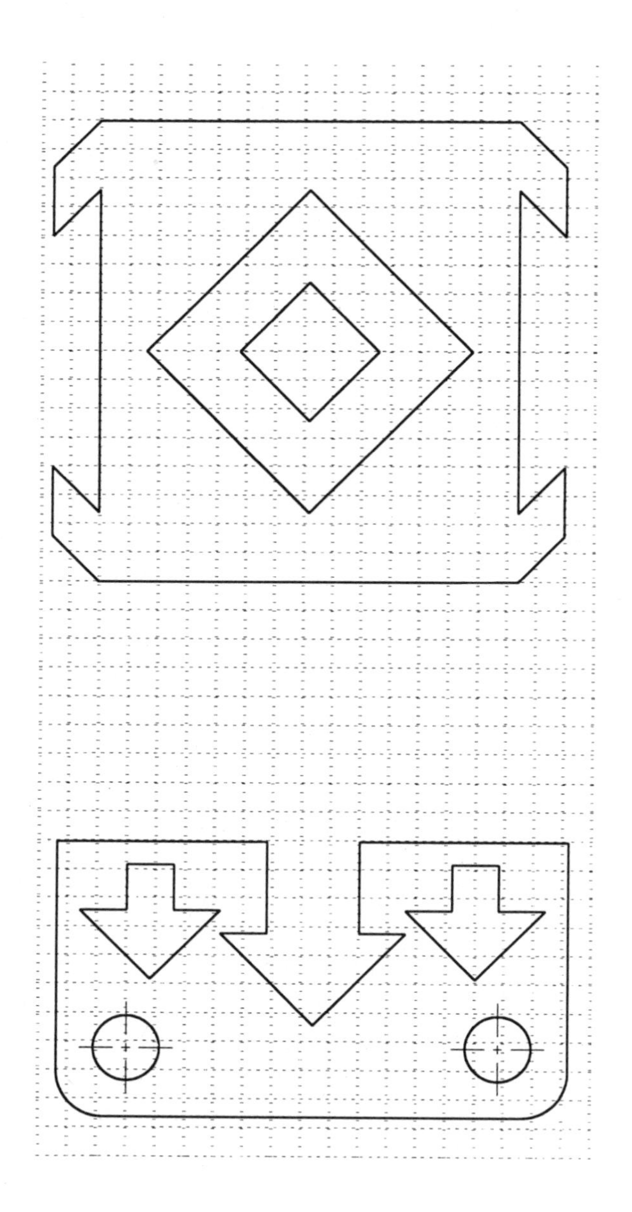

DRAWN BY DATE

PICTORIAL DRAWINGS

Pictorial drawings are similar in appearance to a picture or a photograph. They can be represented in many ways, some of which appear in Figs. 1–3 to 1–5.

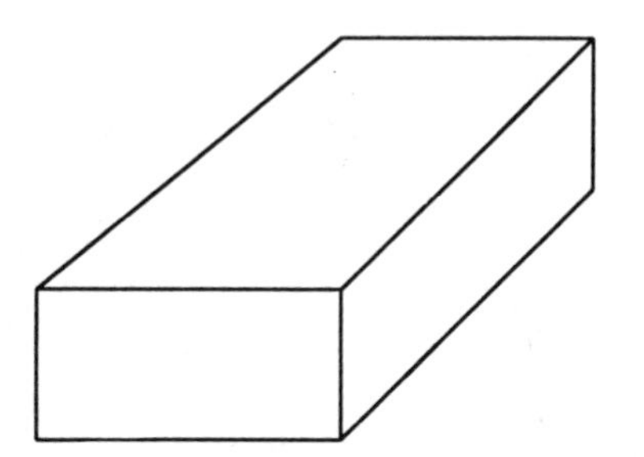

FIGURE 1–3 *Oblique projection.*

Oblique projection: Note that the front of the object in Fig. 1–3 is made up of horizontal and vertical lines. The sloping lines going from the front to the back of the view are usually drawn at a 45° angle to the horizontal, although other angles may be used. This type of pictorial drawing, called *oblique projection*, is not used very often.

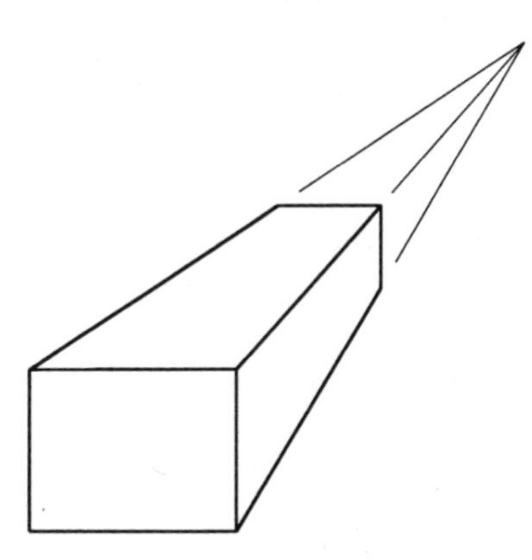

FIGURE 1–4(A) *One-point perspective.*

Perspective projection—One-point: Note that the lines going from the front to the back of the objects in Figs. 1-4(a) and (b) get closer together. In one-point perspective, shown in Fig. 1–4(a), lines going from front to back converge to a single point. Also note that lines representing the front face in one-point perspective appear as horizontal and vertical lines and do not converge.

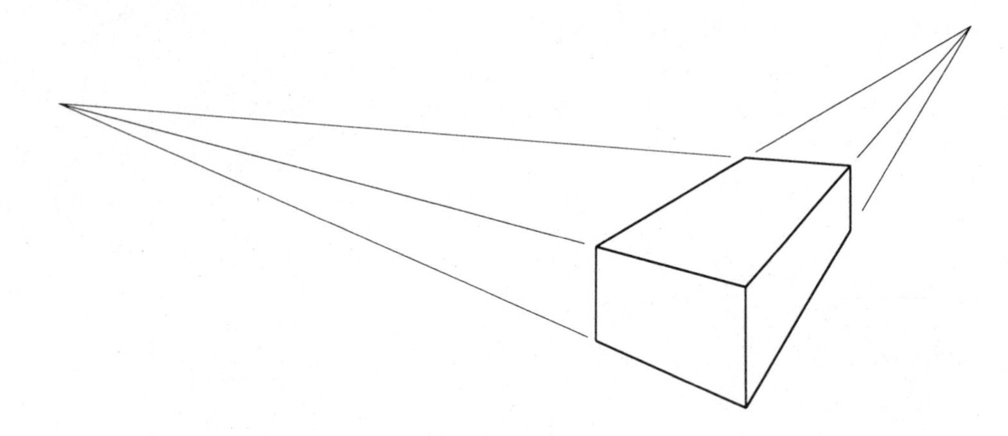

FIGURE 1–4(B) Two-point perspective.

Perspective projection—Two-point: In two-point perspective, shown in Fig. 1–4(b), lines representing vertical edges are vertical. Lines representing edges from the front to the back converge to one point, while lines representing edges from right to left converge to a second point. This type of projection is often used on architectural drawings.

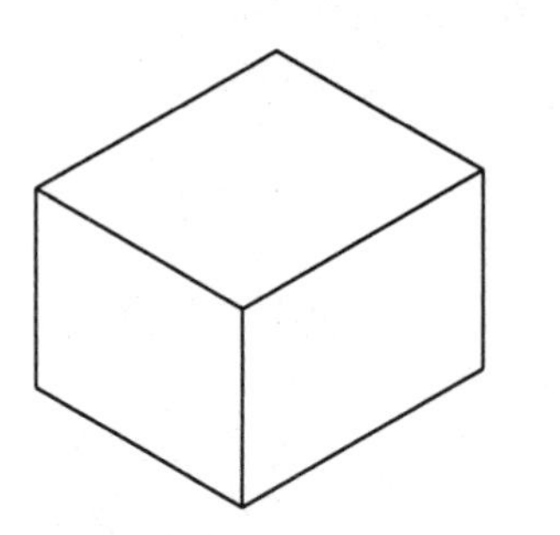

FIGURE 1–5 Isometric projection.

Isometric projection: In isometric projection (Fig. 1–5), note that all lines are either vertical or at a 30° angle to the horizontal and that there are no horizontal lines. We will learn to draw this pictorial projection.

ISOMETRIC VIEWS

RULES ABOUT ISOMETRIC VIEWS

a. Hidden edges (those not visible from your viewpoint), such as the back corner of a box that you cannot see from the front, are usually not drawn. See Fig. 1–6.

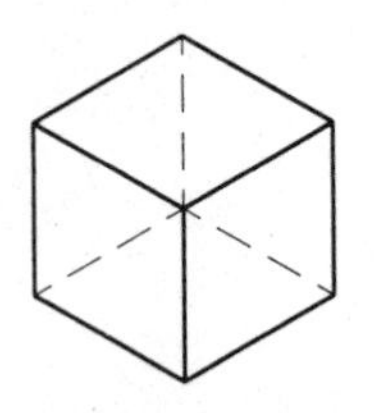

NOT RECOMMENDED

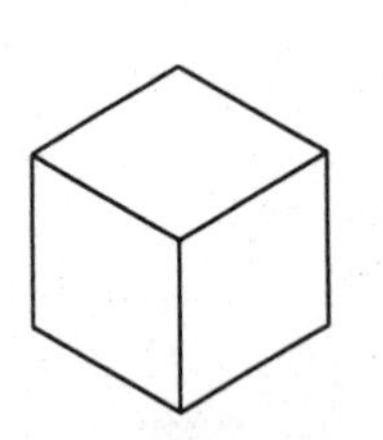

PREFERRED

FIGURE 1–6

b. Vertical edges of an object are represented by vertical lines in an isometric view.

c. Horizontal edges of an object are represented by lines at a 30° angle. See Fig. 1–7.

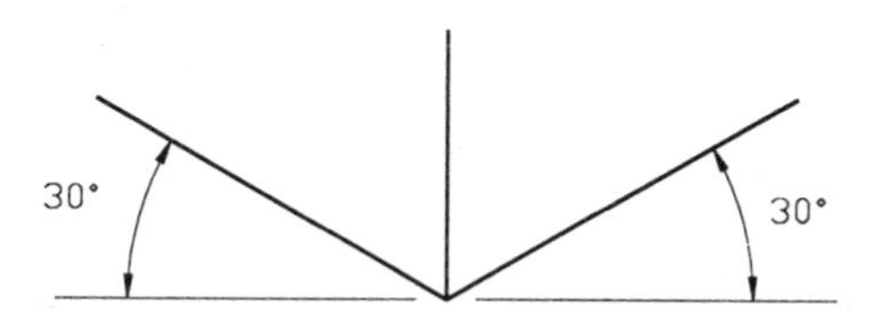

FIGURE 1–7 The isometric axes.

STEPS FOR SKETCHING THE ISOMETRIC AXES

STEP 1 Lightly sketch a horizontal and a vertical line. This gives you two 90° angles. See Fig. 1–8(a).

FIGURE 1–8(A)

STEP 2 Divide each 90° angle into three equal parts by approximating equal angles. This will result in each section being 30°. In future chapters, you will use drafting instruments to draw these lines accurately. See Fig. 1–8(b).

FIGURE 1–8(B)

STEP 3 Note the three lines that make the isometric axes. See Fig. 1–8(c).

FIGURE 1–8(C)

STEP 4 Erase unnecessary lines. See Fig. 1-8(d).

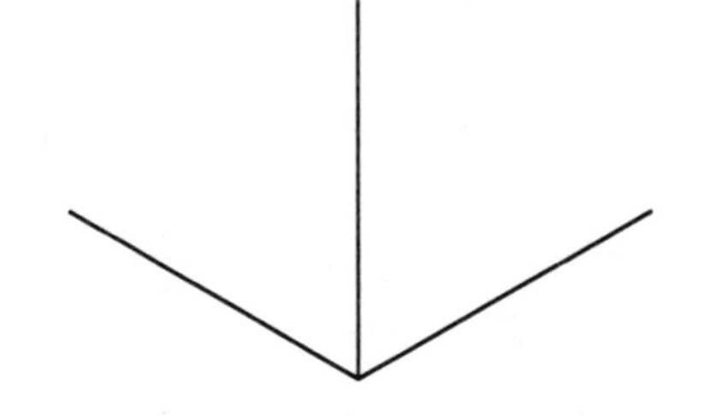

FIGURE 1–8(D)

ACTIVITY: On any blank sheet of paper, sketch the isometric axes by following the steps described above.

STEPS FOR SKETCHING AN ISOMETRIC CUBE

STEP 1 Lightly draw the isometric axes, as described above. The lines may be of any convenient length. See Fig. 1–9(a).

FIGURE 1–9(A)

STEP 2 Mark off approximately equal distances to represent the height, width, and depth of the cube. See Fig. 1–9(b).

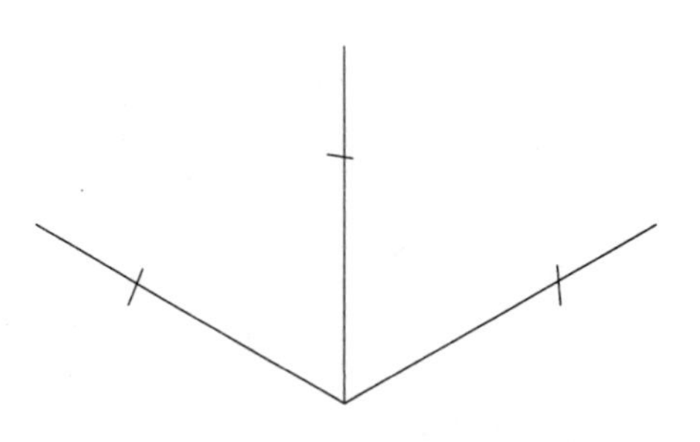

FIGURE 1–9(B)

STEP 3 Add two vertical lines of a convenient length at the appropriate marks, as shown in Fig. 1–9(c).

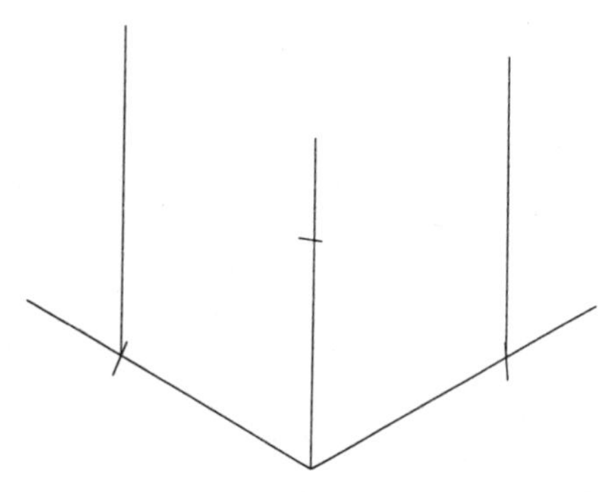

FIGURE 1–9(C)

STEP 4 Add two 30° lines at the appropriate point parallel to two of the *isometric axes*, as shown in Fig. 1–9(d).

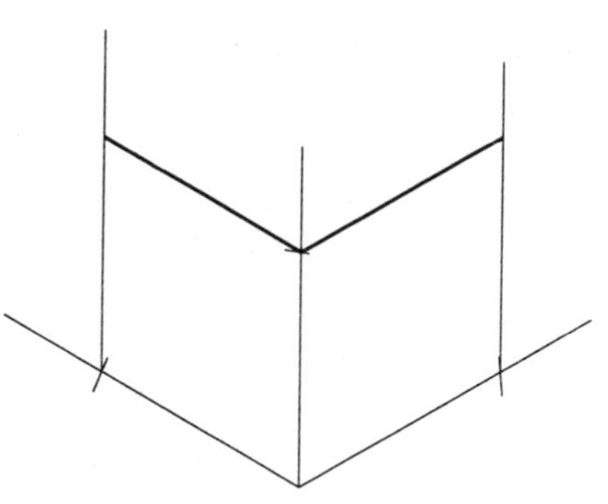

FIGURE 1–9(D)

STEP 5 Add a 30° line (any length) as shown in Fig. 1–9(e).

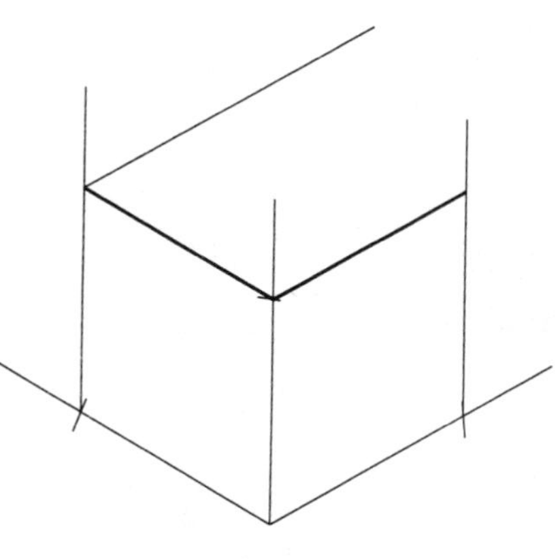

FIGURE 1–9(E)

STEP 6 Complete the cube. See Fig. 1–9(f).

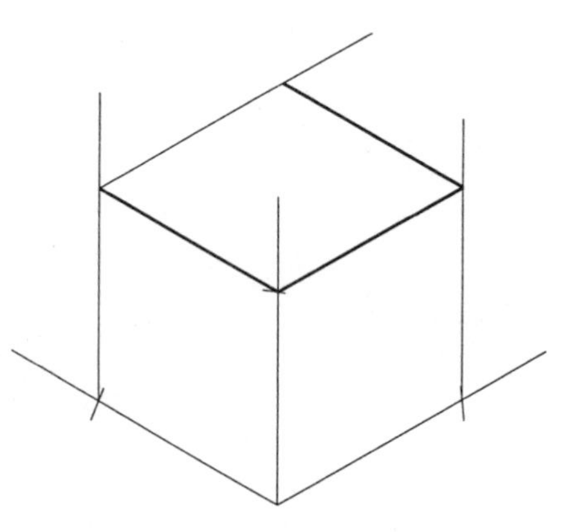

FIGURE 1–9(F)

STEP 7 Erase unnecessary lines if they are too dark (light ones may be left). Darken the remaining lines. See Fig. 1–9(g).

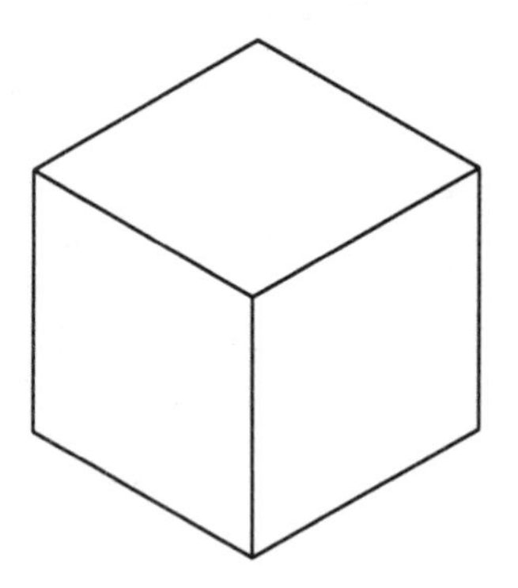

FIGURE 1–9(G)

ACTIVITY: On a blank piece of paper, sketch an isometric cube of approximately the same size as the one on this page.

STEPS FOR SKETCHING ANY ISOMETRIC VIEW

STEP 1 IMPORTANT: *Always* start your isometric drawing, whether sketching or using instruments, by *lightly* drawing a box to completely enclose the object. This box should have the same overall dimensions and shape as the object, in our case, the object shown in Fig. 1–10(a).

Figure 1–10(b) shows the lightly drawn box that we start with.

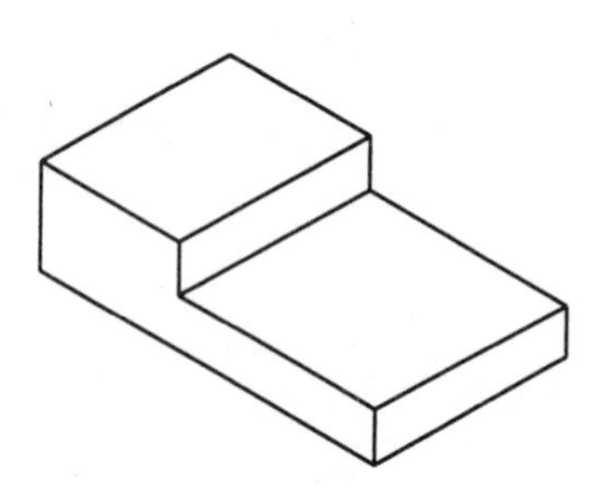

FIGURE 1–10(A)

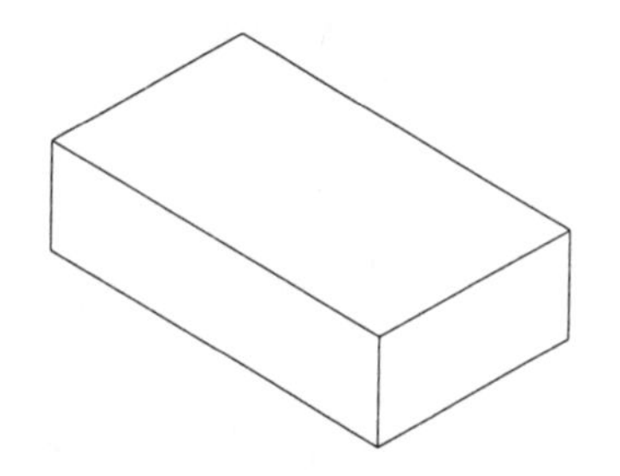

FIGURE 1–10(B)

STEP 2 Add some lines to show the shape of the object. See Fig. 1–10(c). These are *parallel* to the isometric axes (unless they are "nonisometric" lines—explained later in this chapter).

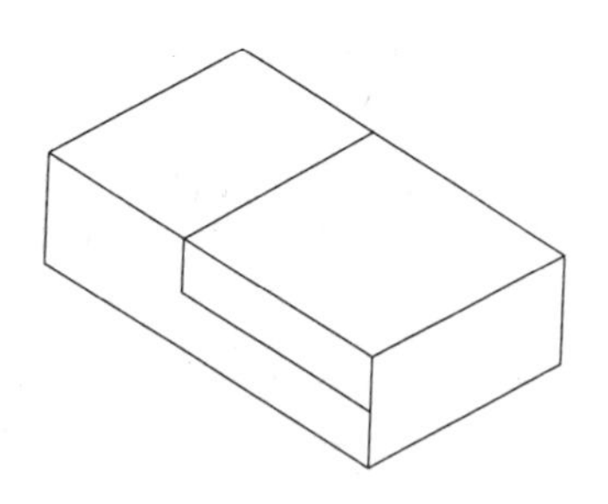

FIGURE 1–10(C)

STEP 3 Add more lines to show the shape of the object. See Fig. 1–10(d).

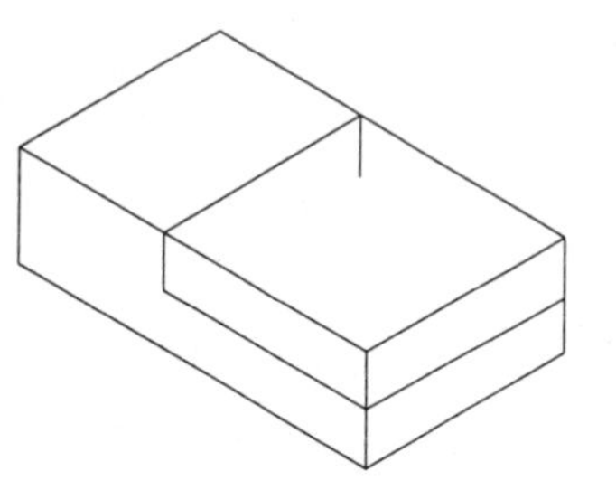

FIGURE 1–10(D)

STEP 4 Finish the shape of the object. See Fig. 1–10(e).

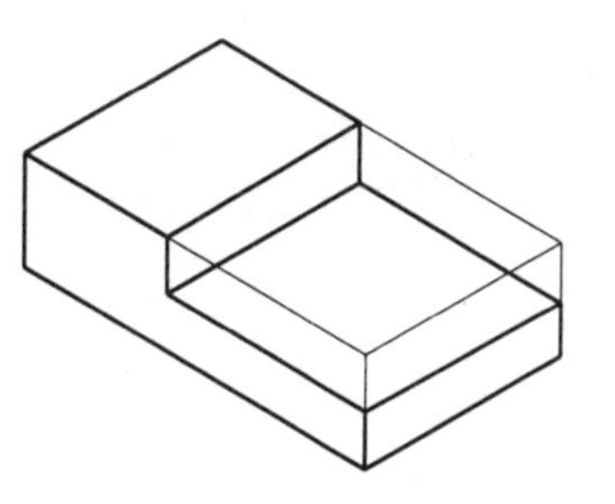

FIGURE 1–10(E)

STEP 5 Erase construction lines if they are too dark (light ones may be left). If you have already purchased your erasing shield, you may use it to cover the lines you are not erasing. See Fig. 1–10(f).

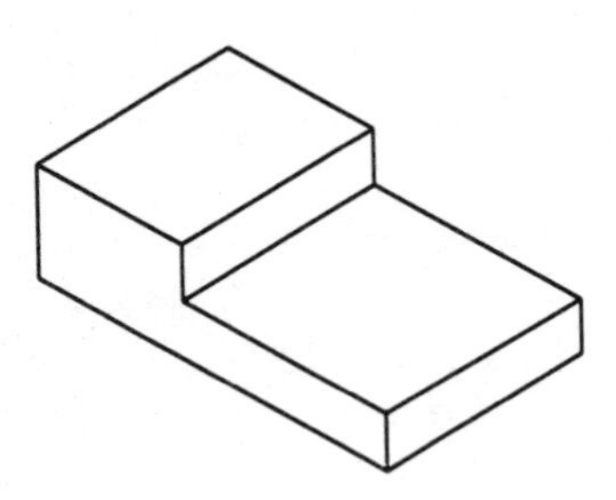

FIGURE 1–10(F)

ACTIVITY: On a blank piece of paper, follow the preceding steps to draw the object shown in Fig. 1–10(f).

NOTE: You may find the isometric grid in the Appendix (page 333) a convenient guide. Place this grid under a blank paper to use as a reference.

An incorrect method of starting your drawing is shown in Fig. 1–11: The isometric box was not drawn first. This *very poor method* will make it difficult to sketch the isometric view.

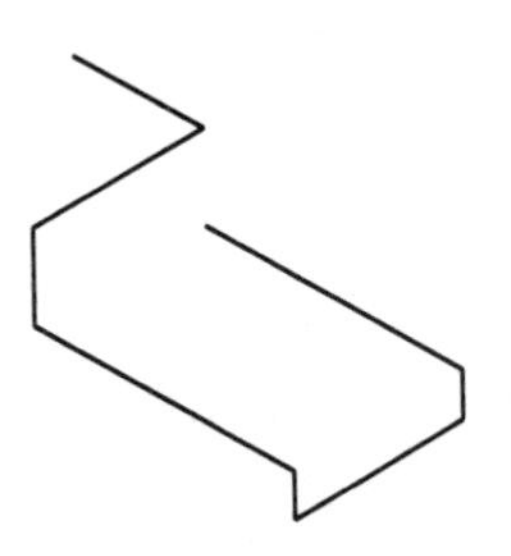

FIGURE 1–11

EXERCISE 1–3

Make freehand sketches of the isometric objects shown below by copying them in the space provided to the right. Do not use a straightedge. If your box or object does not look correct, you probably drew a line that was not parallel to one of the isometric axes. You may find it helpful to use the isometric grid in the Appendix (page 333) by placing it under this sheet. Remember, the isometric box that you first sketch as a guide must have the overall shape of the object to be drawn.

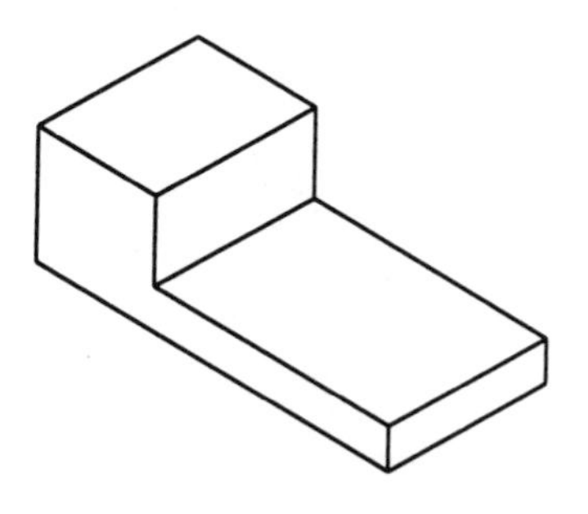

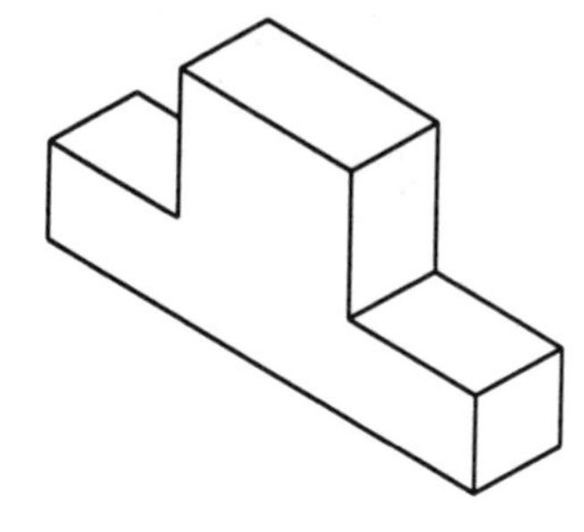

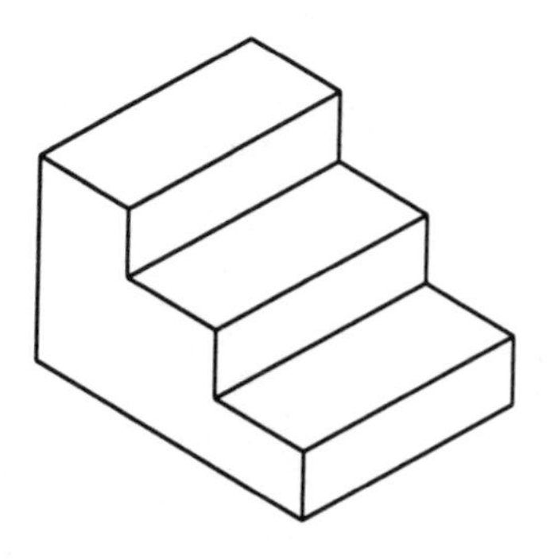

DRAWN BY DATE

NONISOMETRIC (OBLIQUE) LINES

Nonisometric lines (or oblique lines) are lines drawn on an isometric view that are not parallel to any of the three isometric axes. They are used to represent sloping surfaces. See Fig. 1–12(a).

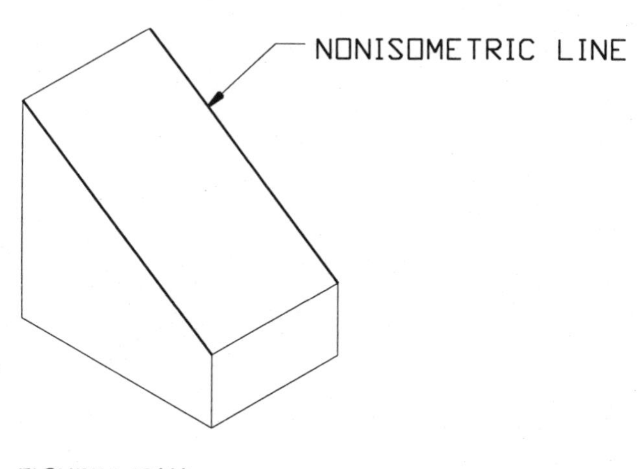

FIGURE 1–12(A)

To draw sloping lines in isometric, locate the *endpoints* of the sloping line and join them with a straight line. In Fig. 1–12(b), endpoints have been designated with the letters A, B, C, and D. In Fig. 1–12(c), points A and B, as well as points C and D, have been joined.

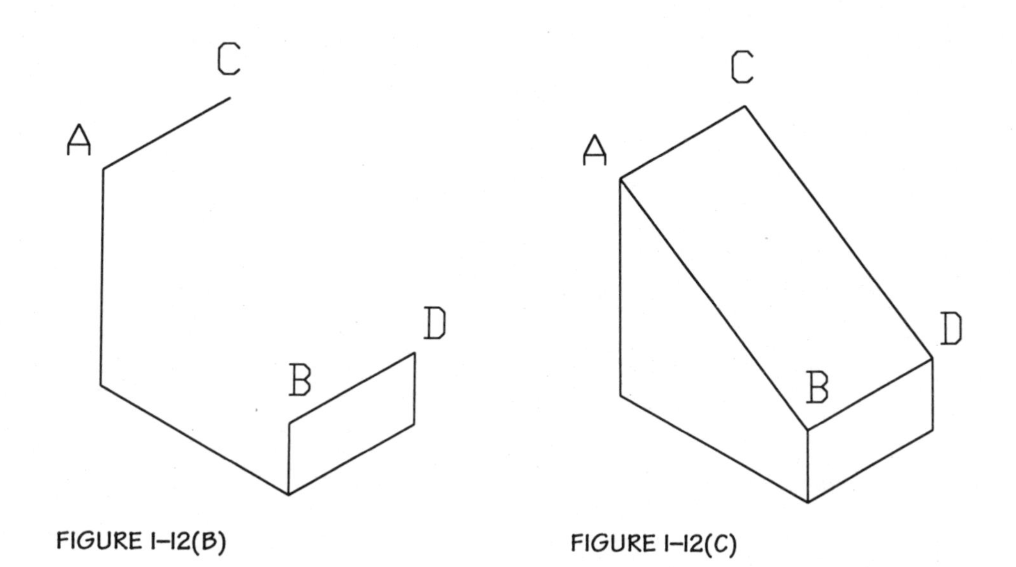

FIGURE 1–12(B)

FIGURE 1–12(C)

ANGLES IN ISOMETRIC

On an isometric view, measure only the linear distances, never the angles. For example, on the drawing in Fig. 1–13, although angle A is 90° on the actual piece, it obviously does not appear as 90° on the isometric drawing (it is actually a 60° angle). All linear distances are the actual lengths, including both isometric and nonisometric lines.

REMEMBER: Angular measures in isometric are not the same as on the actual piece.

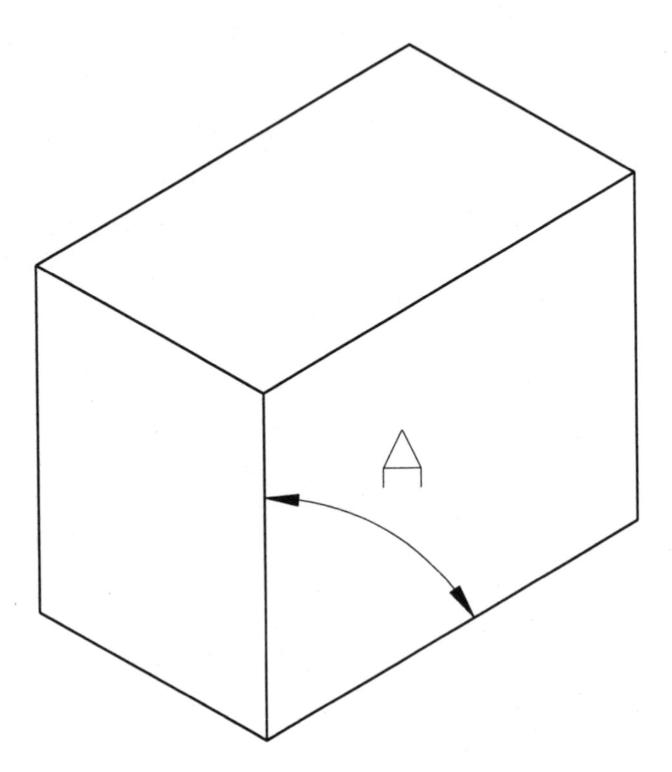

FIGURE 1–13

SAMPLE EXERCISE

The illustrations below (Fig. 1–14) show the sequence of steps in making an isometric sketch. Note that the faces of the view were drawn one at a time, and that each line is parallel to one of the isometric axes.

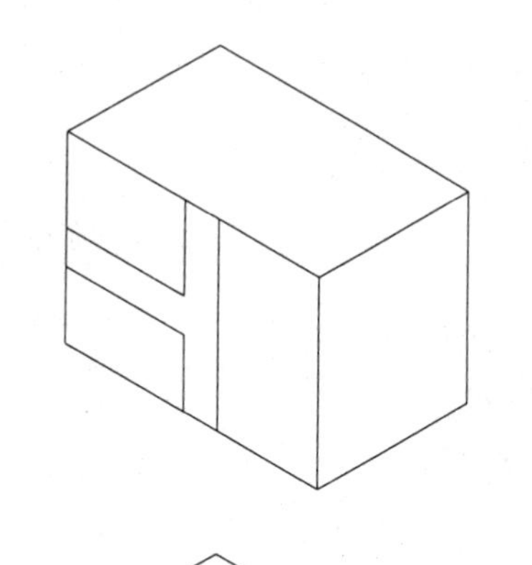

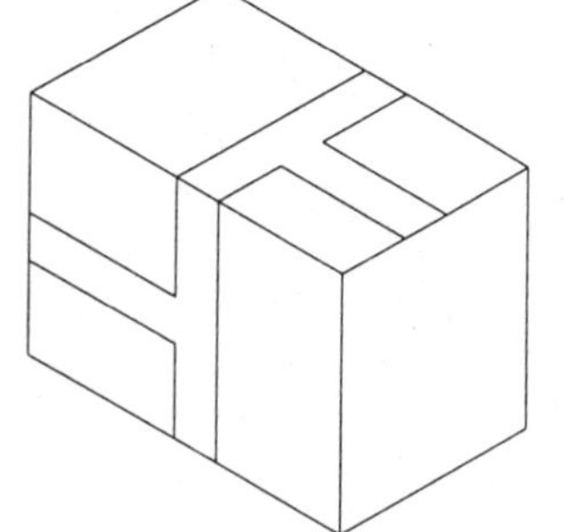

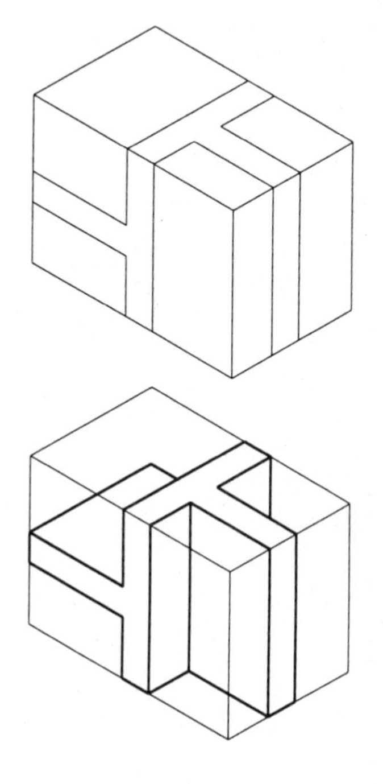

FIGURE 1–14

EXERCISE 1–4

In the space provided on this page, make freehand sketches by copying the isometric objects shown below. The first one was illustrated for you in Fig. 1–14. Do not use a straightedge. If your box or object does not look correct, you may have drawn a line that was not parallel to one of the isometric axes. Remember to start with an isometric box.

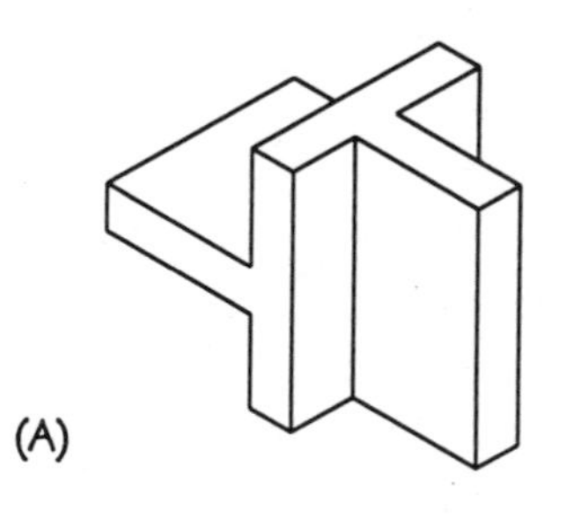

(A)

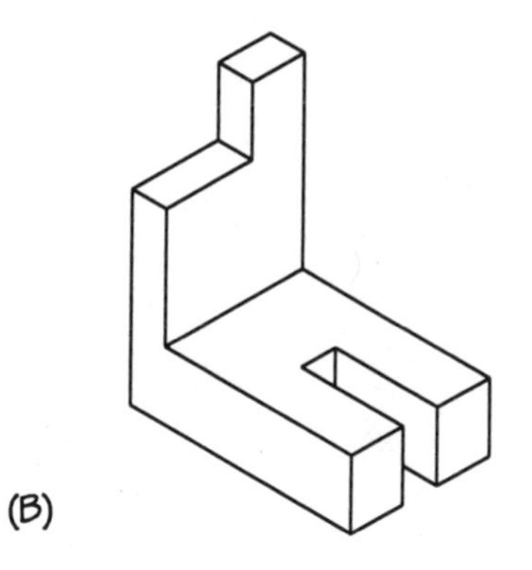

(B)

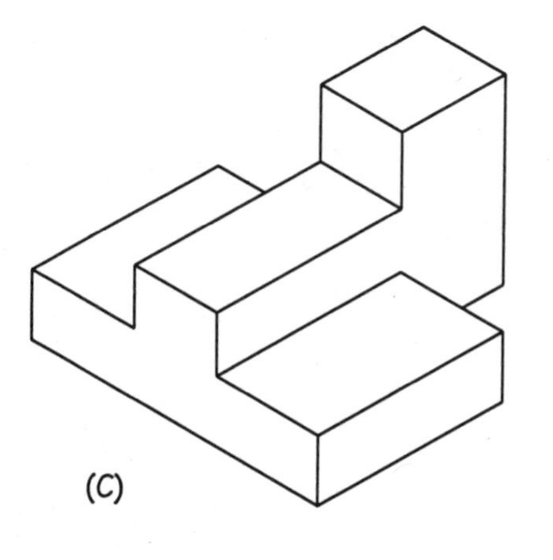

(C)

DRAWN BY DATE

SAMPLE EXERCISE

The illustrations in Fig. 1–15 show the sequence of steps in making an isometric sketch of an object that contains nonisometric lines.

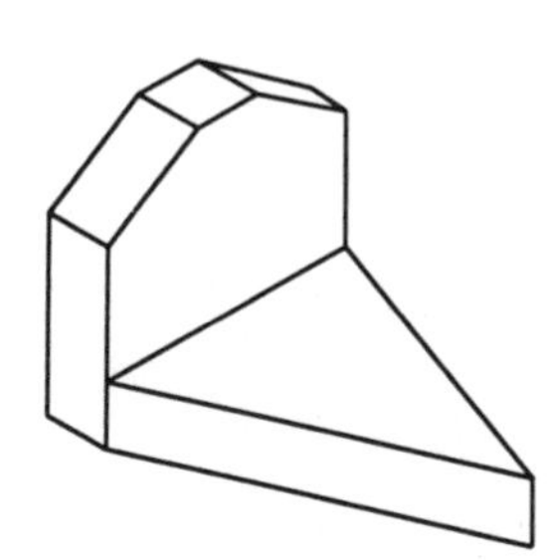

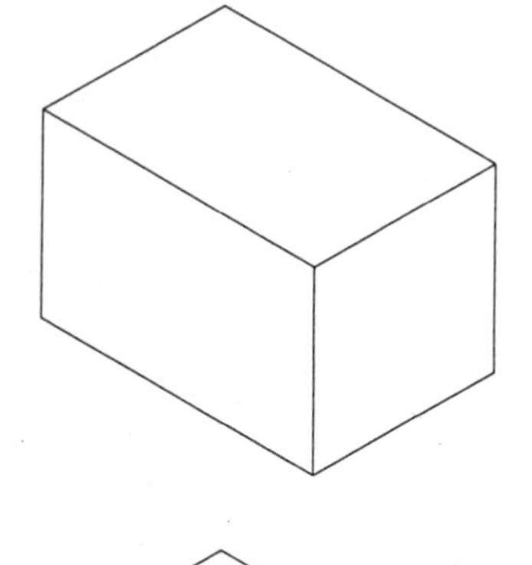

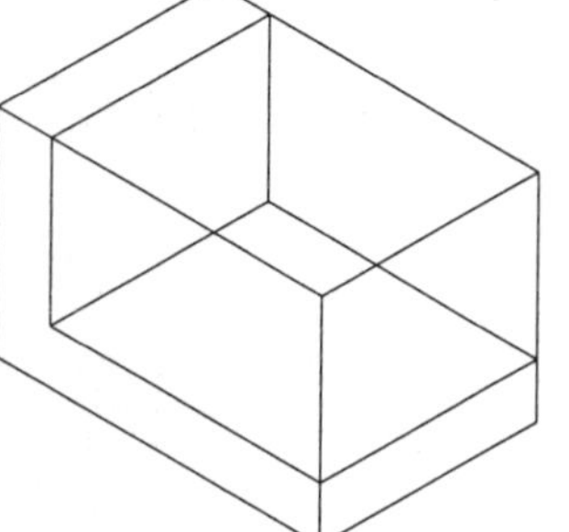

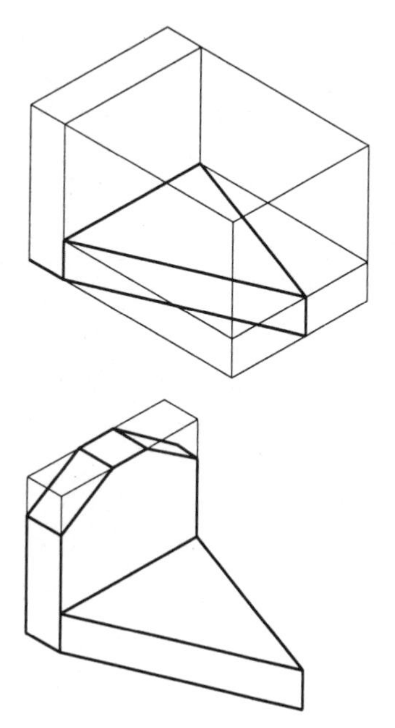

FIGURE 1–15

EXERCISE 1–5

In the space provided on this page, make freehand sketches of the isometric objects shown below. Notice that all these objects contain some nonisometric lines. Remember to start with an isometric box. Do not use a straightedge. Part (c) was illustrated for you in Fig. 1–15.

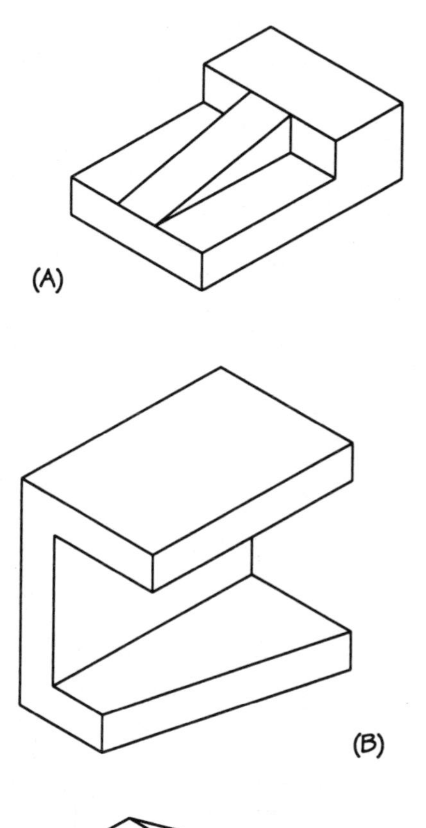

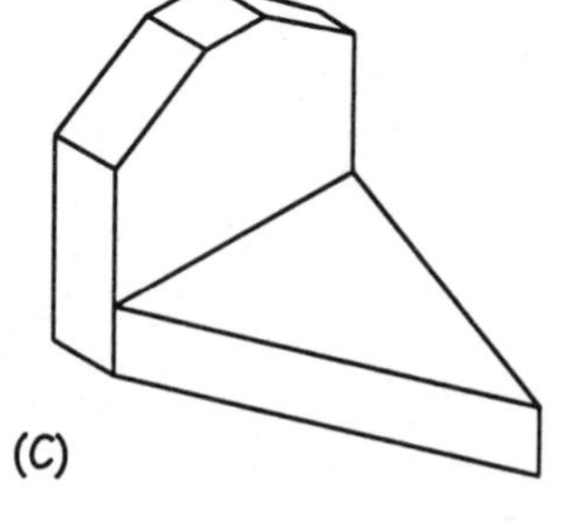

DRAWN BY ______________________ DATE ____________

ORTHOGRAPHIC VIEWS: SKETCHING 2

OBJECTIVE

When you have finished this chapter, you should be able to sketch orthographic views.

EQUIPMENT

To complete this chapter, you should have two technical pencils of different lead thickness or two lead holders with different grades of lead. However, this chapter may be completed with one pencil if necessary.

ORTHOGRAPHIC VIEWS

Most shop drawings used in industry are *orthographic* rather than pictorial drawings. An orthographic view shows one side of an object—it appears as though you are looking directly at that side.

Think of the isometric view shown in Fig. 2–1(a) as an office building. From an airplane, we would view only the roof or the top of the building, surfaces B and D. If we stand facing the front of the building, we would see only surface A. From the side street on the right side of the building, we would see only surfaces C and E.

Figure 2–1(b) illustrates the views we would see from these different directions. The front view shows only surface A. The top view shows only surfaces B and D. The side view shows only surfaces C and E.

The placement of these views is *very important.* The top view is always above the front view, the right side view always to the right of the front view. *Never* put them in any other relative positions.

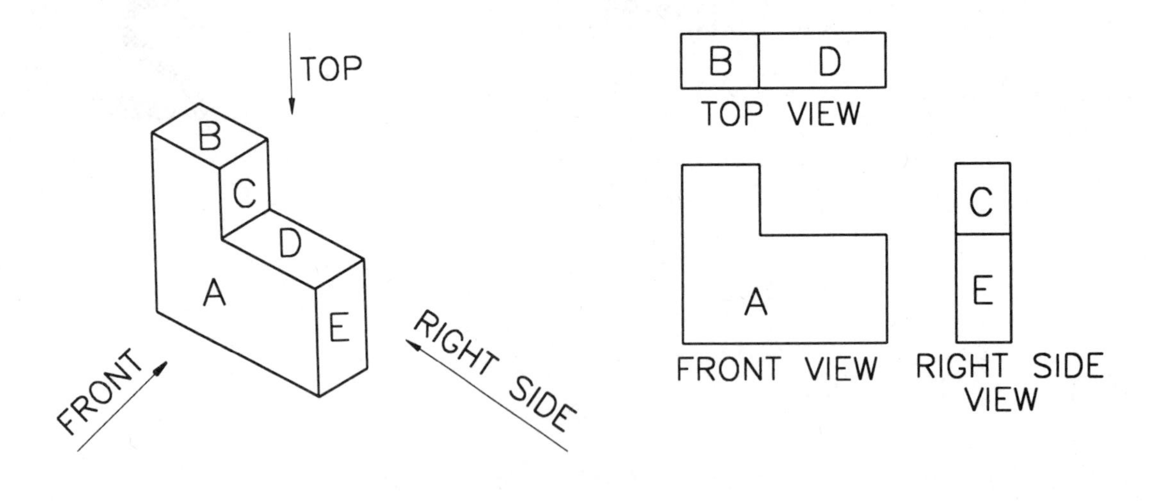

FIGURE 2–1(A) Isometric View

FIGURE 2–1(B) Orthographic Views

The front view is sometimes called the *principal view,* since it usually shows the *characteristic* or *identifiable shape* of the object. In Fig. 2–2, for example, the front view shows that the object is stair-shaped. This would be the characteristic shape, since we would identify this object by its stair shape. We would never think of this object as three rectangles, as shown in the top and right side views.

The top view is sometimes called the *plan* view. From the top view of the orthographic drawing, we see only three rectangles beside each other, as we also do in the right side view. In our particular example, these views do not really identify the object's shape in the same way the front view does.

Note also that distance X in the top view must equal distance X in the side view, since both represent the distance from the front to the back (or the depth) of the object. If you have difficulty with this concept, think of the building analogy described on the previous page.

STEPS IN MAKING ORTHOGRAPHIC VIEWS

STEP 1 It is usually helpful to start with the front view when drawing orthographic views.

STEP 2 Use *construction lines* (very light, use 0.3 mm lead) to project surfaces and sharp corners to the other views.

STEP 3 You do not have to erase the construction lines when you have finished. If they are drawn lightly enough, they may be left.

REMEMBER:

- The top view is above the front view. The right side view is on the right side of the front view.
- We can measure the angles on orthographic views as well as linear distances.

ACTIVITY: Label the orthographic views in the proper locations with the letters that correspond to those on the isometric view.

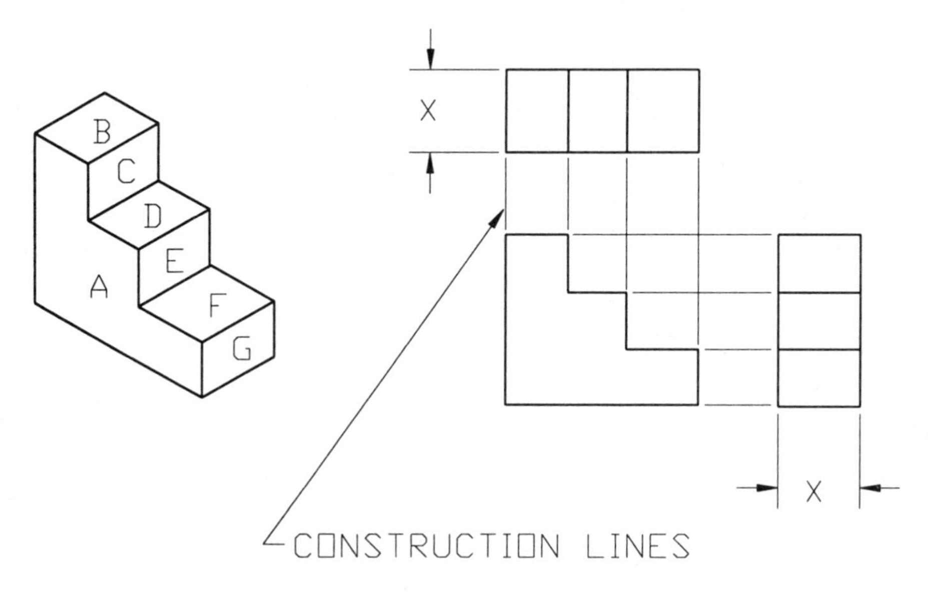

FIGURE 2–2

EXERCISE 2–1

Refer to the isometric drawing below to answer the following questions:

1. What letters do you see on the top view? ____________
2. What letters do you see on the right side view? ____________
3. What letters do you see on the front view? ____________

In the space provided on this page, draw the three main orthographic views of the object (*top, front,* and *side*) shown in the figure. Letters have been added to the different sides of the isometric view to help you analyze the problem. Start by drawing the *front view,* then draw construction lines up for the top view and to the right for the side view.

HINT: For this problem, all your lines on the orthographic views will be horizontal or vertical. To help you solve this problem, refer to the building analogy on page 35 and think of this object as a large building with a small building attached.

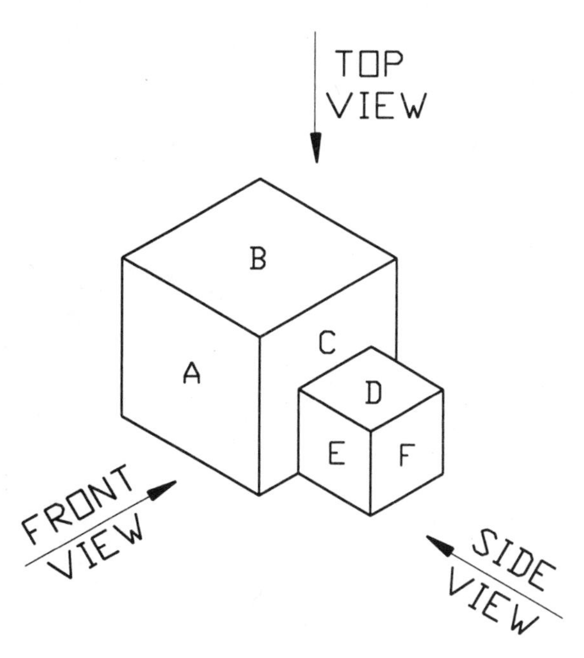

DRAWN BY ____________ DATE ____________

HIDDEN FEATURES

Figures 2–3(a) and (b) illustrate the need for a complete set of orthographic views in order to fully describe an object. In these two examples, the top view is the same in both cases. You cannot tell from the top view whether the middle two lines represent a surface that is higher or one that is lower than the surfaces beside it.

REMEMBER: Always study all views of a drawing when interpreting its shape.

In Fig. 2–3(b), note the use of hidden (dashed) lines to show hidden features on the right side view. The horizontal construction line part way down the front view tells us that there should be a line in the side view. If we think about this object, we realize that we do not see this lower horizontal surface; therefore, the line drawn for this construction line is dashed.

REMEMBER: When you see a dashed line, it is a hidden line representing a corner or surface that we do not see.

ACTIVITY: In the boxes to the right of the orthographic views below, sketch an isometric view of each object.

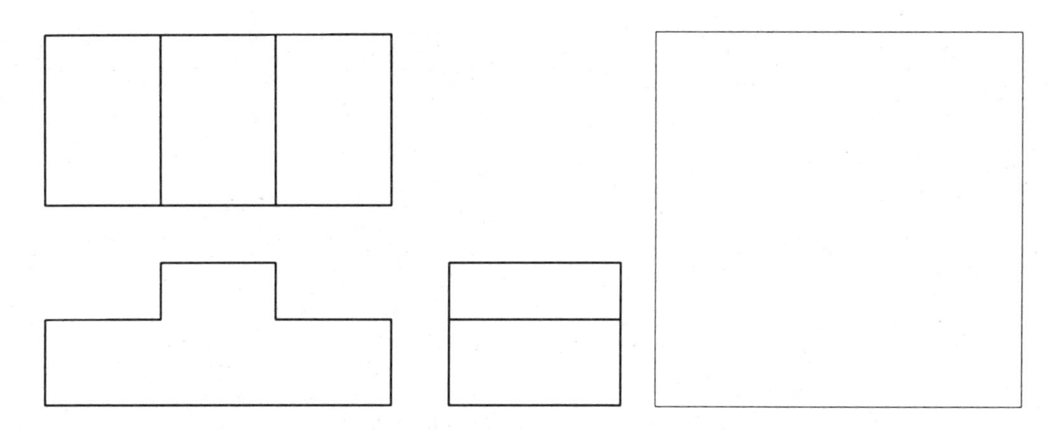

FIGURE 2–3(A)

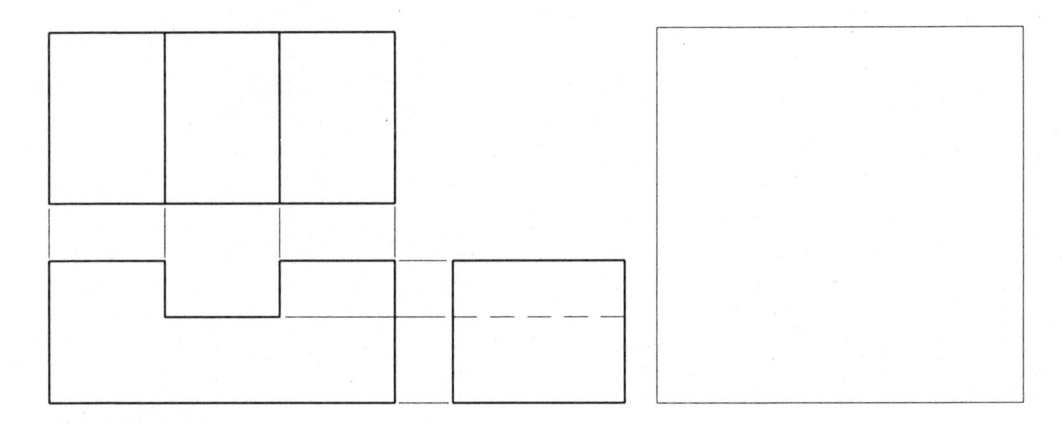

FIGURE 2–3(B)

EXAMPLES OF ORTHOGRAPHIC PROJECTION

Notice that the top view is the same for the three examples shown in Fig. 2–4. The corners seen in all the front views are shown as lines in the other views. Notice in Example 2 the use of the hidden line in the side view to represent the intersection of the two sloping surfaces. In Example 3, the rounded edges shown in the front view are not seen as rounded in the other views.

If you have difficulty with the concept of orthographic projection, lay your pencil down on the desk. If you look directly down on it, it looks like a long rectangle and you only realize that it is round by looking at the end of the pencil or at its shadings. In one orthographic view, you can see that your pencil will look like a circle, and in an adjacent view, it will look like a rectangle. If you have an overhead projector, place a piece of chalk on the glass. You will see that its projection on the screen is a rectangle. This represents one view of the object.

NOTE: Construction lines are not shown below because if they were made very lightly, they would not appear on a copy of the final drawing.

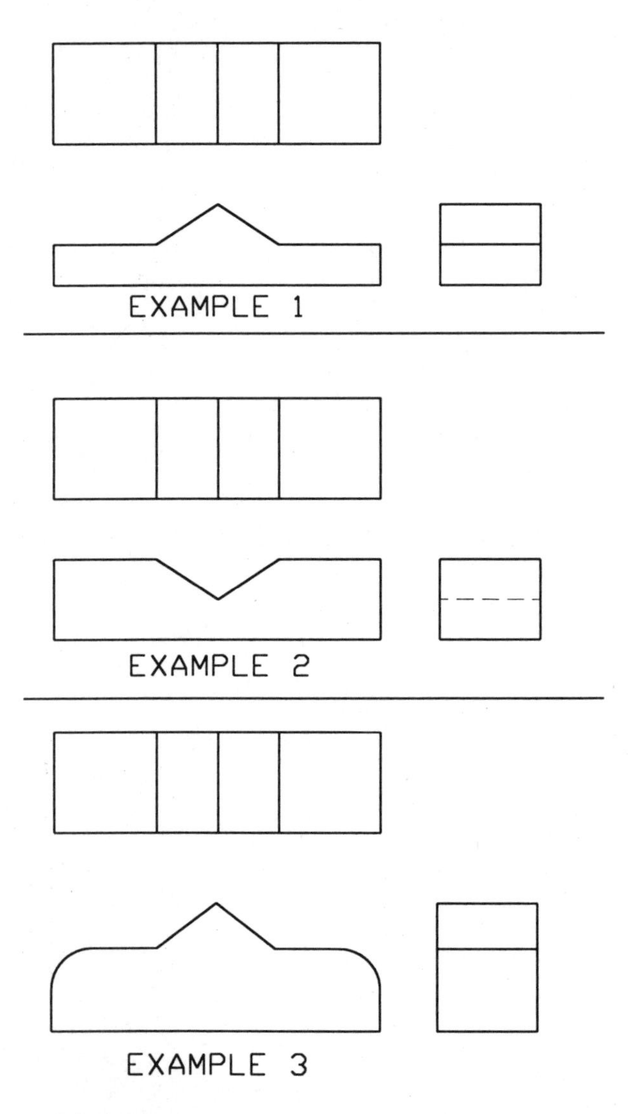

FIGURE 2–4

In the following two examples, notice the projection of curves and surfaces in the top and side views. (Also notice that the views are not labeled top, front, or side. We assume that the reader knows this.) By looking only at the top and side views of Fig. 2–5(a), we do not realize that two corners are rounded. In this example, we can represent the rounded corners in the front view only.

When a solid line and a hidden line are projected by the same construction line, only the solid line is shown. In Fig. 2–5(b), for example, there are no hidden lines in the top and side views, even though lines representing some surfaces are hidden in these views. The isometric views are shown to help you visualize each object.

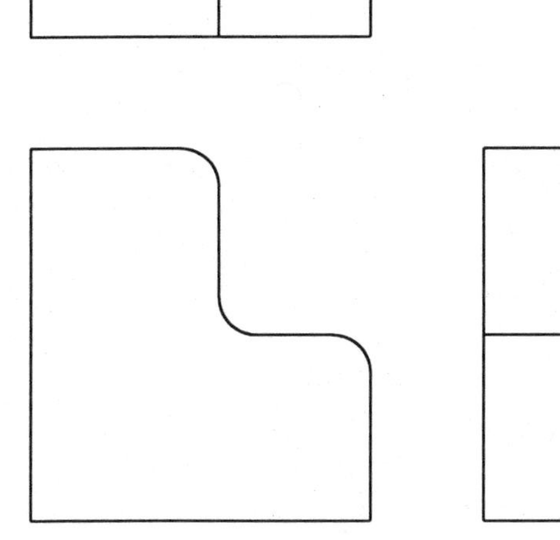

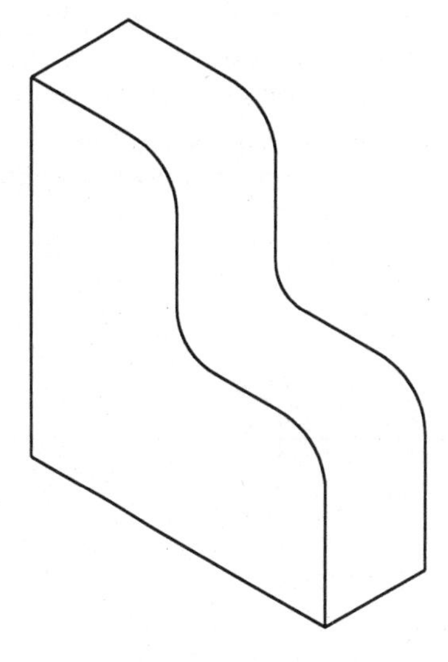

FIGURE 2–5(A)

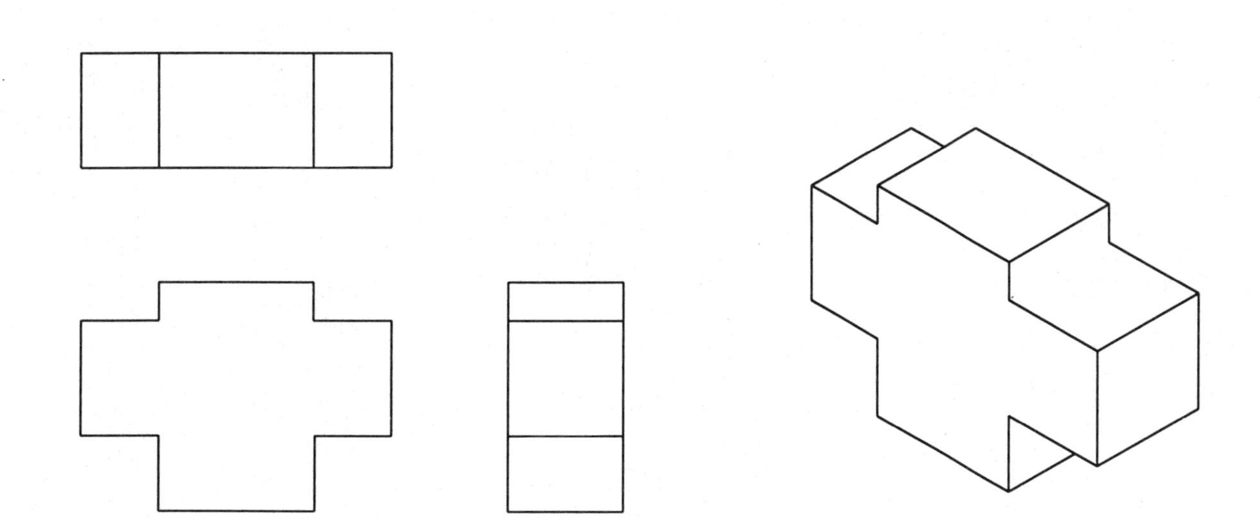

FIGURE 2–5(B)

Figure 2–6 illustrates how a center line and a hole are shown on different views. Three types of lines are used in this drawing. (A complete list of the various types of lines used on a technical drawing is given in the Appendix, page 321.)

Object lines are *thick*, drawn with 0.7 mm lead, and represent the outline or the edge of an object.

Center lines are alternating long and short lines. They are *thin* and *dark*. Use your 0.3 mm lead to draw them. Center lines extend beyond the object line or circle by about 1/4" (7 mm). They are used to represent the center of a hole. Notice that in the center of the circle the center line has two short lines perpendicular to each other. This is illustrated in the front view in Fig. 2–6.

Hidden lines are thin and dark dashed lines and represent hidden features or the edges of a hole. You should use the 0.3 mm lead for hidden lines.

The preceding three lines are all dark, even though they vary in thickness. Remember that only construction lines are light.

Study the three orthographic views below and try to visualize the object. To assist you, an isometric view of this object is shown in Fig. 2–8(c).

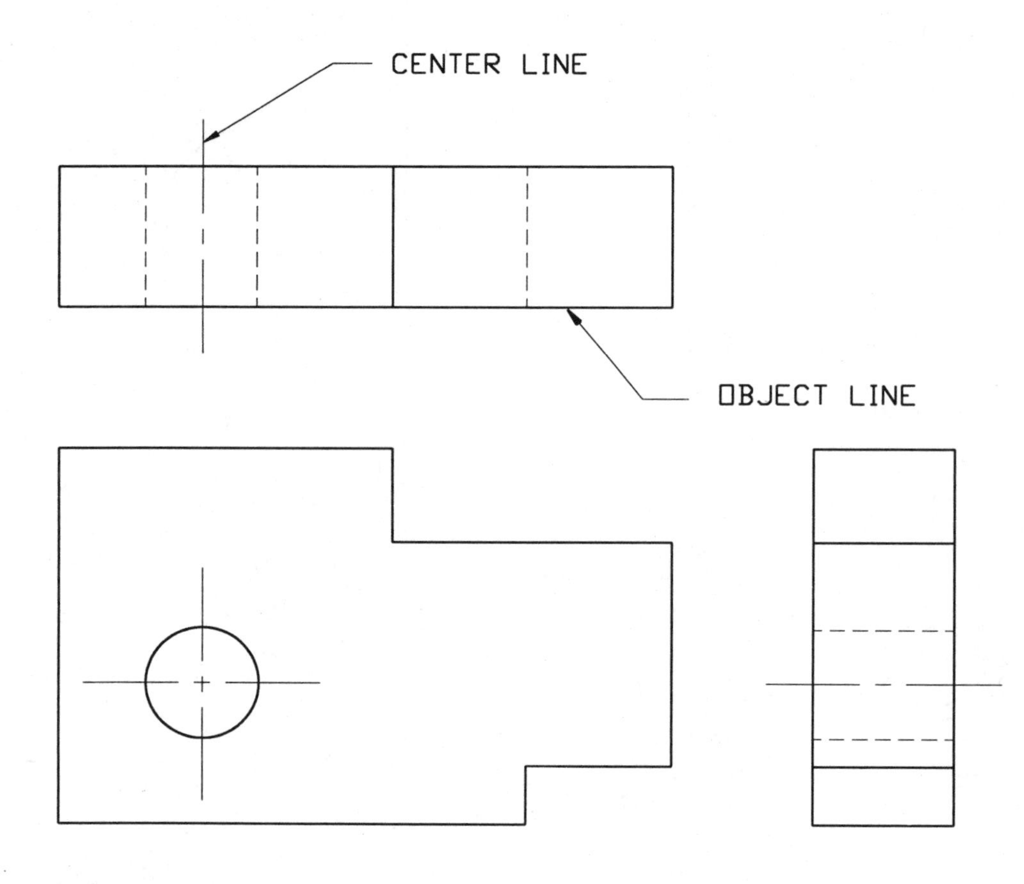

FIGURE 2–6

Notice that the top views are the same in Figs. 2–7(a) and (b), even though the shapes of the objects differ. The isometric views are drawn to the right of the orthographic views so you can visualize each object. You may also think of the office building analogy to assist you.

Figure 2–7(a) also illustrates that distance D in the top view must be the same as distance D in the side view.

A horizontal construction line is drawn in Fig. 2–7(b) to show you how the hidden line in the right side view lines up with the bottom of the curve in the front view.

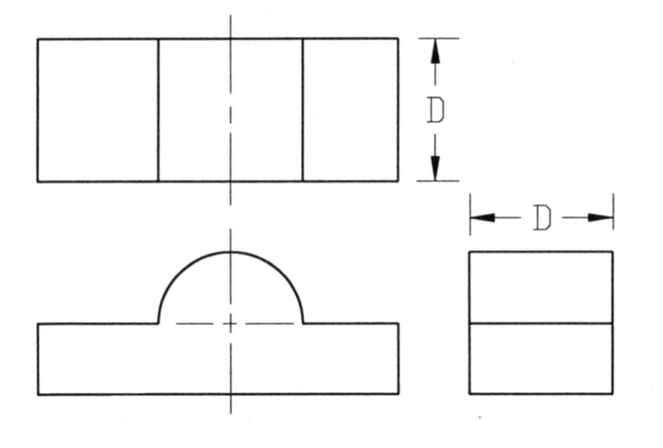

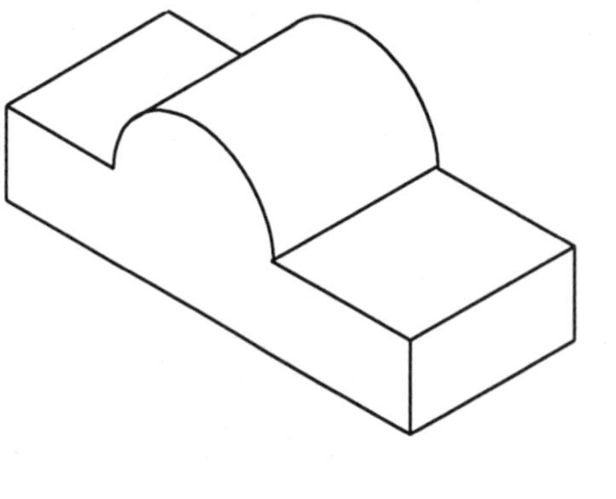

FIGURE 2–7(A)

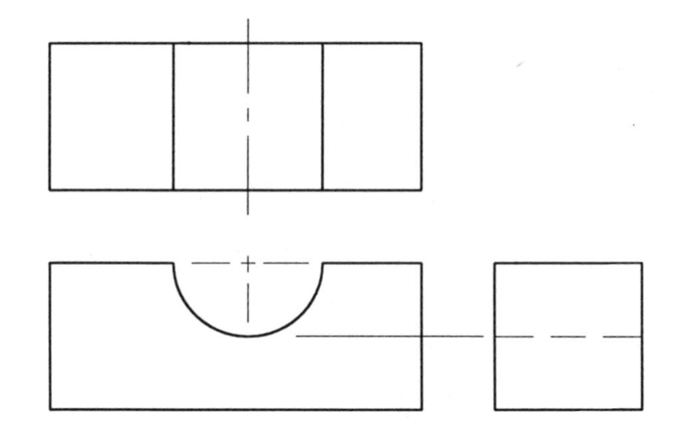
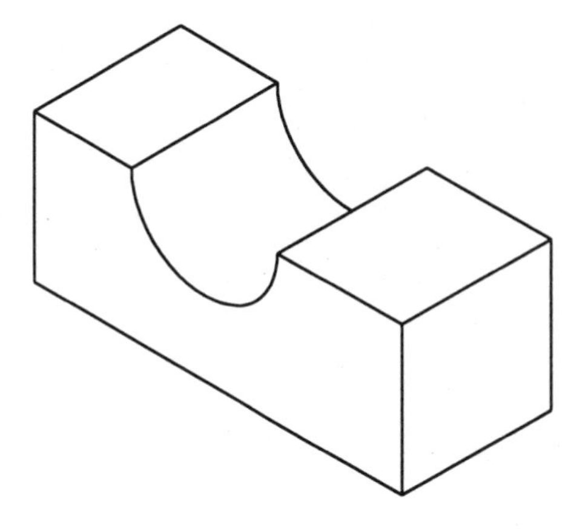

FIGURE 2–7(B)

EXERCISE 2–2

In the top view of the object drawn on this page, we cannot differentiate between a hole and a solid protrusion. Note also that in the front view we see two small holes, one above the other. In the top view, we see only one of these holes because one hole is directly below the other.

Sketch the right side view of this object. Remember to use construction lines to line up the views to ensure that the side view is located directly beside the front view.

HINT: If you are not sure how to begin, start by sketching horizontal construction lines very lightly to the right of the front view.

HINT: In case you need help, an isometric view of this object is drawn in Fig. 2–8(d).

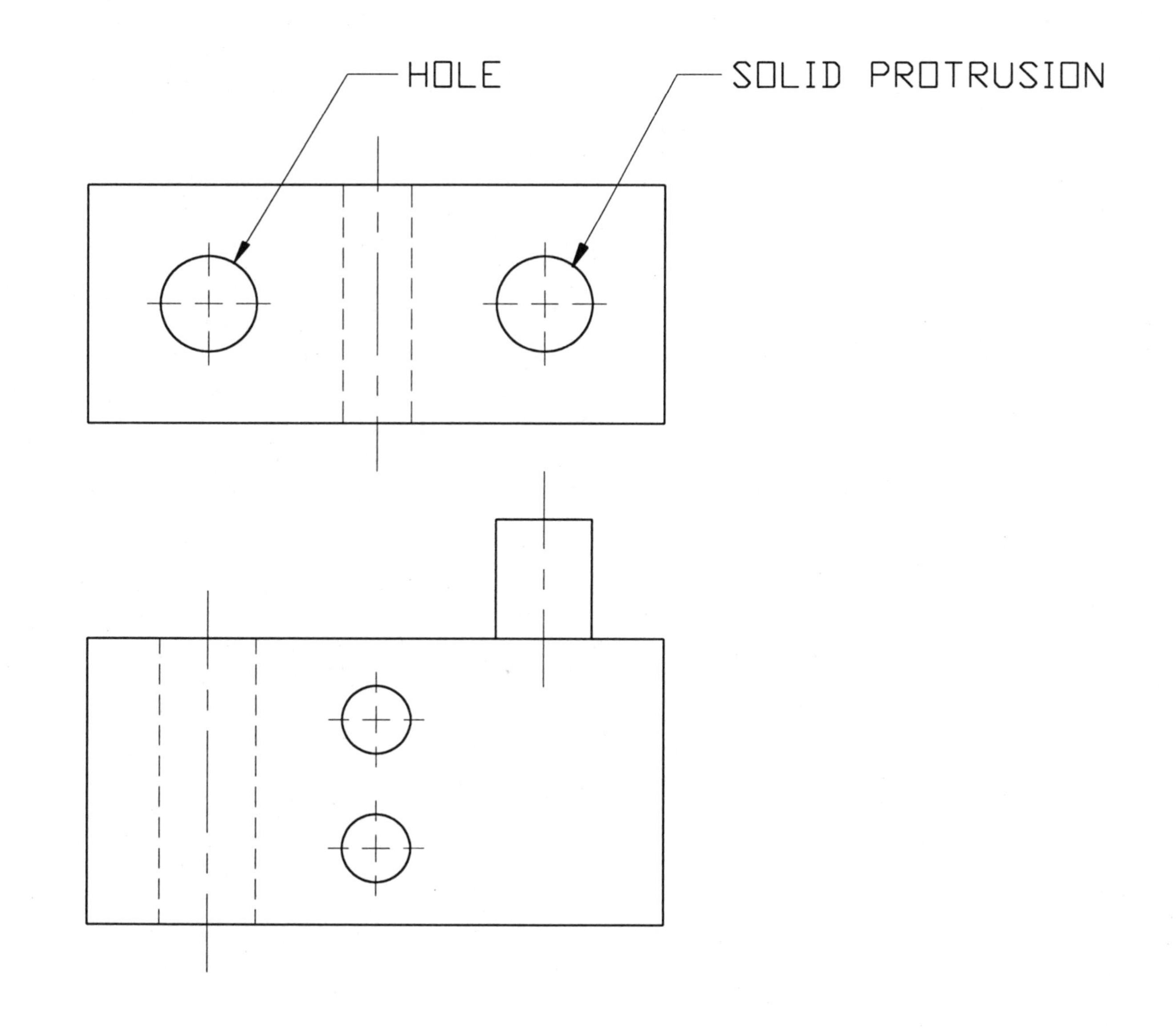

DRAWN BY ______________________ DATE ______________

TWO-VIEW DRAWINGS

Up to now, we have seen orthographic drawings that use three views. Sometimes, however, three views are not necessary to completely describe an object. Figures 2–8(a) and (b) are examples of two-view drawings. Figure 2–8(a) illustrates a hollow object, such as a washer. Sketch the side view. Do you see why it is not needed? Now put a large X through it to remind you that it is not needed.

The drawing in Fig. 2–8(b) shows a solid round bar, or shaft. Notice that the center line is thin and the outline is thick. We shall see on the next page that only one view is required when dimensions are added to this drawing.

Figures 2–8(c) and (d) are isometric views of the objects drawn orthographically in Fig. 2–6 and Exercise 2–2, respectively. They are drawn here to help you visualize the objects.

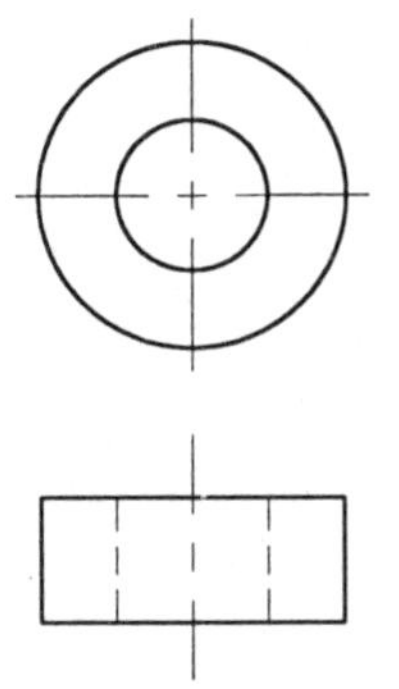

FIGURE 2–8(A)

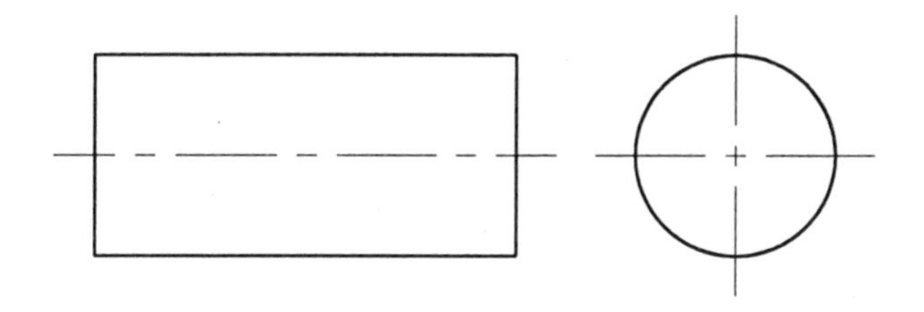

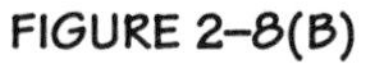

FIGURE 2–8(B)

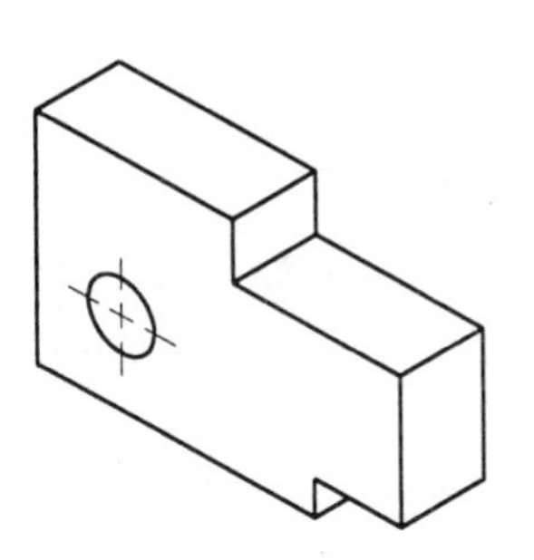

FIGURE 2–8(C)

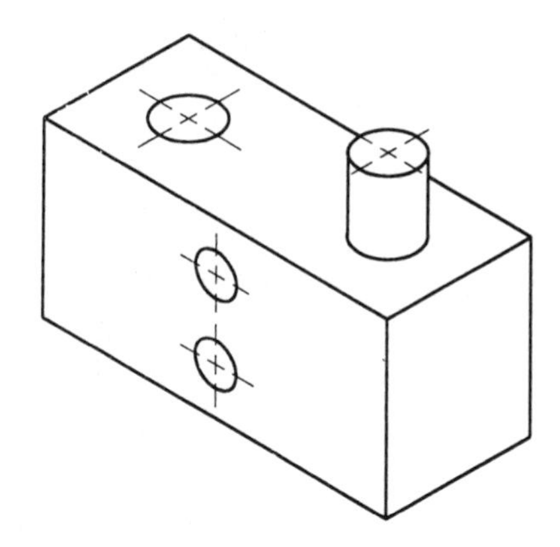

FIGURE 2–8(D)

SINGLE-VIEW DRAWINGS

Figures 2–9(a) and (b) show examples of one-view drawings.

In Fig. 2–9(a), the thickness is given in a note. Usually this type of view would be used for a very thin part, such as a gasket.

Only one view of the shaft is needed in Fig. 2–9(b) because the diameter symbol (Ø) tells you that the object is round.

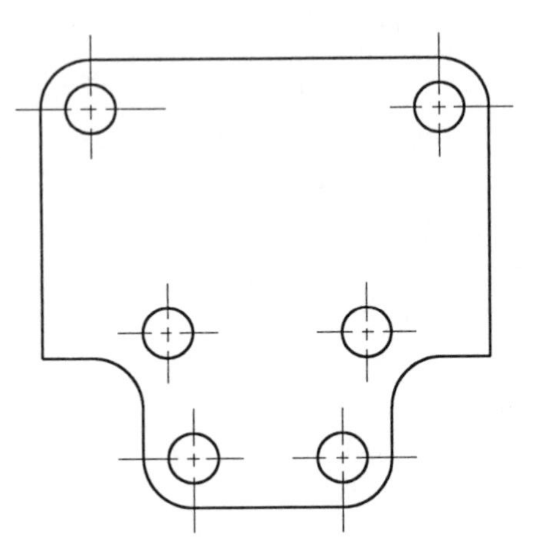

MATERIAL: .125 THICK ALUMINUM

FIGURE 2–9(A)

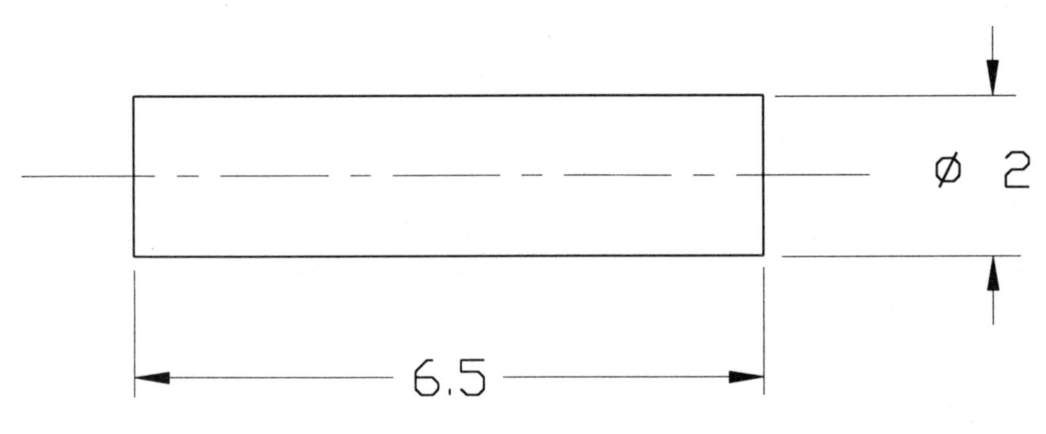

FIGURE 2–9(B)

ARRANGEMENT OF VIEWS

You should plan your view arrangement carefully. Remember that the front view usually has the *characteristic shape* of the object. Figure 2–10 illustrates another factor in placing views: conserving space on the drawing sheet. Figure 2–10(b) is a better arrangement of views than part (a) because the same three views fit into a smaller area.

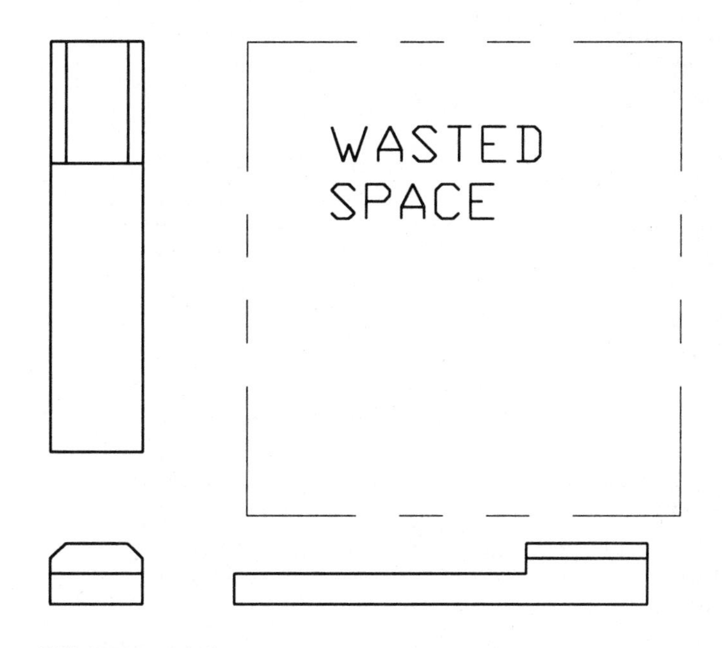

FIGURE 2–10(A)

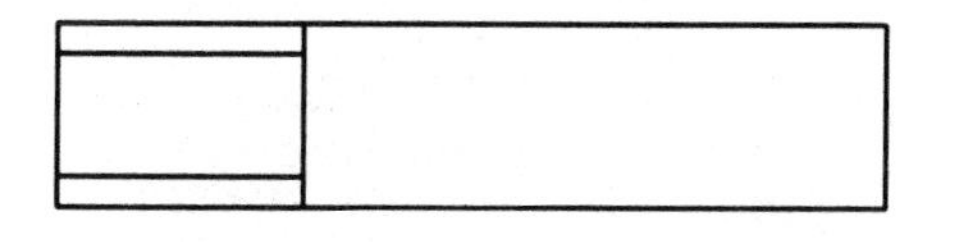

FIGURE 2–10(B)

REVIEW

We have seen examples of two- and three-view orthographic drawings. We have also seen that it is possible to draw a one-view orthographic drawing of a thin part, such as a gasket, by specifying the thickness in a note. Usually, two or three views are needed for most simple parts. Use only the *minimum number* of views to *completely describe* the part so as not to waste drawing time.

Do we need three views of the part drawn in Fig. 2–1? No, because the part can be described completely either by the front *and* side views or by the front *and* top views.

Do we need three views of the part shown in Exercise 2–2? Again, the answer is no. The top and front views are a good choice. The side view is not required.

If you are having difficulty with orthographic projection, think of the object as suspended in a glass box. (A simulated glass box is shown in Fig. 2–12.) Imagine that you are looking through the top of the box, and draw on the glass what you see. This would be the top view. Then look through the front of the box and draw what you see. This is the front view. Similarly, we can draw the side view. Note also that there is a back, left side, and bottom on the glass box. We could draw these views, but they usually would be a repeat of the top, front, or right side views.

Compare the drawing in Fig. 2–1(b) with the one in Fig. 2–5(a). Note that the top and side views correspond in each drawing. The curves in Fig. 2–5(a) are represented only in the front view.

Note that a box has six sides: *top, front, right side, left side, bottom,* and *back.* Normally three views are drawn of an object, but we have seen examples of two- and one-view drawings. We could also draw four, five, or six views of an object if it would help to describe the object. Actually, we can draw many more views when necessary, but we will deal with these *auxiliary* and *section* views in later chapters of this book. Figure 2–11 on the next page shows a six-view drawing of the object shown in Exercise 2–1. Remember that this representation is for teaching purposes only and that two views are all we really need.

THE SIX PRINCIPAL ORTHOGRAPHIC VIEWS

Figure 2–11 illustrates the six principal orthographic views of the object that you first saw in Exercise 2–1. It is assumed that the part is made from one piece, not from two pieces joined together. For this reason we have a hidden line in the *bottom view* rather than a solid line. Note that the hidden line in the bottom view is used to represent the corner A, where the big square meets the small square as seen in the *front view.* The corresponding line in the top view is solid.

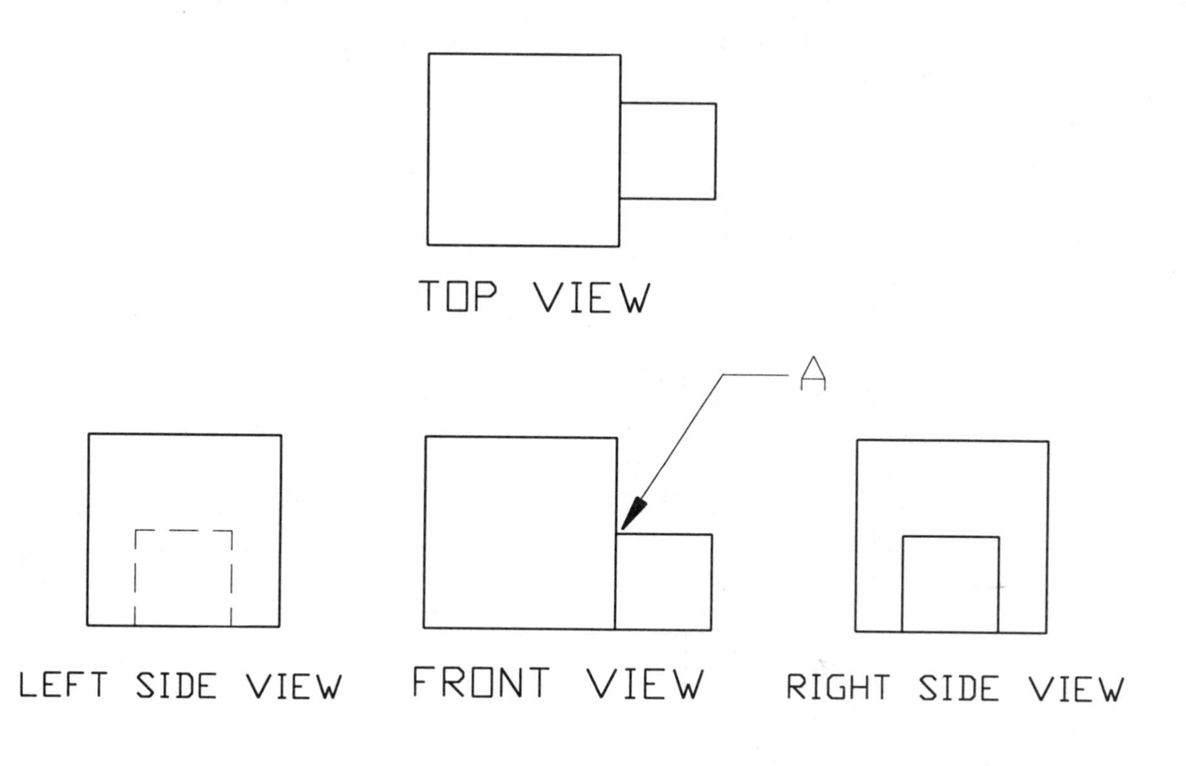

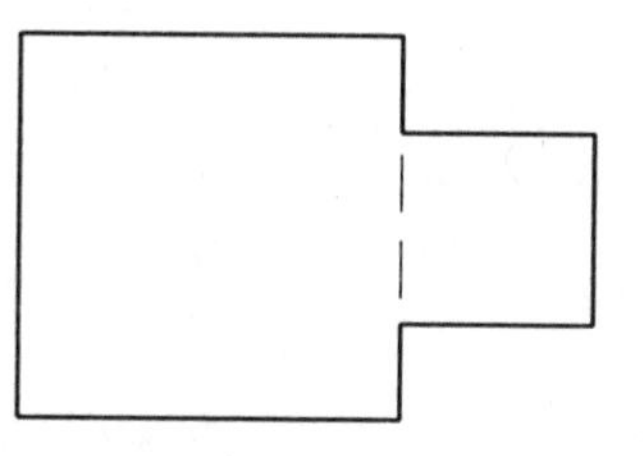

BOTTOM VIEW

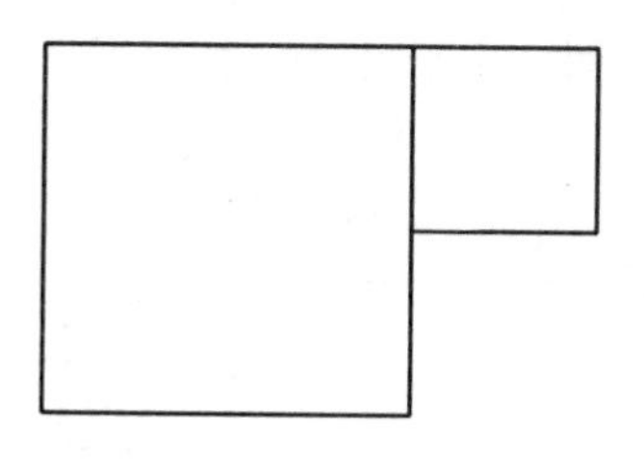

BACK VIEW

FIGURE 2–11

ACTIVITY: To help you visualize the object in Exercise 2–1, cut out the cross in Figure 2–12 on the thick lines and then fold on the dashed lines. This is how the object would look in a glass box.

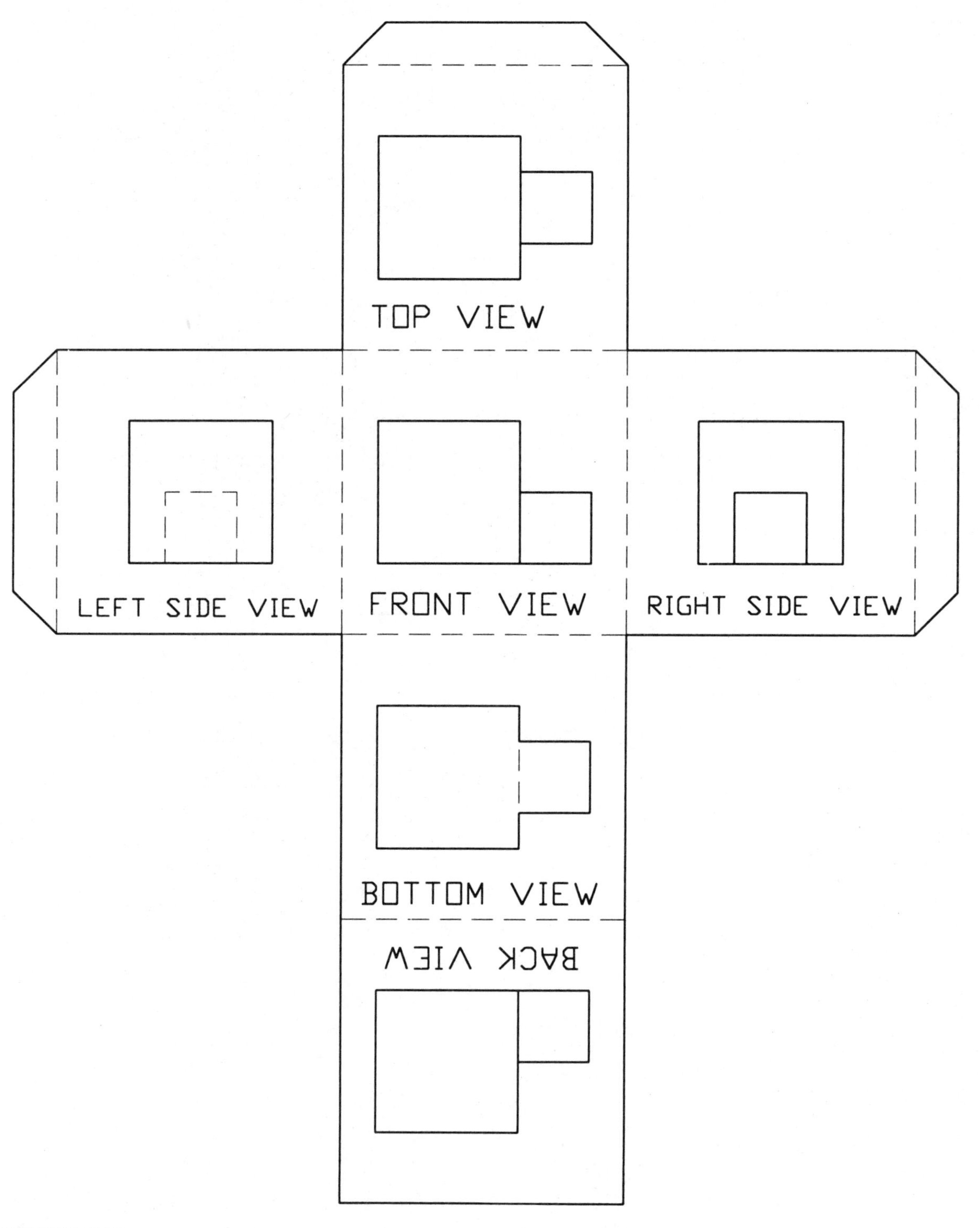

FIGURE 2–12

EXERCISE 2–3

Refer to the drawing below to answer the following questions:

1. What letters do you see on the top view? ______________________
2. What letters do you see on the right side view? ______________________
3. What letters do you see on the front view? ______________________

In the space provided, sketch the three orthographic views (top, front, right side) of the object. It is usually easier to start with the front view. Then use construction lines to line up the top and right side views. Letters have been placed on the isometric views to help you. If you are having difficulty, try putting the corresponding letters on the surfaces of the orthographic views. Use them *only* to help solve this problem. The letters should *not* appear on a good industrial drawing.

Think of this object as an interesting office building. From an airplane, we would see the roofs, surfaces B and D. From the front, we would see surfaces A and F. From the side street, we would see surfaces C, E, and G.

NOTE: *Construction lines* are drawn from the front view to help you to draw the top and side views. Some beginners draw construction lines from the isometric view as an aid to drawing the orthographic views. *This is wrong. Never* draw construction lines from an isometric view to draw the orthographic.

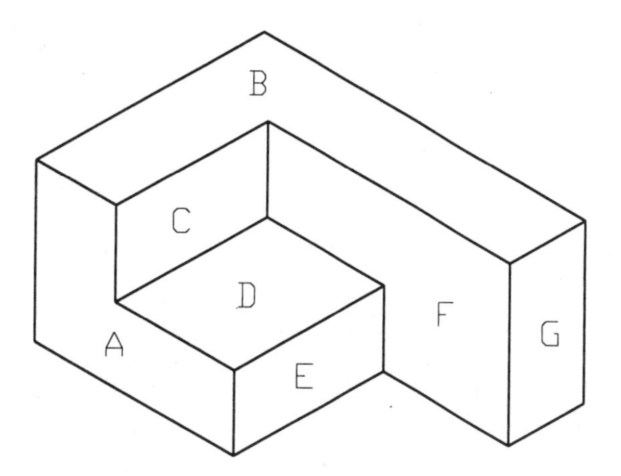

DRAWN BY ____ DATE ______________

EXERCISE 2–4

On blank paper, sketch three orthographic views (top, front, right side) of each of the objects shown below.

HINT: Start with the front view.

For part (c), choose your own front view so that there are no hidden lines in the front view and only one hidden line in the right side view.

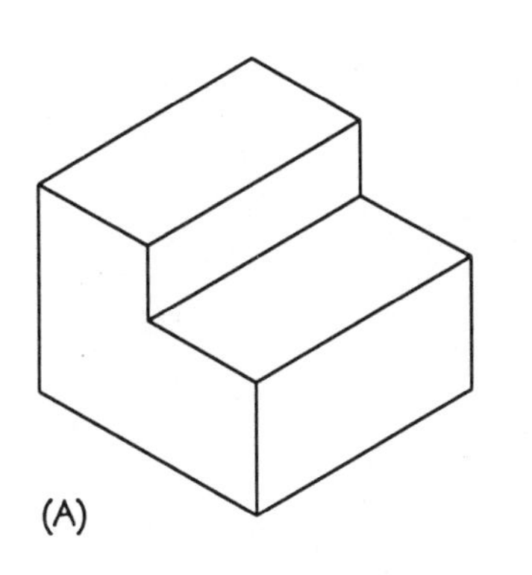

(A)

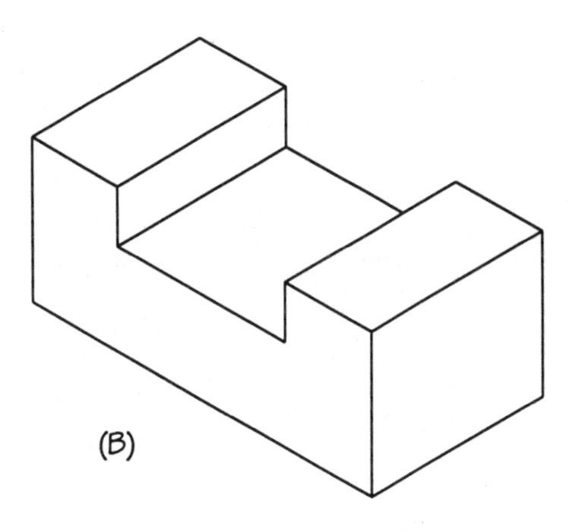

(B)

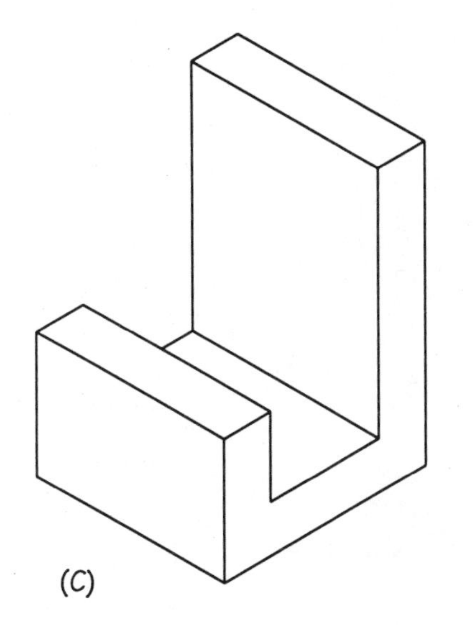

(C)

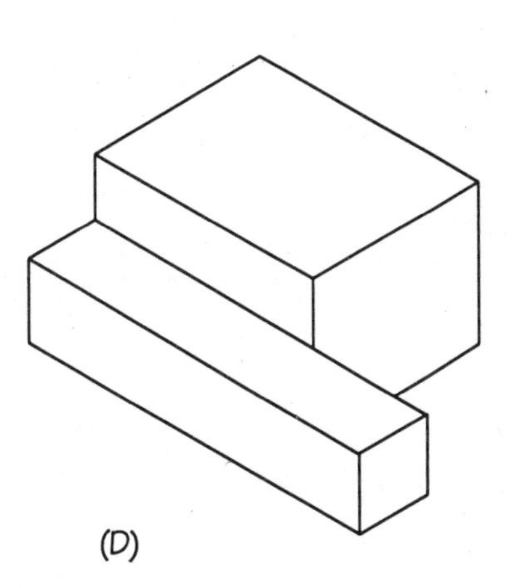

(D)

EXERCISE 2–5

In the space provided, sketch three orthographic views (top, front, right side) of the object shown below. The letters on the surfaces serve only to help you. It is usually easier to start with the front view. Use construction lines to line up the views.

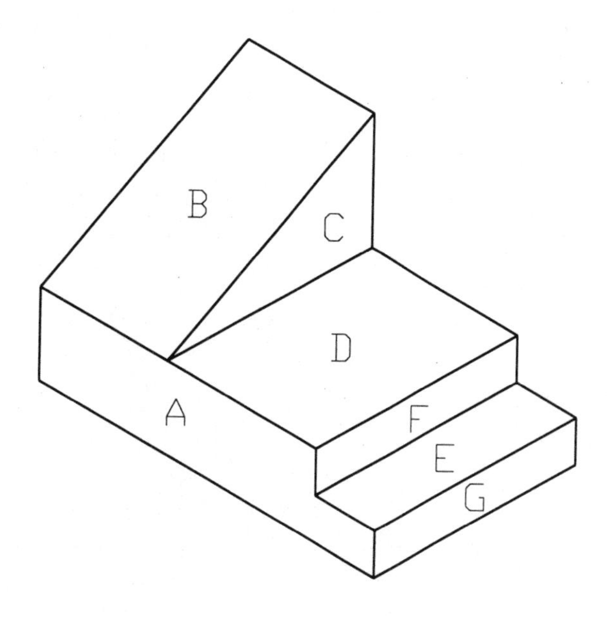

DRAWN BY DATE

EXERCISE 2–6

On blank paper, sketch orthographic views (top, front, right side) of each of the figures below.

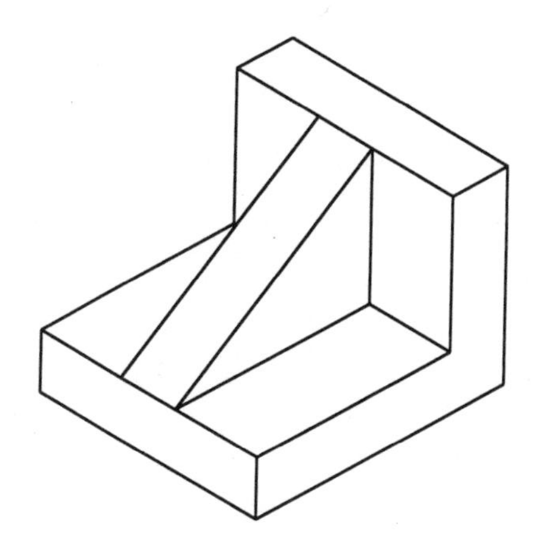

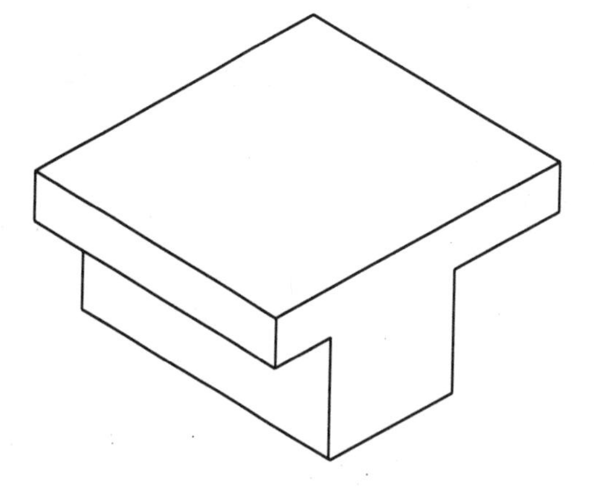

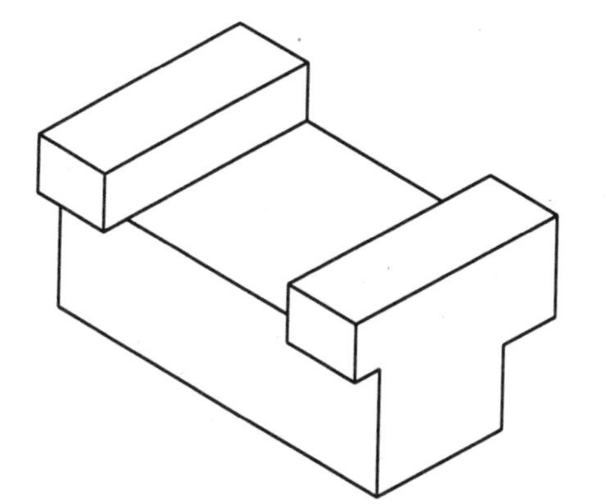

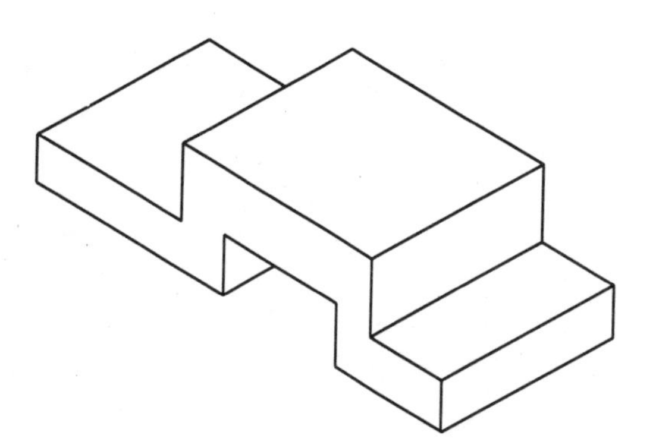

DRAWN BY ______________________ DATE ______________

EXERCISE 2–7

In the space provided, sketch three orthographic views (top, front, right side) of the object shown below. The letters on the surfaces serve only to help you. It is usually easier to start with the front view. Use construction lines to line up the views.

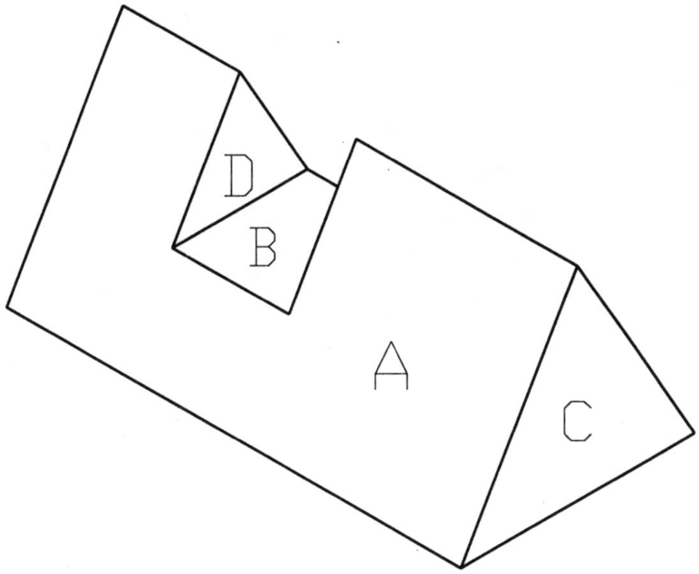

D

B

A

C

DRAWN BY DATE

EXERCISE 2–8

In the space provided, sketch three orthographic views (top, front, right side) of the object shown below. The letters on the surfaces serve only to help you. It is usually easier to start with the front view. Use construction lines to line up the views. If you are having difficulty with this problem, take a piece of cheese and take a cut out of it so that it looks like the object below. You will then have a three-dimensional model to help you solve your problem.

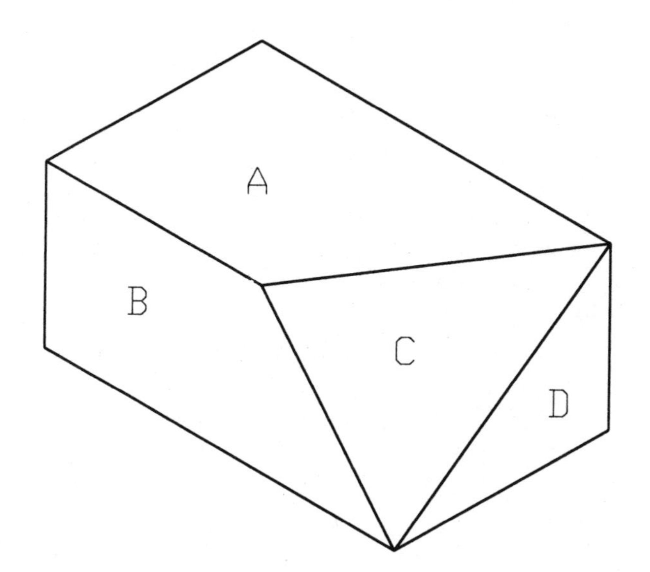

DRAWN BY DATE

EXERCISE 2–9

On blank paper, sketch orthographic views (top, front, right side) of each of the figures shown below.

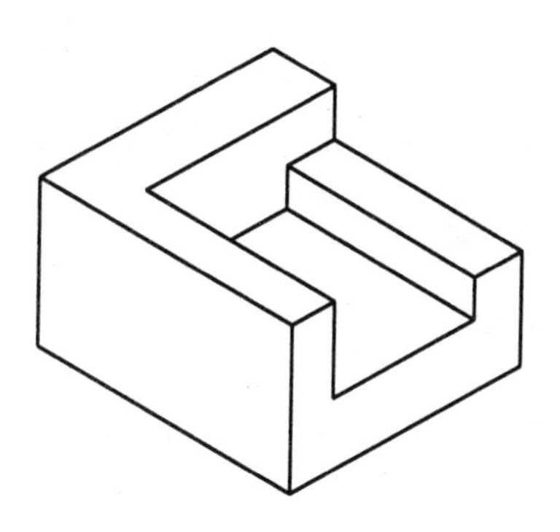

(A)

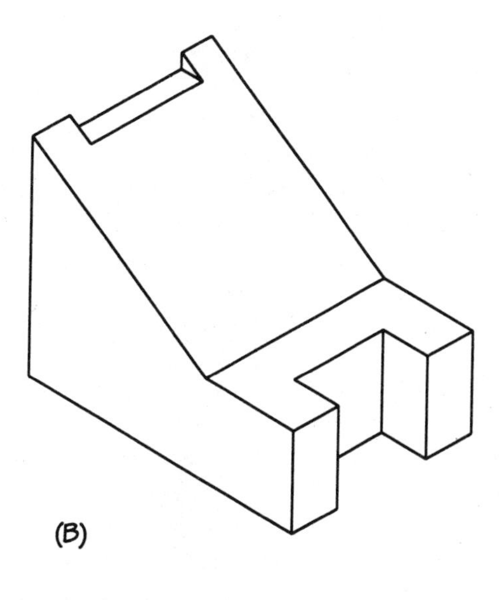

(B)

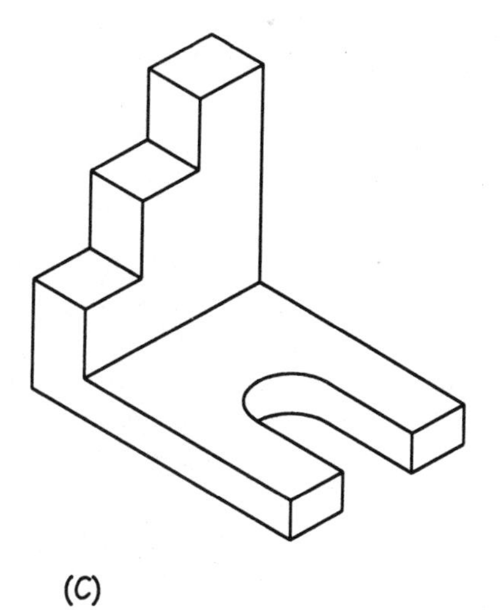

(C)

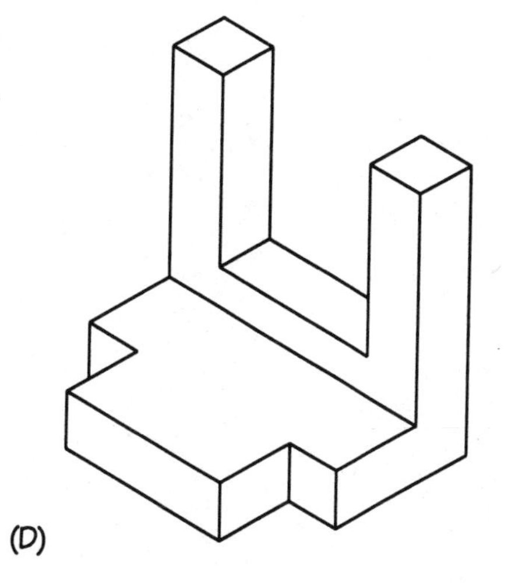

(D)

DRAWN BY ______________________ DATE ____________

SCALES

3

OBJECTIVE

When you have finished this chapter, you should be able to use the architect's scale, the metric scale, and the engineer's scale to make accurate measurements.

EQUIPMENT

You should have your technical pencils as well as your architect's, metric, and engineer's scales.

INTRODUCTION

We all have used a ruler at one time or another to measure, say, the size of a table, a room, or our height. In drafting, several special "rulers" (scales) are used; this chapter will discuss three specific ones.

Suppose we want to draw a large object such as the floor plan of a room, or the engine of a car. We would need an unreasonably large sheet of paper to draw it the same size as the actual part, obviously an impossible task. We solve this problem by drawing the object to a *reduced scale*, using the architect's scale, the engineer's scale, or the metric scale. These three scales are discussed in this chapter.

Because the plastic scale has engraved markings, it feels rough, like a saw, if you run your finger along the edge of the scale. For this reason, you should never use the scale as a straightedge to draw lines. The rough edge will wear down the lead in your pencil, dirty the scale and your fingers, and consequently dirty the drawing. Use the scale only for measuring distances, and draw lines only with the triangles or parallel straightedge.

From now on, when referring to these kinds of measuring instruments, we will use the term *scale* rather than ruler.

Figure 3–1 shows a sample of each of the scales described in this chapter.

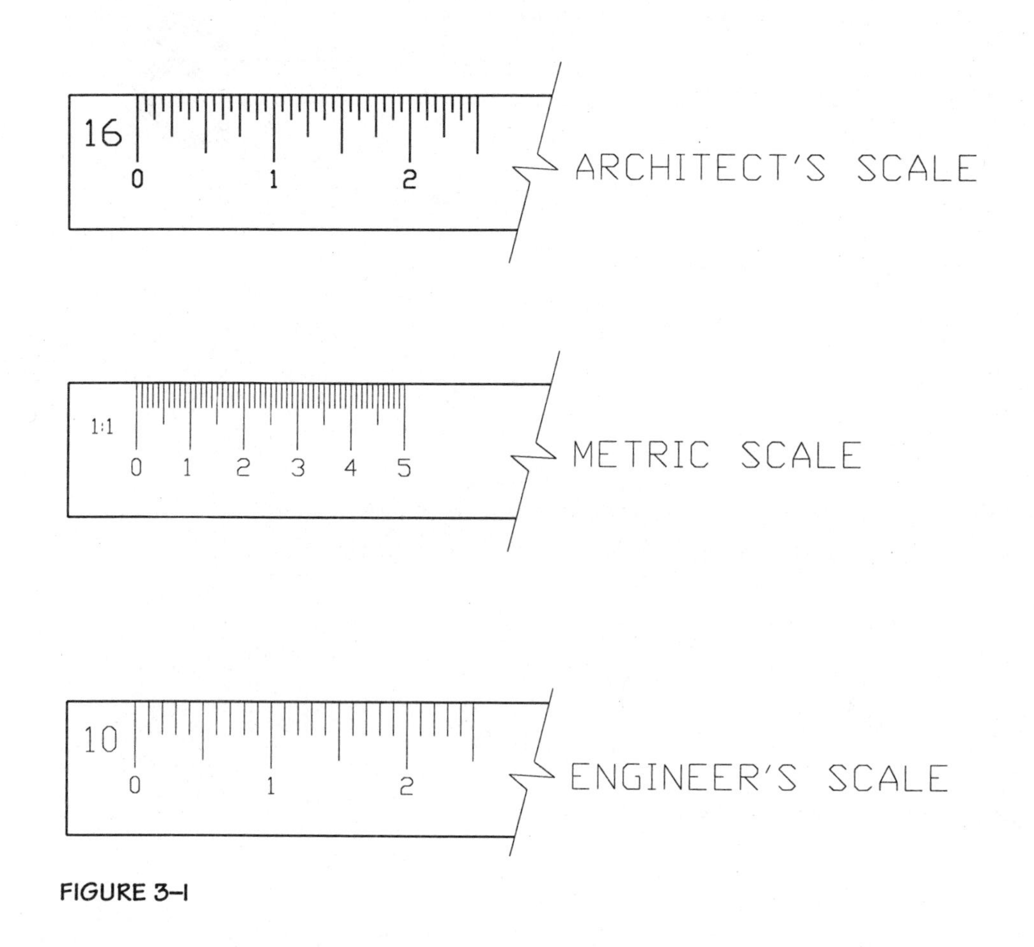

FIGURE 3–1

THE ARCHITECT'S SCALE

Study your architect's scale and notice that it has 11 different scales. They are marked on the various sides as follows: 3/8, 3/4, 1 1/2 (read "one and a half"), 3, 1/8, 1/4, 1/2, 1, 3/32, 3/16, 16. An example of how to specify a particular scale is 3/8" = 1'-0". This is read as "3/8 of an inch equals one foot, zero inches" and means that 3/8" on the drawing represents 12" (one foot) on the actual part.

The 11 scales are as follows:

3/8" = 1'-0" (3/8 inch represents 12 inches)
3/4" = 1'-0" (3/4 inch represents 12 inches)
1 1/2" = 1'-0" (1 1/2 inches represents 12 inches)
3" = 1'-0" (3 inches represents 12 inches)
1/8" = 1'-0" (1/8 inch represents 12 inches)
1/4" = 1'-0" (1/4 inch represents 12 inches)
1/2" = 1'-0" (1/2 inch represents 12 inches)
1" = 1'-0" (1 inch represents 12 inches)
3/32" = 1'-0" (3/32 inch represents 12 inches)
3/16" = 1'-0" (3/16 inch represents 12 inches)
12" = 12", also called the *full scale.* The scale is labeled with 16 to indicate that each inch is divided into 16 equal parts, each equal to 1/16". It may be written as 1'-0" = 1'-0", 1"=1", 1:1, or *full scale.*

Full scale means that the drawing is drawn the same size as the actual part. Try to use full scale whenever possible when making a drawing. This way your drawing will be the same size as the actual part.

On the 1" = 1'-0" scale (1" = 12"), each full unit (one inch) represents 12", or one foot. You would use two 1" units, or 2 inches, to represent 2 feet. The actual drawing would be 1/12 of the real size (or 1:12). On the 1" = 1'-0" scale, each inch representing 1 foot is usually divided into 24 equal parts. That is, the smallest unit on the 1" = 1'-0" scale represents 1/2". Depending on the make of the scale, the smallest unit could represent 1/4" (each inch having 48 divisions), since not all manufacturers divide the scale into the same number of parts. The following table gives similar information for all 11 scales.

SCALE	RATIO OF SIZE OF DRAWING TO ACTUAL OBJECT	SMALLEST UNIT ON SCALE*
1" = 1'-0"	1:12	1/2" (or 1/4")
1/2" = 1'-0"	1:24	1/2"
1/8" = 1'-0"	1:96	2"
1/4" = 1'-0"	1:48	1"
1 1/2" = 1'-0"	1:8	1/4"
3" = 1'-0"	1:4	1/8"
3/8" = 1'-0"	1:32	1"
3/4" = 1'-0"	1:16	1/2"
3/32" = 1'-0"	1:128	2"
3/16" = 1'-0"	1:64	1"
12" = 1'-0"	1:1	1/16"

***** *This may vary depending on the make of the scale.*

NOTE: 1/2" = 1'-0" is *not* read as half-scale! A half-scale drawing would be drawn half the full size of the actual part. 1/2" = 1'-0" scale would be 1/24 of the actual part.

REMEMBER: 1" = 1'-0' is not the same scale as 1' = 1' (or full scale).

Use the scale only *once* for each dimension. The error that some beginners make is to pick up the scale, measure the number of full feet, then move the scale and measure the inches. To use the scale properly, in one operation measure the full feet on one side of the 0 marking and then, without moving the scale, measure the inches on the other side of the 0 marking.

On a completed drawing that includes dimensions, do not measure with your scale to determine the size of a feature—because the accuracy of your measurements will depend on the accuracy of the drawing. Instead, rely on the dimensions (numbers) that appear on the drawing.

ARCHITECT'S SCALE: FULL SCALE

Figure 3–2 illustrates the full scale of the architect's scale. Use this figure for reference only, for the process used to reproduce this book may have distorted the sizes of the scales, rendering them inaccurate as measuring instruments. Study this figure carefully.

In the inch system, one foot is divided into 12 inches. On your architect's scale, each inch on the full scale is divided into 16 units, each of which represents 1/16". Two 1/16 units represent 1/8" (2/16). Four 1/16 units represent 1/4" (4/16). Six 1/16 units represent 3/8" (6/16). Eight 1/16 units represent 1/2" (8/16), and so forth.

Notice that the architect's full scale differs from all other scales (which are reduced scales). The full scale is divided by the same number of units for its entire length, whereas reduced scales label the number of feet to the right of the zero marking and the number of inches to the left of the zero marking.

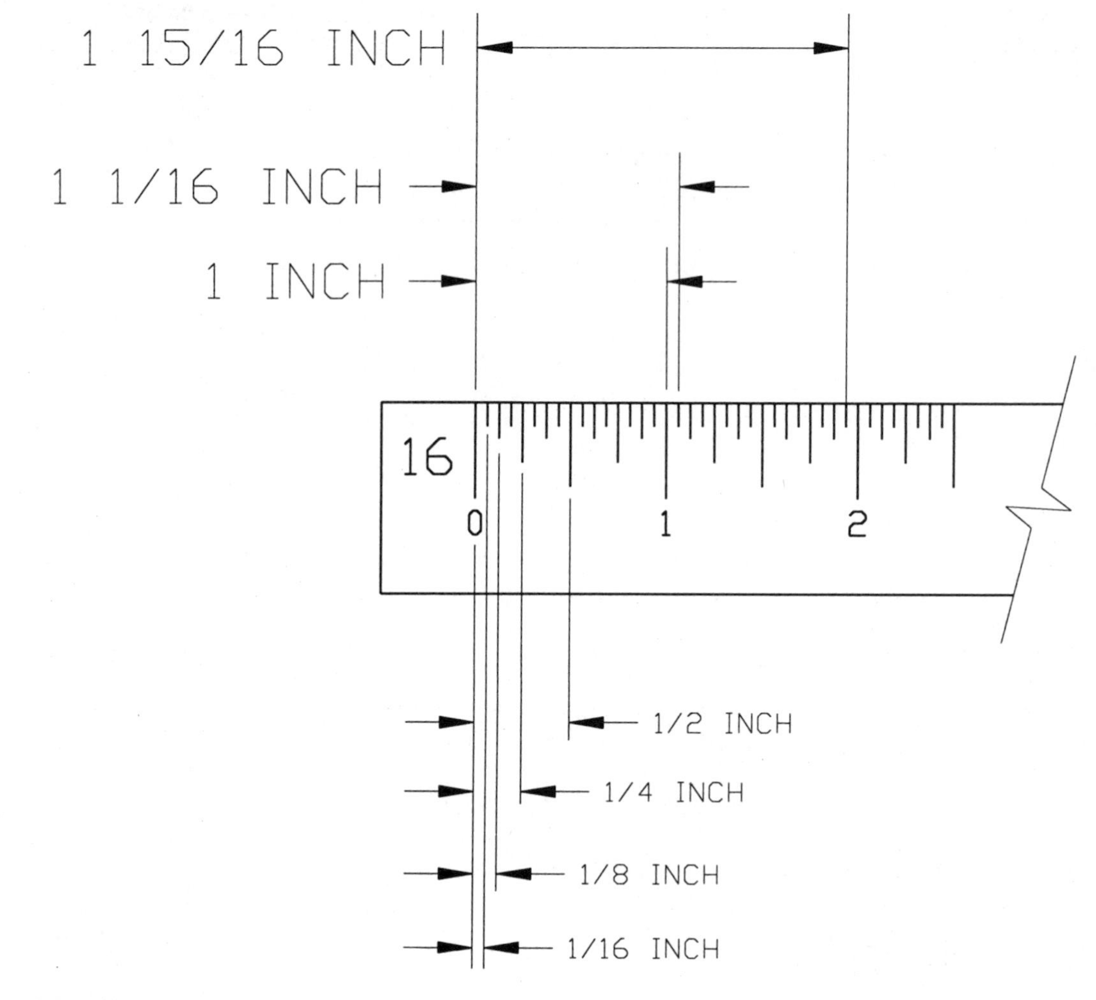

FIGURE 3–2 Architect's full scale

EXAMPLES OF HOW TO READ THE ARCHITECT'S SCALE

Use Fig. 3–3 as a study guide, but do not try to measure the examples with your scale (for the reason given on the previous page).

NOTE: On the architect's scale, all the edges (except for the full scale) represent two scales: one read from the left and one read from the right. For example, the scales marked 1 1/2 and 3 appear at opposite ends of the same edge. This could lead to confusion when reading the scale, since by looking at the 1 1/2" = 1'-0" scale in Fig. 3–3, we see the number 2 representing 1 foot. The number 2 is actually for the 3" = 1'-0" scale, which is read from the right. In general, the left and right scales on each edge have a ratio of 1 to 2; that is, one scale is double the other. The markings are used for scales read from both the left and the right, although the actual numbers will apply to only one scale. One row of numbers is placed higher than the other to help you identify which row you are reading.

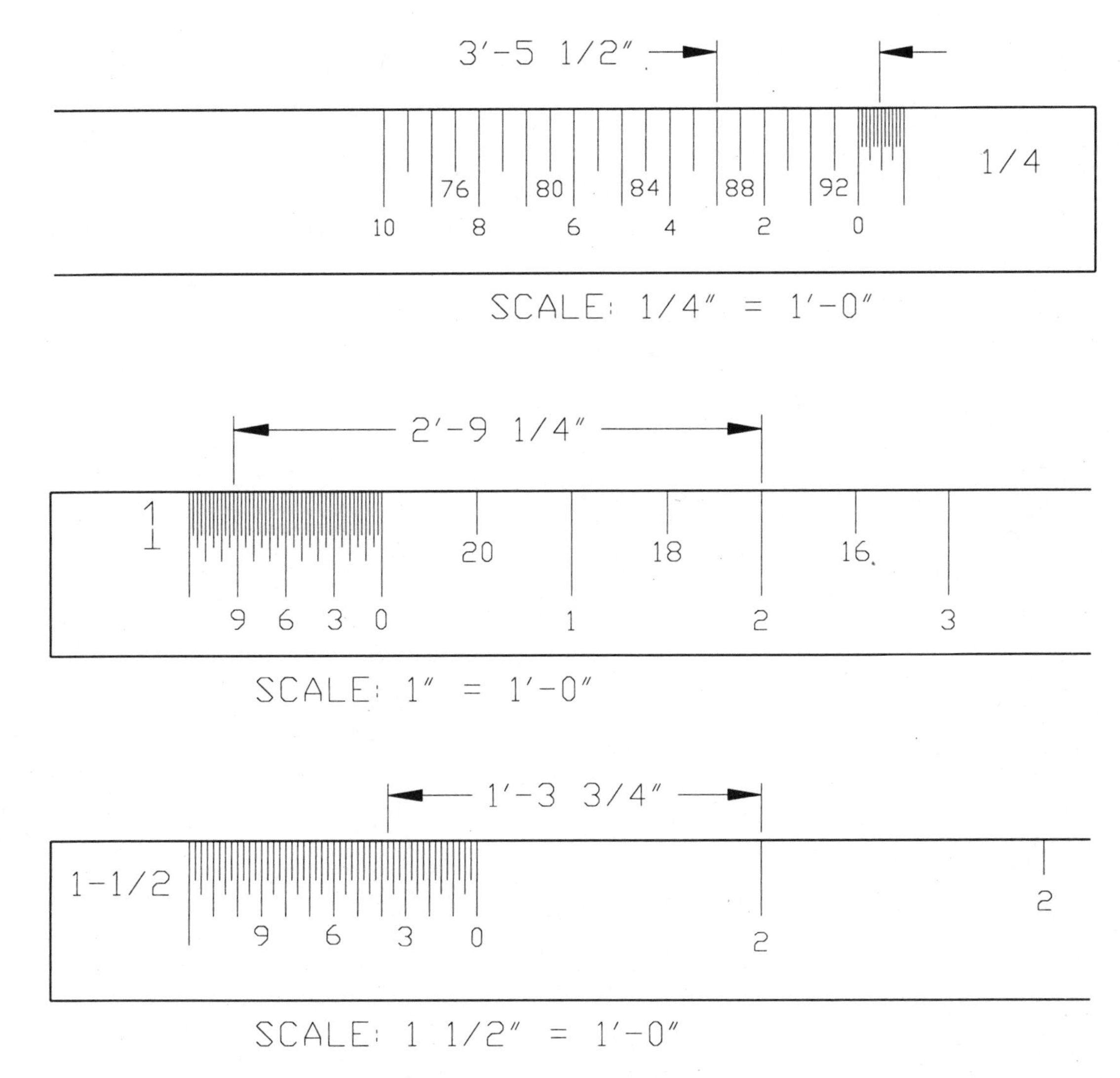

FIGURE 3–3

ACTIVITY: Measure the size of this page from top to bottom with the full scale. What do you get? Measure it again with the 1" = 1'-0" scale. Now measure the paper from top to bottom with the 1/8" = 1'-0" scale.

EXERCISE 3–1

On the lines provided, and using the scales given below, mark off the lengths indicated. The starting points are given.

(a) SCALE: 3/4" = 1'-0", 4'-6 1/2"

(b) SCALE: 3/8" = 1'-0", 10'-4 1/2"

(c) SCALE: 1-1/2"= 1'-0", 2'-3 1/4"

(d) SCALE: 3" = 1'-0", 1'-1 1/8"

(e) SCALE: 1/8" = 1'-0", 28'-4"

(f) SCALE: 1/4" = 1'-0", 11'-7 1/2"

(g) SCALE: 1" = 1'-0", 3'-2 1/4"

(h) SCALE: 1/2" = 1'-0", 5'-5"

(i) SCALE: 3/32" = 1'-0", 36'-4"

(j) SCALE: 3/16" = 1'-0", 8'-5"

(k) SCALE: 12" = 12", 3"

DRAWN BY DATE

THE METRIC SCALE

The basic metric unit used in technical drawing is the millimeter. One millimeter (mm) is one thousandth of a meter.

10 mm = 1 centimeter (cm)
100 mm = 1 decimeter (dm)
1000 mm = 1 meter (m)
1/1000 mm = 1 micrometer (μm)

Typical metric scales include the following examples:

1:1 (full scale, 1 mm represents 1 mm)
1:2 (1 mm on the drawing represents 2 mm on the actual part)
1:5 (1 mm on the drawing represents 5 mm on the actual part)
1:10 (1 mm on the drawing represents 10 mm on the actual part)
1:20 (1 mm on the drawing represents 20 mm on the actual part)
1:50 (1 mm on the drawing represents 50 mm on the actual part)

Many other scales are used, and they are read in a similar method to those discussed.

The International Organization for Standardization (ISO) and the Canadian Standards Association (CSA) prescribe that the metric system be used. The American National Standards Institute (ANSI) states that the metric or decimal inch may be used but recommends the metric system. See page 329 in the Appendix for a discussion of these standards.

NOTE: When the dimension is less than 1 mm, a zero (0) is placed in front of the number (for example, 0.3 mm).

EXAMPLES OF HOW TO READ THE METRIC SCALE

Use this page as a study guide, but do not measure the examples in Fig. 3–4 with your scale.

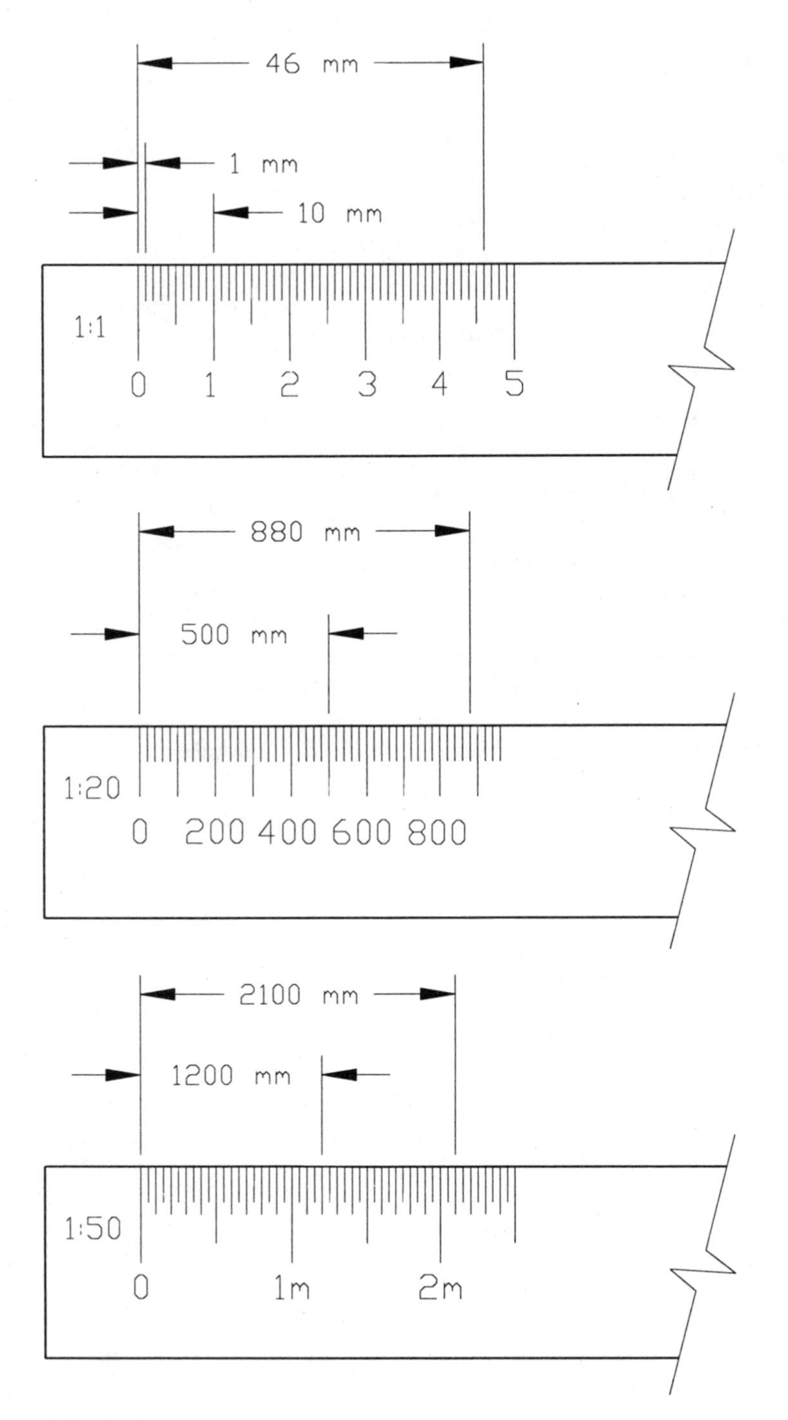

FIGURE 3–4

ACTIVITY: Using the metric scale, measure the size of this page from top to bottom with the full scale. What do you get?

EXERCISE 3–2

On the lines provided, and using the scales given below, mark off the lengths indicated. The starting points are given.

(a) 1:1 80mm

(b) 1:10 750mm

(c) 1:50 4800mm

(d) 1:20 1850mm

(e) 1:5 350mm

(f) 1:2 135mm

(g) 1:1 65mm

(h) 1:10 350mm

(i) 1:50 3750mm

(j) 1:20 1120mm

(k) 1:1 26mm

DRAWN BY DATE

THE ENGINEER'S SCALE

The engineer's scale is sometimes called the civil engineer's scale because it is often used on outdoor engineering projects such as roads and structures. It is a decimal-based scale with divisions in multiples of 10 units.

Typical scales include 10, 20, 30, 40, 50, 60.

The number on each scale indicates the number of units representing an inch. Hence, in the 10 scale, each unit is divided into 10 smaller units and each smaller unit is one-tenth (1/10) of an inch. Similarly, for the 60 scale, each inch is divided into 60 smaller units, and each of these smaller units is one-sixtieth (1/60) of an inch. The scale is written as 1 = 60 or 1:60, meaning that one inch represents 60 units. The units can be feet, inches, or other units. For example, using the 10 scale, we could say that the scale is 1" = 100', meaning that one inch on the drawing represents 100 feet on the actual structure and, since each inch is divided into 10 equal parts, each of these parts represents 10 feet on the actual structure.

Examples of how metric scales may be used:

SCALE CHOSEN	POSSIBLE USES
10	1" = 1" (full scale)
10	1" = 100' (as explained above)
20	1" = 20"
30	1" = 3'
40	1" = 400'
50	1" = 5'
60	1" = 600'

EXAMPLES OF HOW TO READ THE ENGINEER'S SCALE

Use this page as a study guide, but do not measure the examples in Fig. 3–5 with your scale.

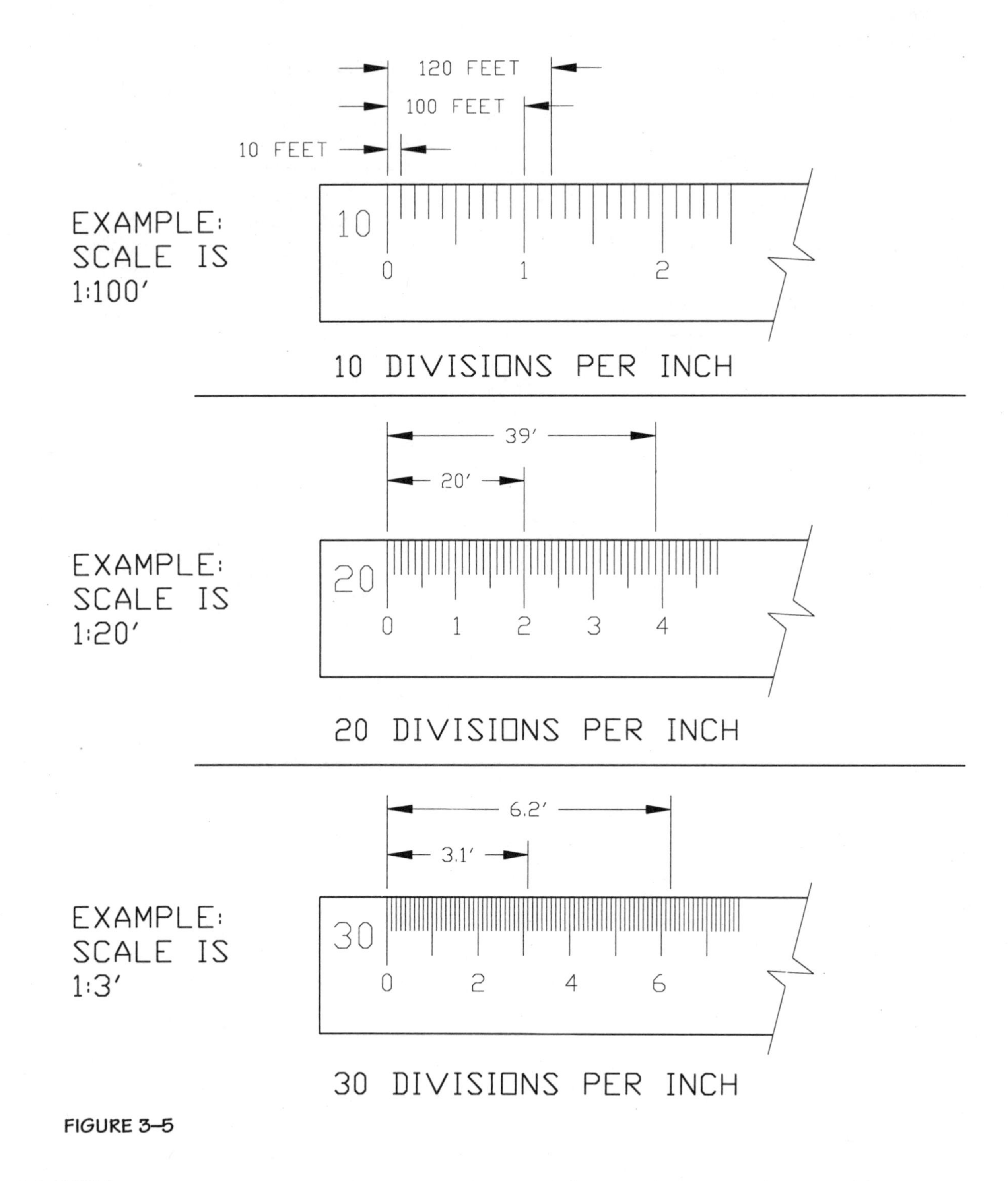

FIGURE 3–5

ACTIVITY: Using the engineer's scale, measure the size of this page from top to bottom with the full scale. What do you get?

EXERCISE 3–3

On the lines provided, and using the scales given below, mark off the lengths indicated. The starting points are given.

(a) FULL SCALE, 4.2"

(b) 1"=10'-0" , 23'-0"

(c) 1"=20'-0" , 58'-0"

(d) 1"=30'-0" , 95'-0"

(e) 1"=40'-0" , 136'-0"

(f) 1"=50'-0" , 178'-0"

(g) 1"=60'-0" , 215'-0"

(h) 1"=10'-0" , 15'-0"

(i) 1"=20'-0" , 75'-0"

(j) 1"=30'-0" , 81'-0"

(k) 1"=10'-0" , 9'-0"

DRAWN BY ____________ DATE ____________

ADDITIONAL NOTES ABOUT SCALES

a. Because the original drawing is a valuable document representing many hours of work, it is kept in the drawing office and only copies of the original are sent to the tradespeople to make the part or structure. After a drawing has been issued from the drawing office and sent to the fabrication shop or construction site, certain changes to the drawing may be required. All changes are done in the drawing office on the original drawing only. If the drawing is to be changed in any way, a "revision note" is placed on the drawing so that everyone will know which copy is the most recent and what changes have been made. Usually this is done by adding REV A (B, C, ...) after the drawing number. A revision box may be used to give details about the particulars of the revision. If a dimension was changed but the lines were not erased and redrawn to represent the revised dimension, some companies put NTS below the dimension to indicate that the feature is not to scale. Some companies underline the dimension to remind the reader that it is not to scale.

b. The decimal equivalents table found in the Appendix (page 326) is useful for converting fractions to decimals and vice versa. For example, 3 15/16 inches equates to 3.9375 inches.

GEOMETRIC CONSTRUCTIONS

4

OBJECTIVE

At the end of this chapter, you should be able to use drafting instruments to perfom various geometric constructions.

EQUIPMENT

For this and all subsequent chapters, you are required to have all of your drafting instruments.

This chapter gives detailed, step-by-step instructions on how to perform certain geometric constructions. These constructions are used on technical drawings to help draw certain shapes or to graphically solve problems.

You should be able to follow most instructions without seeking assistance. This chapter is based on the saying, "Give people a fish and they eat for a day, teach them to fish and they eat for the rest of their lives." You will learn (in case you have not already realized it) that for many new challenges you can succeed without someone watching over you every minute. This simple ability is what most employers seek: people who have the initiative and resourcefulness to work on their own.

Before studying some geometric constructions, you should practice using the protractor, one of your drafting instruments. We will begin the chapter by describing how to use this instrument, since you may wish to verify some of the constructions that you will be drawing.

HOW TO USE A PROTRACTOR

A protractor is used to measure angles in units called *degrees*. (A degree is divided into 60 *minutes*, and each minute into 60 *seconds*. We will not be using these smaller units in this book.) Refer to Fig. 4–1 as you follow these steps.

STEP 1 Place the protractor's center mark on the vertex of the angle, that is, on the corner where the two lines meet.

STEP 2 Align one side of the angle with one of the zero markings on the protractor.

STEP 3 Note which mark on the protractor is closest to the other side of the angle. Read the number of degrees off the protractor, using the row of numbers in which your chosen zero is located.

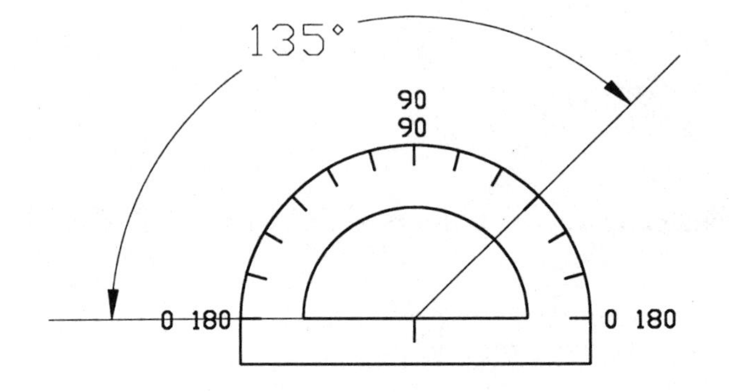

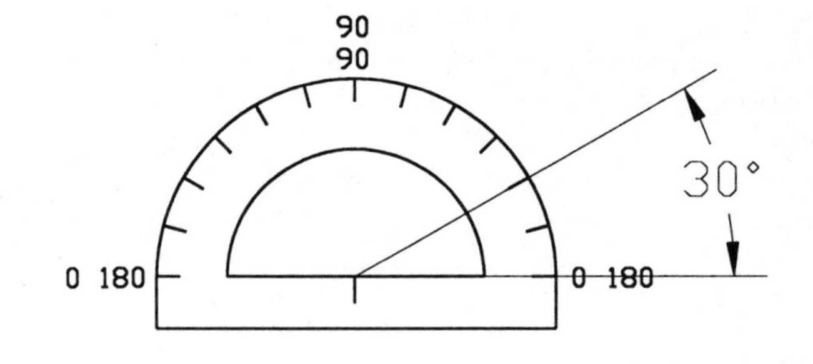

FIGURE 4–1

EXERCISE 4–1

Using your protractor, measure the first row of angles below and verify that the measurements are correct. Then measure the second and third rows and write your answers on each view. You will probably have to extend the lines representing each side of the angle.

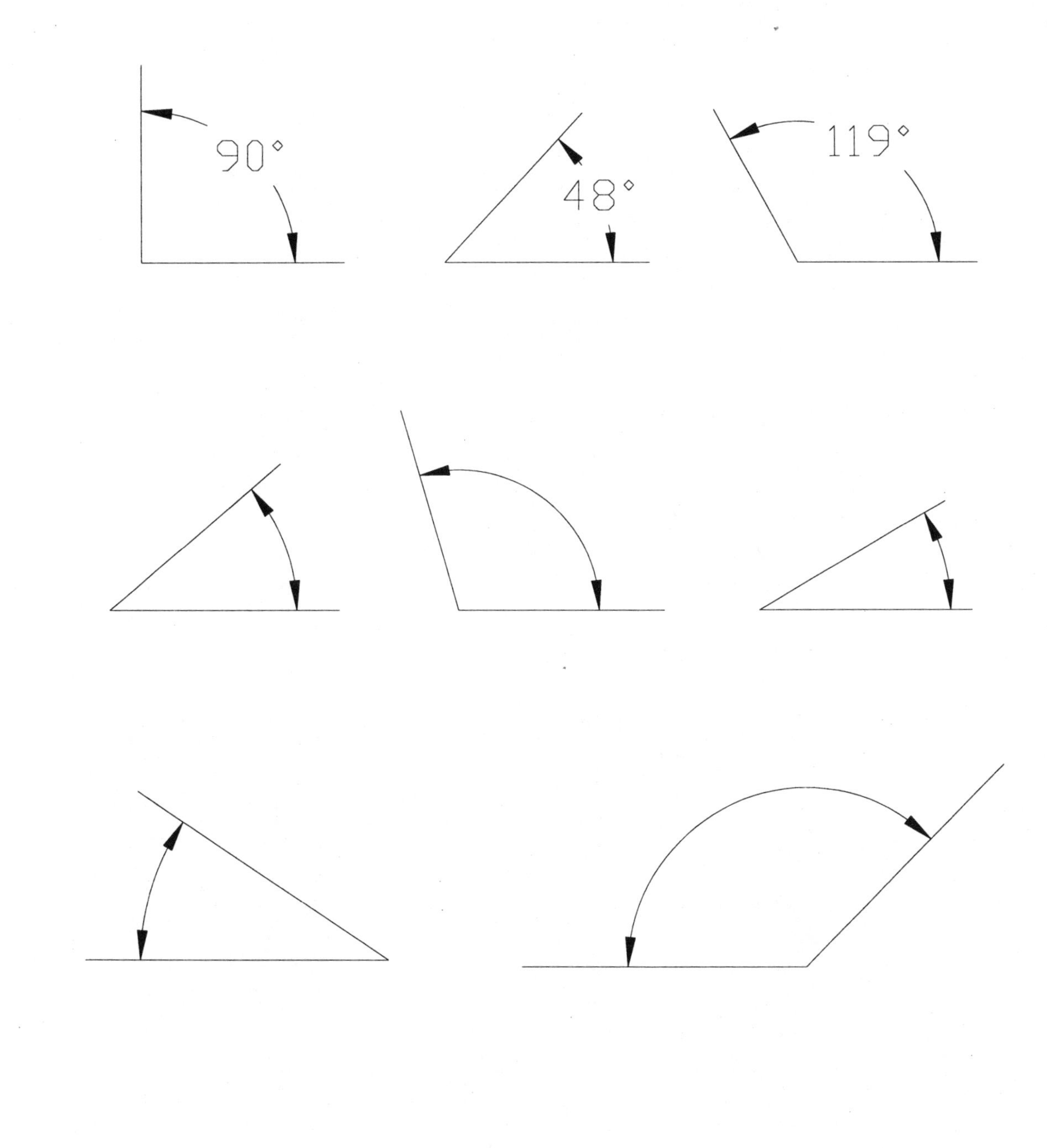

DRAWN BY DATE

CONSTRUCTION 1: To draw a line perpendicular to a given line at a certain point on the given line (Fig. 4–2)

STEP 1 Place your two triangles so that the hypotenuse of one triangle (the side opposite the right angle, or the longest side) is touching the hypotenuse of the other one.

STEP 2 Align one side of one triangle (triangle X in Fig. 4–2a) along given line AB.

STEP 3 Holding the other triangle (Y) firmly, gently slide triangle X along the hypotenuses until a side of triangle X passes through given point C on AB. See Fig. 4–2(b).

STEP 4 Using your triangle as a guide, draw a straight line DE through point C. Line DE is perpendicular to line AB.

FIGURE 4–2(A)

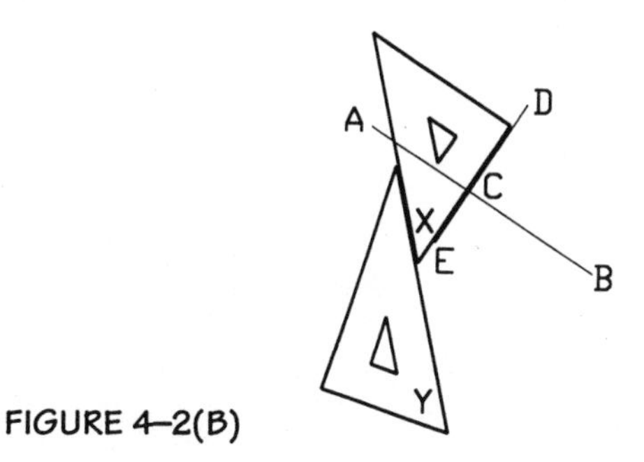

FIGURE 4–2(B)

CONSTRUCTION 2: To draw a line parallel to a given line at a given distance from it (Fig. 4–3)

NOTE: Parallel lines never intersect, no matter how far they are extended.

STEP 1 Choose a convenient point C on given line AB. With center C and radius set to given distance R, draw an arc on the side of line AB where you want the parallel line to be. See Fig. 4–3(a).

STEP 2 Using Construction 1 (making sure that the hypotenuses of your triangles are together), lightly draw a line perpendicular to line AB through point C and cutting the arc at D. See Fig. 4–3(b).

STEP 3 Still holding triangle Y firmly, gently slide triangle X so that its other side passes through D. Draw a line EF through D. Line EF is parallel to AB and is distance R away from it.

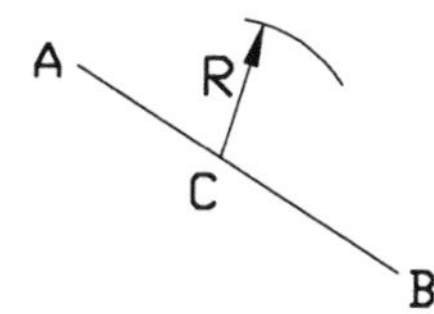

FIGURE 4–3(A)

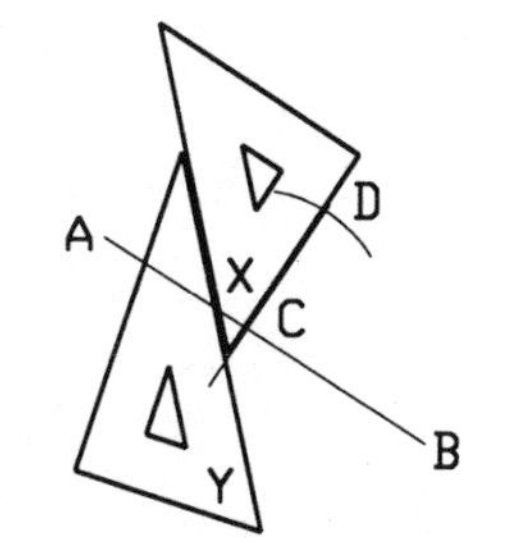

FIGURE 4–3(B)

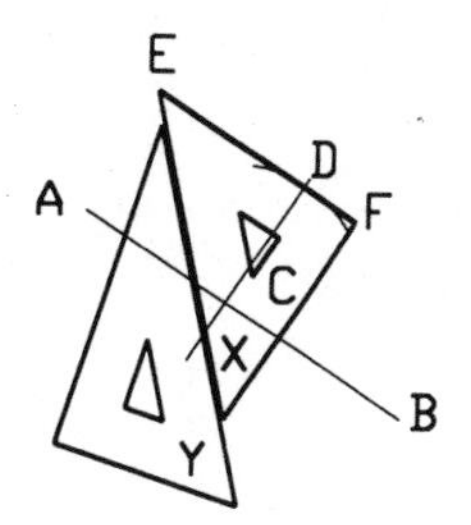

FIGURE 4–3(C)

CONSTRUCTION 3: To divide a straight line into a given number of equal parts (Fig. 4–4)

STEP 1 Given line AB and the number of equal divisions required (eight in our example), draw a convenient acute angle ABC. (An acute angle is smaller than a right angle. See the Appendix, page 325, for more on angles.)

STEP 2 With any small, convenient radius, cut eight successive arcs on line BC, with your first center at B, and each successive center at the intersection of the previous arc. Call the eighth intersection point D.

STEP 3 Join DA. Draw lines parallel to DA from each of the points of intersection along BC to intersect line AB. The intersection points on AB have divided the line into eight equal parts.

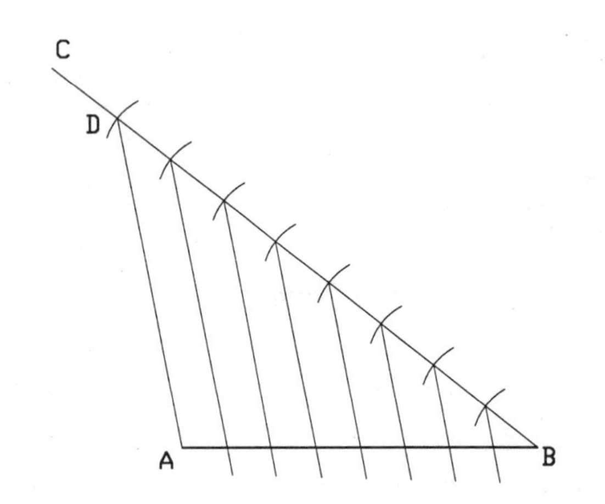

FIGURE 4–4

CONSTRUCTION 4: To draw a perpendicular bisector of a straight line (Fig. 4–5)

NOTE: Line XY will bisect (cut exactly in half) line AB and will be perpendicular (at 90°) to it.

STEP 1 Given line AB, open your compass to a radius greater than half the length of AB.

STEP 2 Put the point of the compass at A and draw an arc above and below the line AB.

STEP 3 Put the compass point at B and, with the same radius, draw arcs that intersect the arcs made in step 2, calling the points of intersection X and Y.

STEP 4 Join X and Y. Line XY divides AB into two equal parts, and XY is perpendicular to AB.

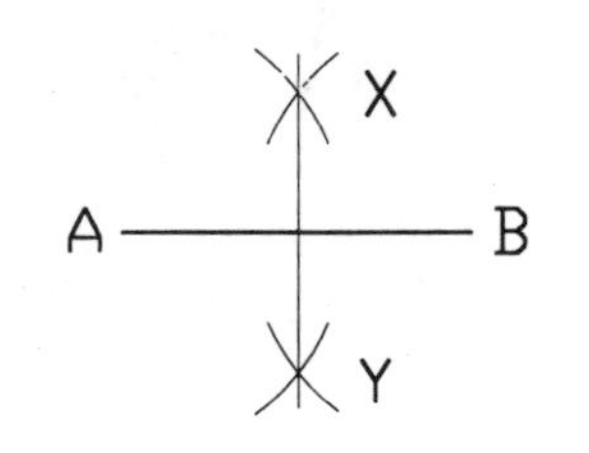

FIGURE 4–5

CONSTRUCTION 5: To bisect an arc (Fig. 4–6)

STEP 1 Given an arc AB, open your compass to a radius greater than half the distance AB.

STEP 2 With your compass point at A, the endpoint of the arc, and then at B, the other endpoint, draw arcs that intersect at X and Y above and below the arc.

STEP 3 Join the points where these arcs intersect to draw line XY. Line XY bisects the arc.

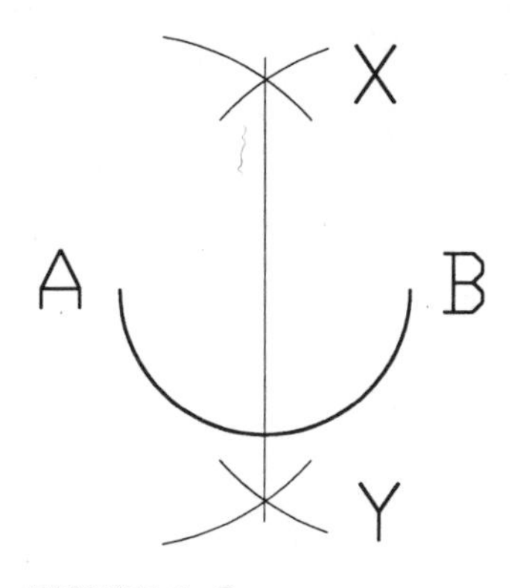

FIGURE 4–6

CONSTRUCTION 6: To bisect an angle (Fig. 4–7)

STEP 1 Given an angle ABC, draw an arc of suitable radius with compass point at B to cut the two sides of the angle at X and Y.

STEP 2 With compass point at X and Y, draw two arcs of equal radius that intersect at Z.

STEP 3 Join Z and B. Line BZ bisects the angle.

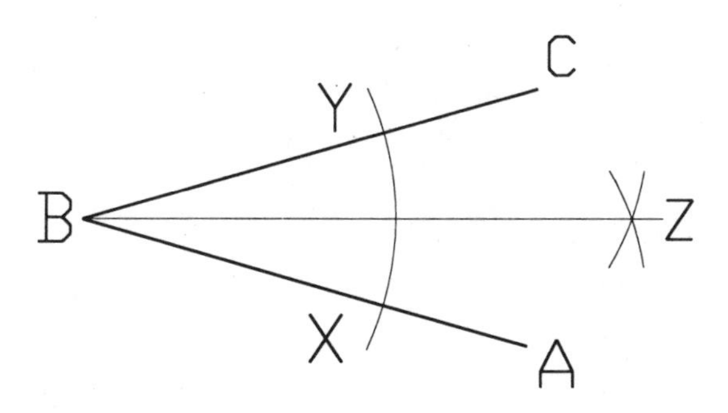

FIGURE 4–7

TANGENTS

A *tangent* is a line that meets another line at a single point but does not intersect it, even if the tangent is extended to infinity. Figure 4–8 shows several examples of tangents. Some lines are thick and others are thin to help you differentiate them; either line may be called tangent to the other one. You can see from these examples that a tangent may be a straight line meeting a curved line, a curved line meeting a straight line, or a curved line meeting another curved line. In Fig. 4–9(a), the straight line is not tangent to the circle, because if it were extended as shown in part (b), it would intersect the circle in two places. (A tangent may also be defined as a straight or curved surface meeting another straight or curved surface.)

QUESTION: If you were given two circles that were not touching each other, how many straight line tangents could you draw to them? How many curved tangents could be drawn? (The answers to these questions are found at the end of the chapter.)

FIGURE 4–8

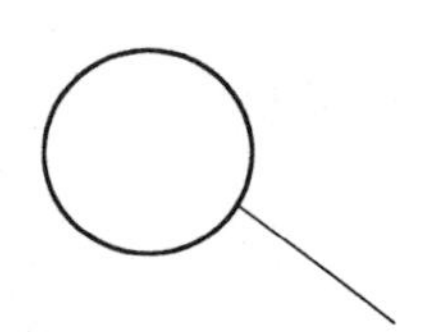

FIGURE 4–9(A)

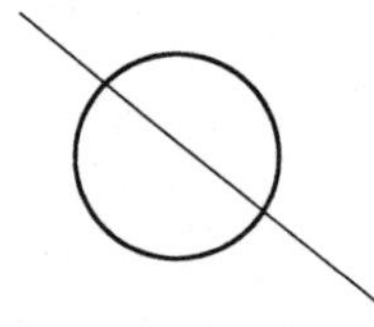

FIGURE 4–9(B)

CONSTRUCTION 7: To draw a line tangent to a circle at a given point on the circumference (Fig. 4–10)

NOTE: For the purposes of this chapter, the *circumference* of a circle is the line that bounds it. (Circumference may also be defined as the length of that line.)

STEP 1 Join center A to given point B on the circumference.

STEP 2 Through point B, construct line CD perpendicular to AB (Construction 1). CD is tangent to the circle.

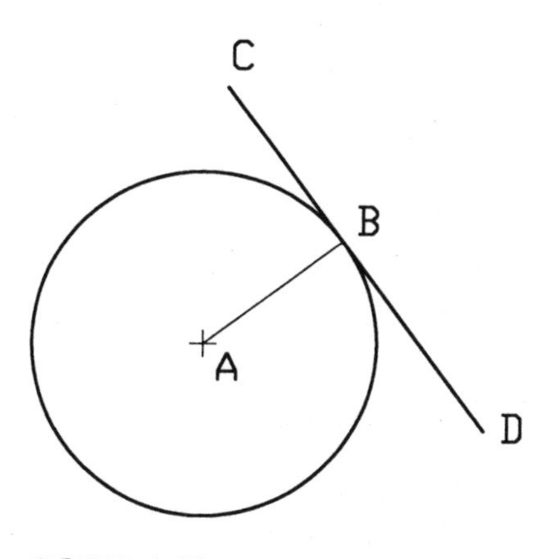

FIGURE 4–10

CONSTRUCTION 8: To draw an arc tangent to a right (90°) angle (Fig. 4–11)

STEP 1 Given radius R of the arc, draw an arc with center (the compass point) at B, and radius R cutting lines AB and BC at X and Y, respectively.

STEP 2 With centers of X and Y, and again with radius R, draw arcs intersecting at P.

STEP 3 With center P, draw the required arc with radius R. This arc is tangent to lines AB and BC.

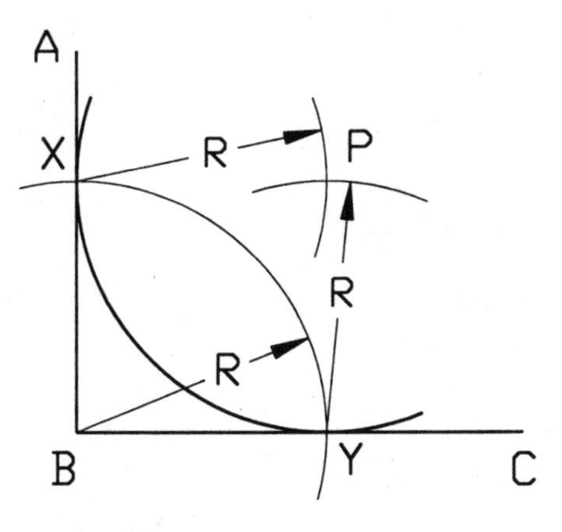

FIGURE 4–11

CONSTRUCTION 9: To draw an arc tangent to the sides of an acute or an obtuse angle (Fig. 4–12)

NOTE: An acute angle measures less than 90° (Fig. 4–12a). An obtuse angle measures between 90° and 180° (Fig. 4–12b).

STEP 1 With the given radius R, draw lines inside the angle parallel to the given lines at R distance away (Construction 2). Where these lines intersect will be the center X of the arc.

STEP 2 With radius R and the compass point at X, draw your arc tangent to the sides of the angle. (In our example, the vertex of each angle has been erased.)

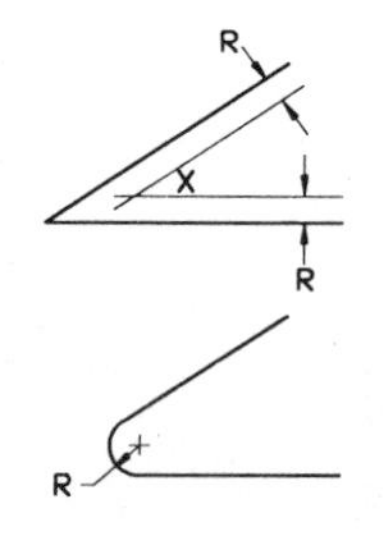

FIGURE 4-12(A) Acute angle

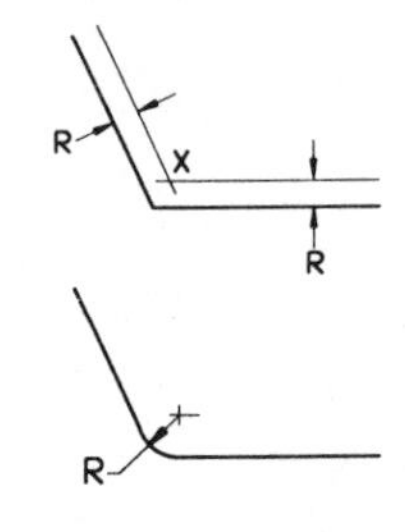

FIGURE 4-12(B) Obtuse angle

CONSTRUCTION 10: To draw an arc tangent to a circle and a straight line (Fig. 4–13)

STEP 1 Given radius R, the radius of the required tangent arc, draw a line parallel to the given line, at a distance R from the given line, on the side of the line closer to the circle (Construction 2).

STEP 2 Put the point of the compass in the center of the circle and, using radius X (the radius of the circle plus R, or X = Y + R), draw an arc to cut the drawn parallel line at point Z. Refer to Fig. 4–13(a).

STEP 3 With the center at Z and radius R, draw the required arc tangent to the circle and the straight line. Refer to Fig. 4–13(b).

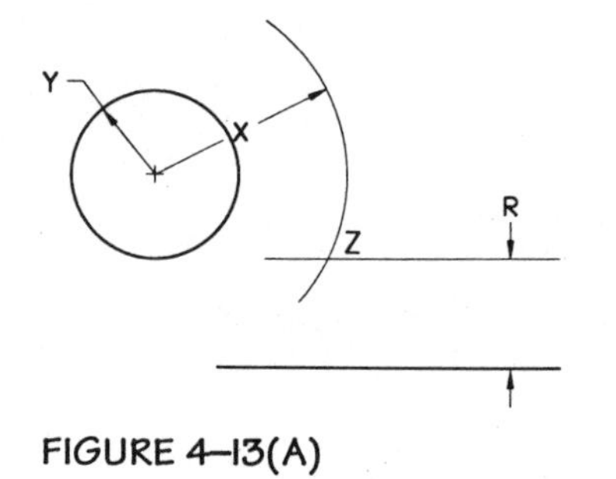

FIGURE 4–13(A)

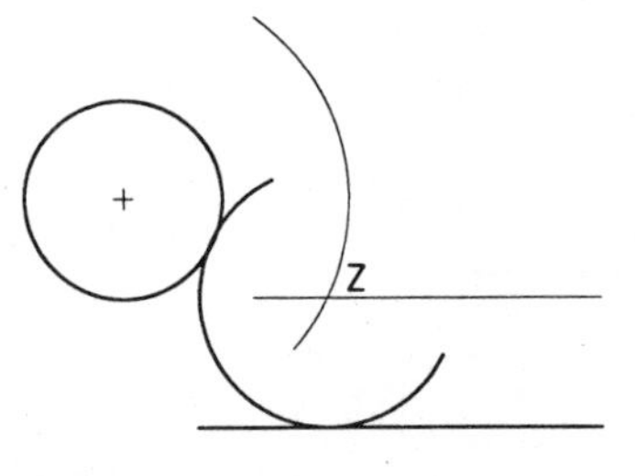

FIGURE 4–13(B)

CONSTRUCTION 11: To draw an arc tangent to two circles (Fig. 4–14)

STEP 1 Given: Radius A of the required tangent arc, and two circles of radius B and C. With the compass point at the center of the circle having radius B, open the compass to radius D (D = A + B), and draw an arc between the circles. See Fig. 4–14(a).

STEP 2 Similarly, open the compass to radius E (E = A + C). With the compass point at the center of the circle having radius C, draw an arc to intersect the first arc at X. See Fig. 4–14(a).

STEP 3 With center X and radius A, draw the required arc tangent to the given circles. See Fig. 4–14(b).

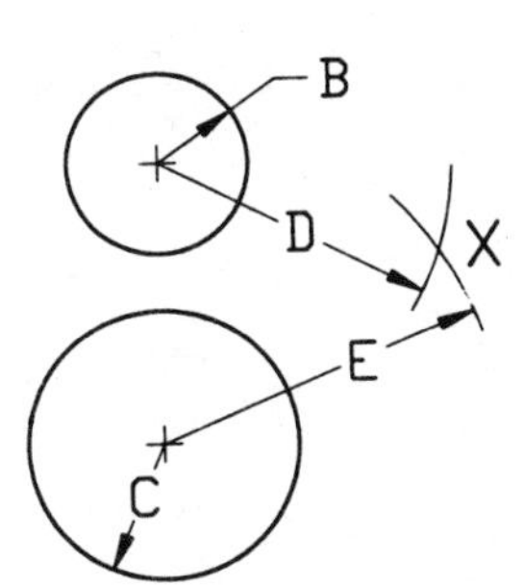

FIGURE 4–14(A)

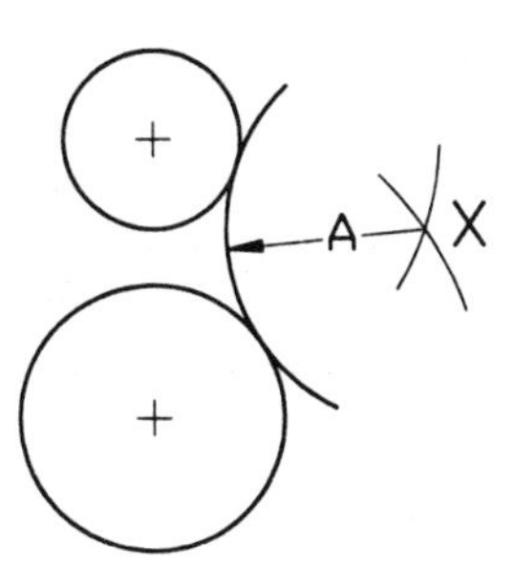

FIGURE 4–14(B)

CONSTRUCTION 12: To draw an arc tangent to two circles and to enclose the circles (Fig. 4–15)

STEP 1 Given: Radius A of the required tangent arc, and two circles of radius B and C. Open the compass to radius D (D = A – B) and, with the compass point at the center of the circle having radius B, draw an arc between the circles (Fig. 4–15a).

STEP 2 Similarly, open the compass to radius E (E = A – C) and, with the compass point at the center of the circle with radius C, draw an arc to intersect the first arc at X (Fig. 4–15a).

STEP 3 With center X and radius A, draw the required arc tangent to the given circles and enclosing the circles (Fig. 4–15b).

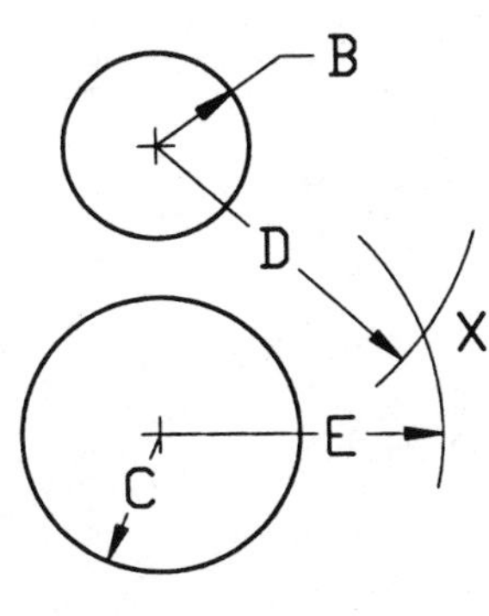

FIGURE 4–15(A)

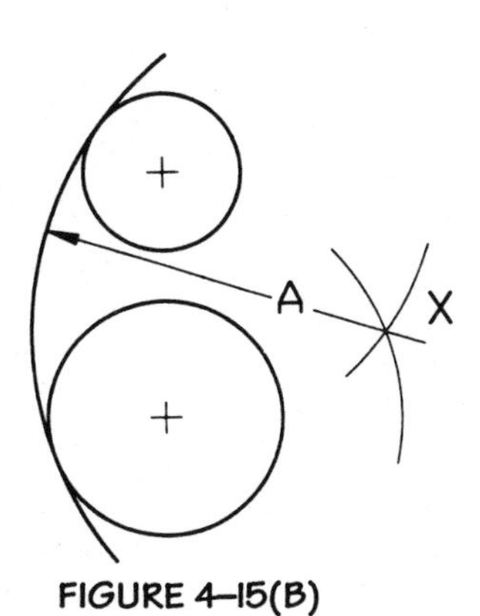

FIGURE 4–15(B)

CONSTRUCTION 13: To draw a hexagon given the distance across flats (Fig. 4–16)

NOTE: The solution to this problem is to *circumscribe a hexagon about a circle* whose diameter is equal to the distance across the flats. That is, the hexagon will be on the outside of the circle and will be tangent to it.

STEP 1 Draw a circle with a radius equal to half the distance across flats.

STEP 2 Use the 30°–60° triangle to make tangents to the circle, as shown in Fig. 4–16. Ensure that the 30°–60° triangle is held firmly against your parallel straightedge.

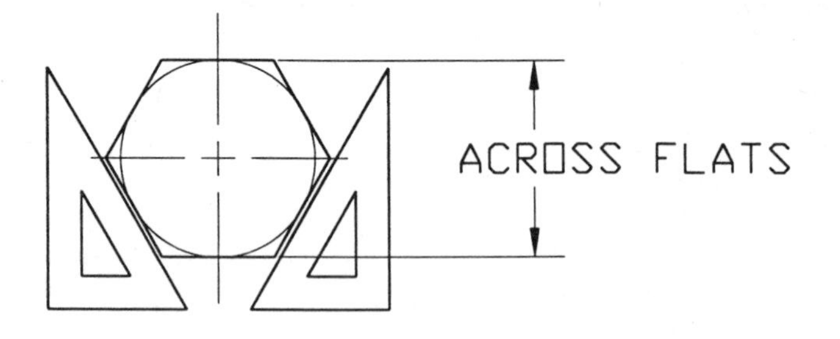

FIGURE 4–16

CONSTRUCTION 14: To draw a hexagon given the distance across corners (Fig. 4–17)

NOTE: The solution to this problem is to *inscribe a hexagon in a circle* whose diameter is equal to the distance across the corners. That is, the hexagon will be inside the circle and touch the circle at its vertices (corners).

STEP 1 Draw a circle with radius equal to half the distance across corners.

STEP 2 With the 30°–60° triangle, draw lines through the center of the circle to establish points on the circumference 60° apart (Fig. 4–17a).

STEP 3 Join these points (Fig. 4–17b).

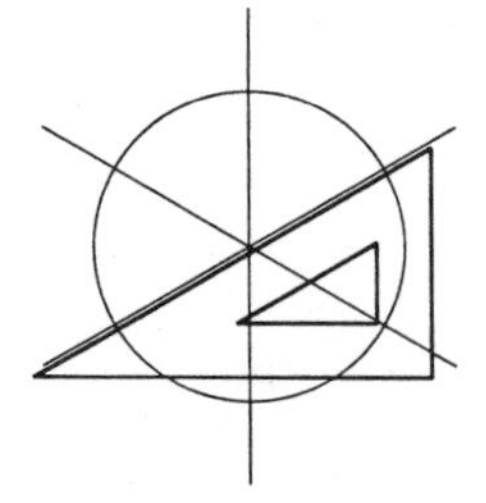

FIGURE 4–17(A)

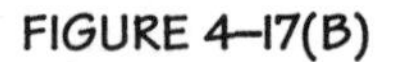
FIGURE 4–17(B)

CONSTRUCTION 15: To draw an octagon given the distance across flats (Fig. 4–18)

NOTE: We will *circumscribe an octagon about a circle* whose diameter is equal to the distance across flats.

STEP 1 Draw a circle with a radius equal to half the distance across flats.

STEP 2 Using one of your triangles, draw vertical lines tangent to the circle. Use the parallel straightedge for horizontal lines. See Fig. 4–18(a).

STEP 3 Use the 45° triangle to draw lines tangent to the circle at 45° to the horizontal (Fig. 4–18b).

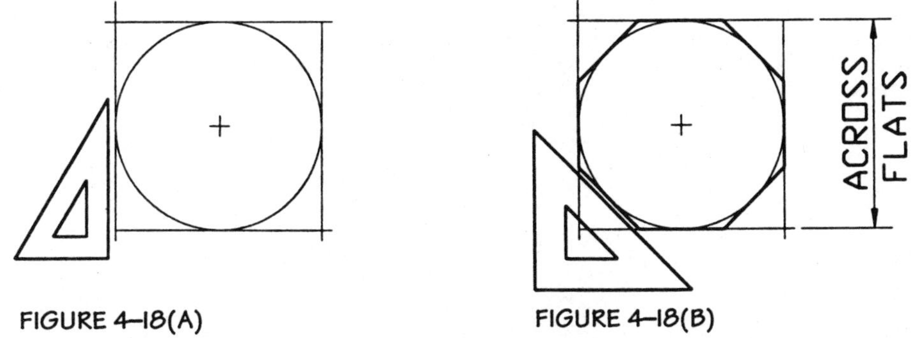

FIGURE 4–18(A)

FIGURE 4–18(B)

CONSTRUCTION 16: To draw an octagon given the distance across corners (Fig. 4–19)

NOTE: We will *inscribe an octagon in a circle* whose diameter is equal to the distance across corners.

STEP 1 Draw a circle equal to half the distance across the corners.

STEP 2 Draw vertical and horizontal center lines.

STEP 3 Holding the 45° triangle firmly against the parallel straightedge, align the hypotenuse of the triangle with the center of the circle. Draw short lines intersecting the circumference (Fig. 4–19a).

STEP 4 Join the points on the circumference as shown in Fig. 4–19(b).

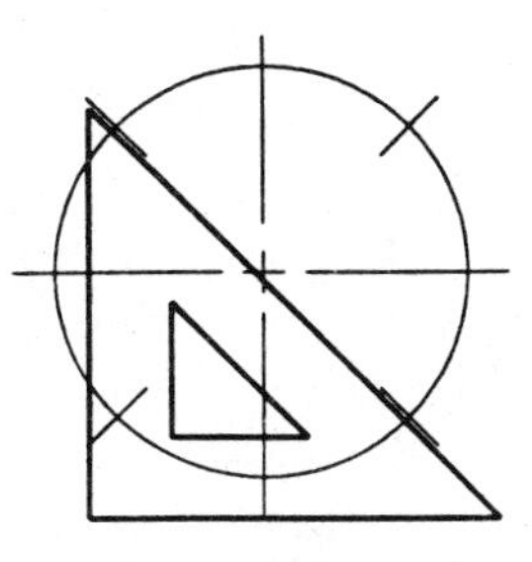

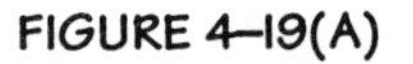

FIGURE 4–19(A)

ACROSS CORNERS

FIGURE 4–19(B)

CONSTRUCTION 17: To draw a reverse (ogee) curve connecting two parallel lines (Fig. 4–20)

STEP 1 Given two parallel lines AB and CD, and distances X and Y, join points B and C with a line.

STEP 2 Draw a perpendicular to AB at B and a perpendicular to CD at C.

STEP 3 Select a point Z on BC where the curves are to meet. If Z is the midpoint of BC, you will draw an even ogee curve (Fig. 4–20a). If Z is not the midpoint of BC, then you will draw an uneven ogee curve (Fig. 4–20b).

STEP 4 Draw the perpendicular bisectors of BZ and ZC (Construction 4).

STEP 5 Points M and N where the perpendiculars from steps 2 and 4 meet are the centers of the arcs forming the ogee curve.

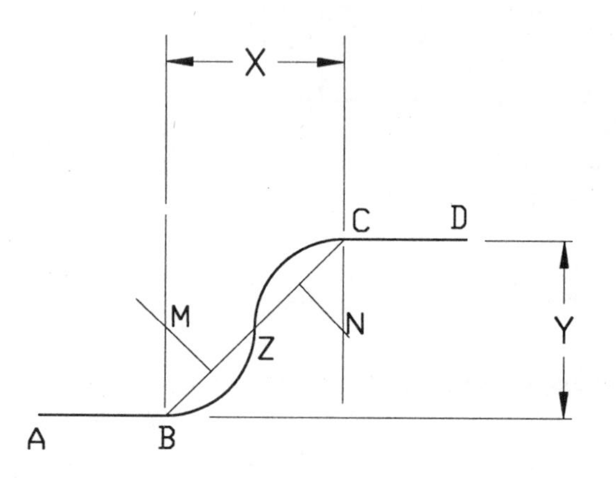

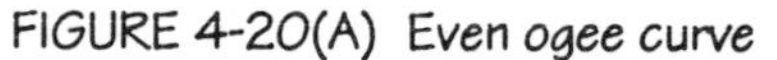

FIGURE 4-20(A) Even ogee curve

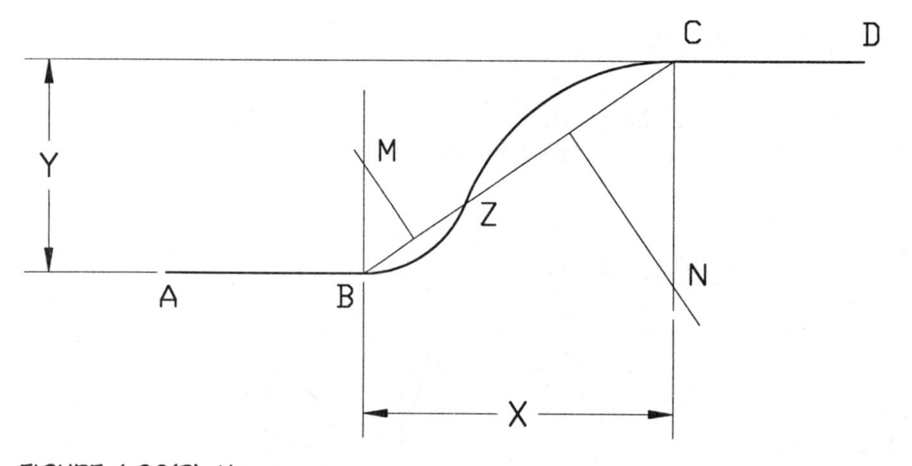

FIGURE 4-20(B) Uneven ogee curve

HINTS ABOUT USING DRAFTING INSTRUMENTS

Study these hints before proceeding to the exercises.

a. Tape down an 8 1/2" by 11" (metric A4) sheet of paper on your drawing board by lining up the top edge of the parallel straightedge with the top of your paper (the wide side), using about 1 1/2" (about 40 mm) of drafting tape at each of the four corners. Place the drawing sheet near the middle of the drawing board, not at the bottom. (See Fig. 4–21.)

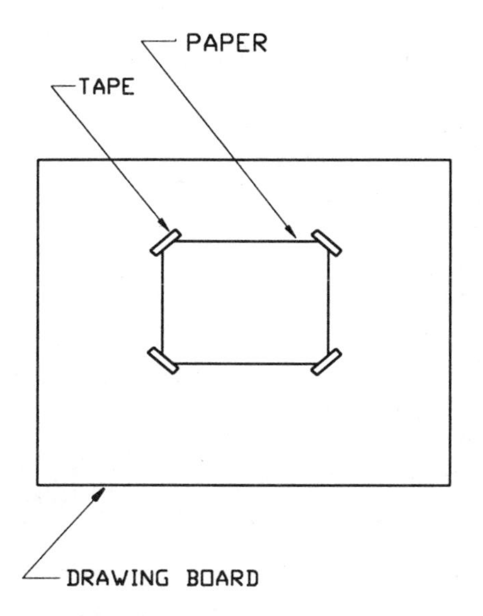

FIGURE 4–21

b. If you are using a lead holder and *not* a 0.3 or 0.7 mm technical pencil: To achieve an even line quality on an instrument drawing, roll the pencil between your thumb and finger as you draw a line. With this method, the lead will wear evenly and the line will not get wider as you draw. This method is not necessary for sketching and that is the reason we are introducing it only now.

If you are using a 0.3 or 0.7 mm technical pencil: You simply hold the pencil in a vertical position—the thickness of the lead will maintain the line thickness. You do not have to rotate this type of pencil.

c. Use the scale for MEASURING ONLY, *not* for drawing lines. Use the triangle or parallel straightedge to draw lines. If you run your fingertip along the edge of the scale, you will feel that it has indentations for each marking or graduation. This edge will quickly wear your lead, can dirty your scale, and consequently can dirty your drawing.

d. Use only the TOP of your parallel straightedge. Avoid using the bottom edge for drawing, and avoid putting your triangle on the bottom edge. The reason is that parallel straightedges move up and down to form parallel lines, but if you use the top and bottom edges, you are assuming that these two edges are parallel to each other. This assumption may or may not be true, depending upon the quality of the straightedge.

e. After you have used your instruments to make a drawing, do not go over the lines freehand.

f. DECIMAL EQUIVALENTS TABLE: Refer to page 326 in the Appendix. This table converts a fraction of an inch to a decimal and vice versa. It also converts a decimal or a fraction of an inch to the metric equivalent in millimeters.

Example: 1/64" equals .015625 in decimal form; 1/64" also equals 0.39688 mm. Similarly, 1/8" equals .125" in decimal form, which is also equal to 3.175 mm.

g. When using triangles, avoid using the tip when drawing a line. Figure 4–22(a) illustrates a poor way of using the triangle. Figure 4–22(b) shows a better way.

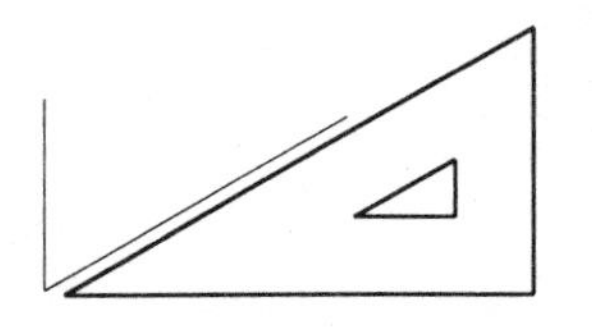

NO

FIGURE 4–22(A)

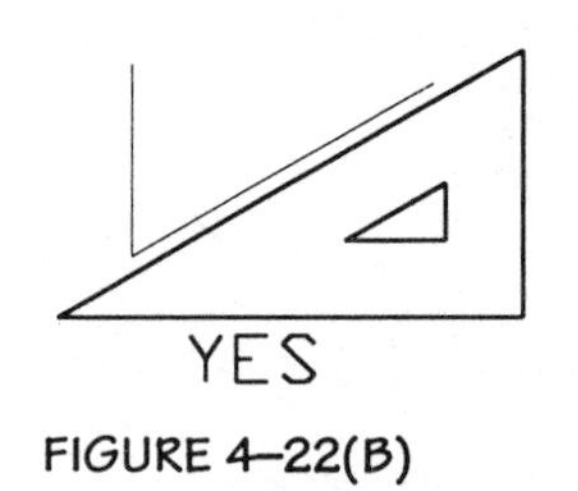

YES

FIGURE 4–22(B)

h. Use your triangles, rather than your protractor, for drawing 30°, 60°, 90°, or 45° angles because they will give you a more accurate angle. See page 323 of the Appendix for information on triangles.

i. You will be using the compass in this chapter. The lead in your compass should protrude about 1/2" (13 mm). Over a waste basket, sharpen your compass lead with the sandpaper pad to a flat chisel edge. The metal point in your compass should protrude about 1/16" (2 mm) beyond the lead, since some of the point will penetrate the paper and soft drawing surface. When purchased, your compass included a piece of lead and may have had some extra lead in a small tube. You probably do not know the grade of this lead. If you do not like the line quality of this lead, discard it and buy H lead to fit your compass. If you are using a lead holder, break off a piece of lead from your H lead holder and sharpen it to a flat chisel edge with your sandpaper pad. You should now be able to make a circle or arc with the same thickness and darkness as a straight line drawn with a pencil. To keep your drawing clear and clean, use your compass only once for each arc or circle. Avoid drawing a light arc or circle and darkening it later.

EXERCISES FOR GEOMETRIC CONSTRUCTIONS

On the following pages, perform the constructions as directed. Do not use a trial-and-error approach. For example, do not move your compass point around to search for the desired result. Rather, follow the steps of the constructions shown on the previous pages.

If you encounter difficulty solving any of these problems, try sketching the solution freehand. This will help you to visualize the answer and then to solve the problem with instruments. Be sure to remove the page you will work on from this book and tape it down on your drawing board. A 0.3 mm technical pencil will assist you in locating your lines accurately. You should:

a. Show *all* construction lines. Do not erase any of them.

b. Show *all* tangent points by extending arcs and lines past the tangent point.

c. Indicate clearly with a center line the center of *all* circles and arcs. Refer to page 44.

d. Maintain consistent line quality for all arcs, circles, and straight lines.

EXERCISE 4–2

Measure the angles below and write the answers on this page. (This is a measurement exercise, not a construction exercise.)

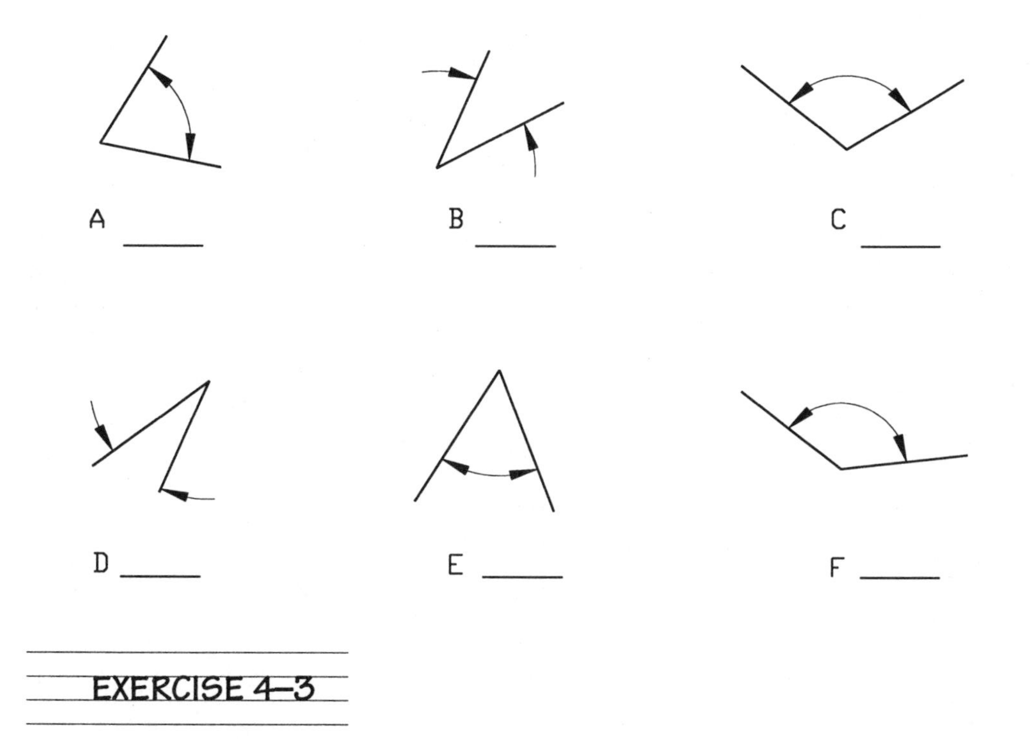

EXERCISE 4–3

Draw a 2" (50 mm) line perpendicular to line AB at point X.

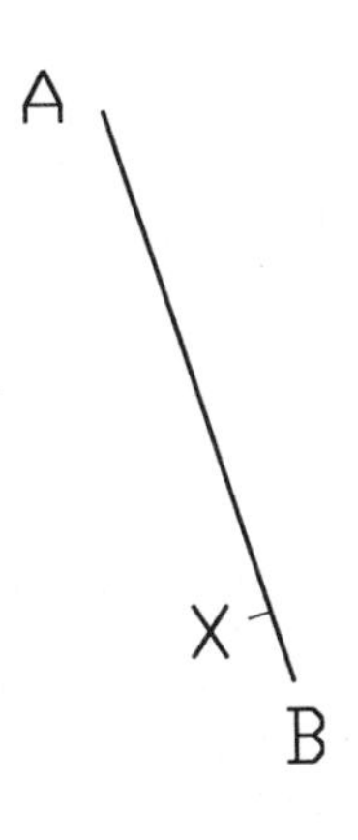

DRAWN BY ______________________________ DATE ______

EXERCISE 4–4

Draw a line parallel to line CD at a distance 2" (50 mm) to the right of line CD.

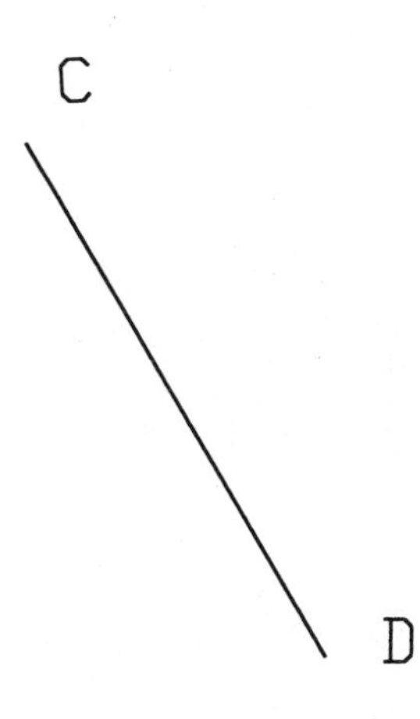

EXERCISE 4–5

Divide line AB into five equal parts.

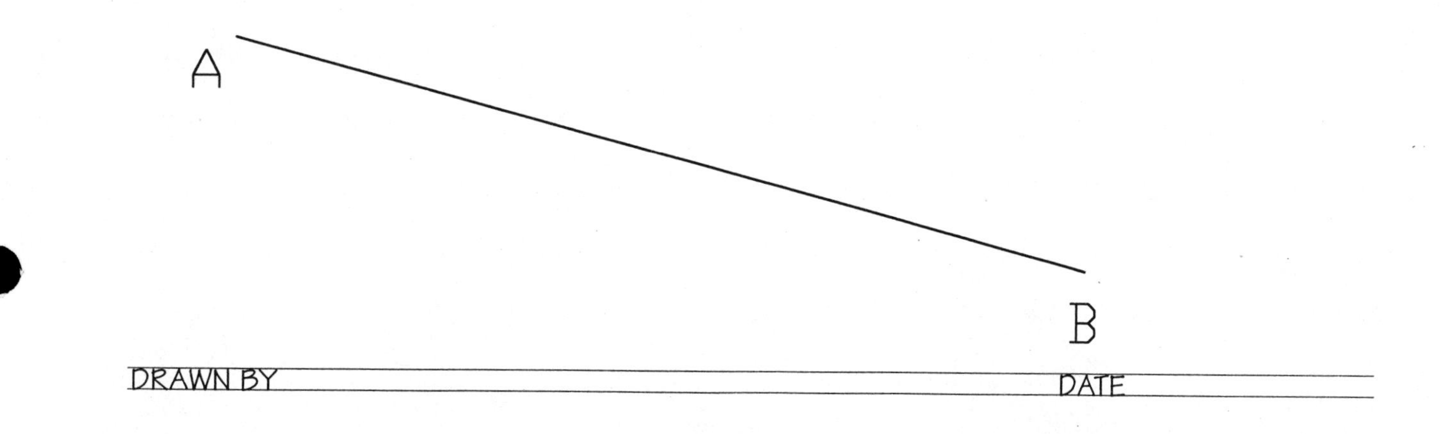

EXERCISE 4–6

Draw a perpendicular bisector of line CD.

C

D

EXERCISE 4–7

Bisect the arc shown below.

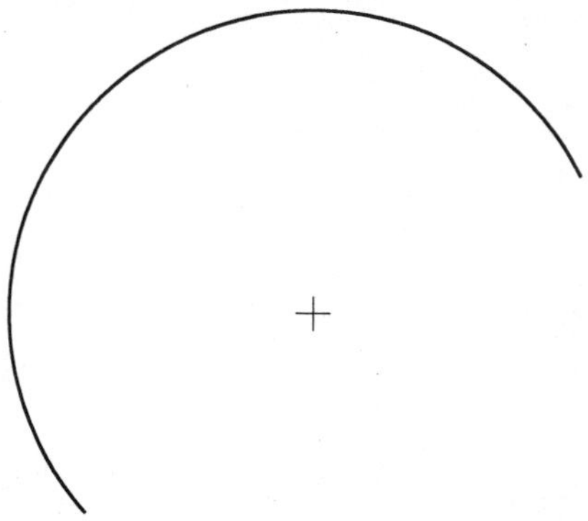

DRAWN BY DATE

EXERCISE 4–8

Bisect angle ABC.

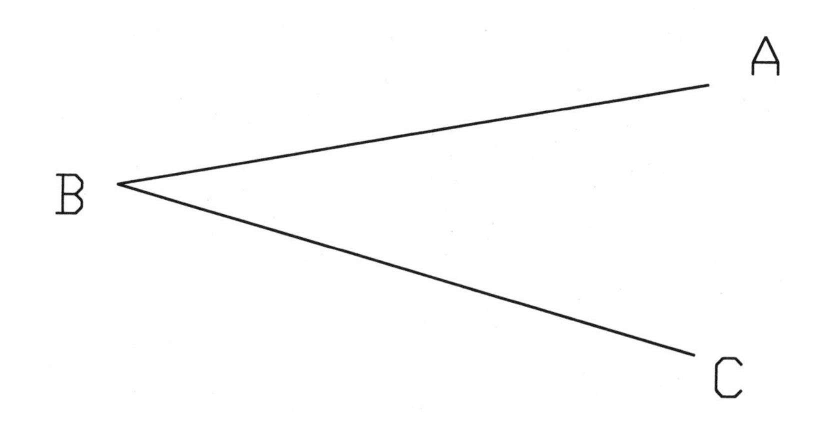

EXERCISE 4–9

Draw a line tangent to the circle below at point X.

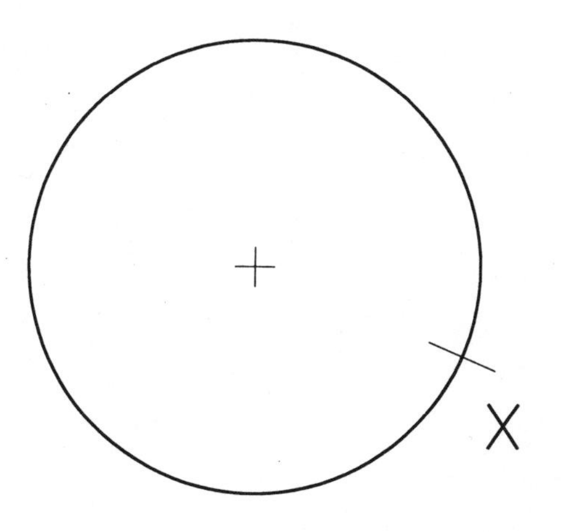

DRAWN BY DATE

EXERCISE 4–10

Draw an arc of radius 1.5" (38 mm) tangent to the right angle ABC.

A

B

C

EXERCISE 4–11

Draw an arc of radius 1" (25 mm) tangent to angle ABC.

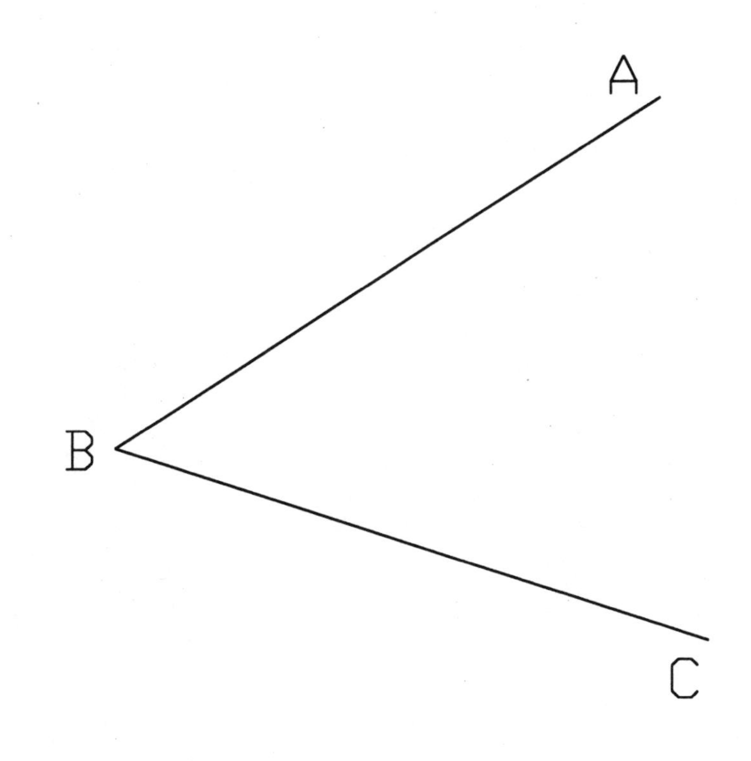

DRAWN BY DATE

EXERCISE 4–12

Draw an arc of radius 1" (25 mm) tangent to angle DEF.

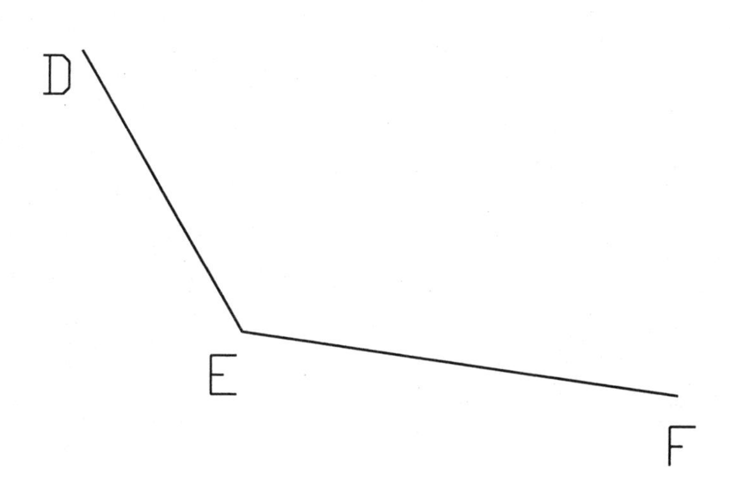

EXERCISE 4–13

Construct an arc of 1.5" (38 mm) radius tangent to the given arc and line. Refer to Construction 10.

DRAWN BY ______________________________ DATE ______________

EXERCISE 4–14

Construct an arc of 1.5" (38 mm) radius tangent to the two given arcs. Refer to Construction 11.

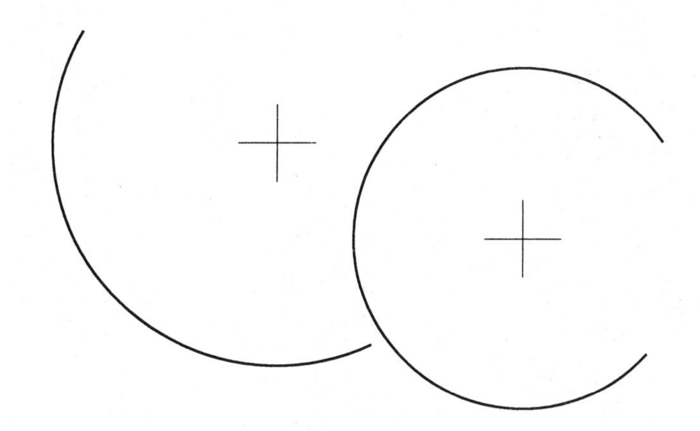

EXERCISE 4–15

Construct an arc of .75" (19 mm) radius tangent to the two given arcs. Refer to Construction 11.

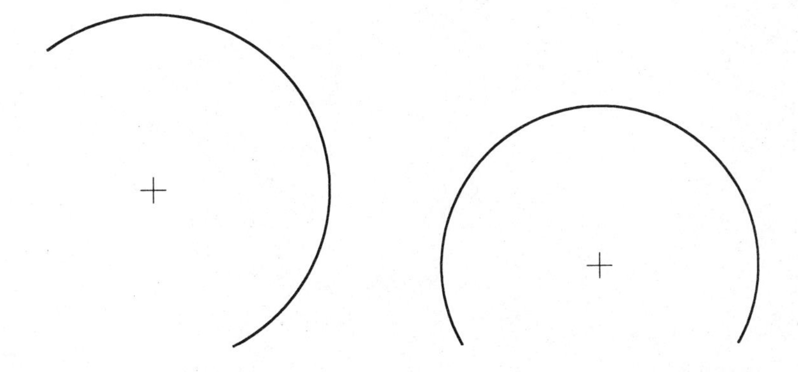

DRAWN BY DATE

EXERCISE 4–16

Construct an arc of 1.5" (38 mm) radius passing through point P and tangent to the given line below. HINT: Draw a line parallel to and 1.5" (38 mm) from the given line. Draw an arc of 1.5" (38 mm) radius with center at P. You figure out the rest!

P

EXERCISE 4–17

Construct a circle of 1" (25 mm) diameter tangent to the given arcs. Refer to Construction 11.

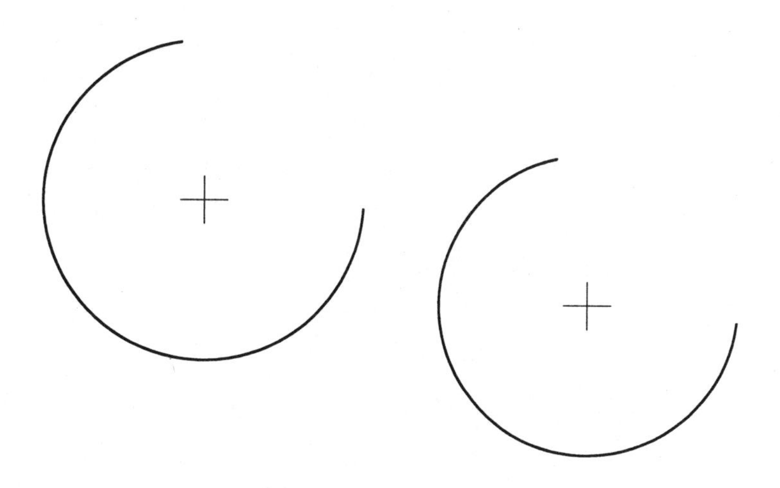

DRAWN BY DATE

EXERCISE 4–18

Construct an arc of 1.5" (38 mm) radius tangent to the two given circles and containing both. Refer to Construction 12.

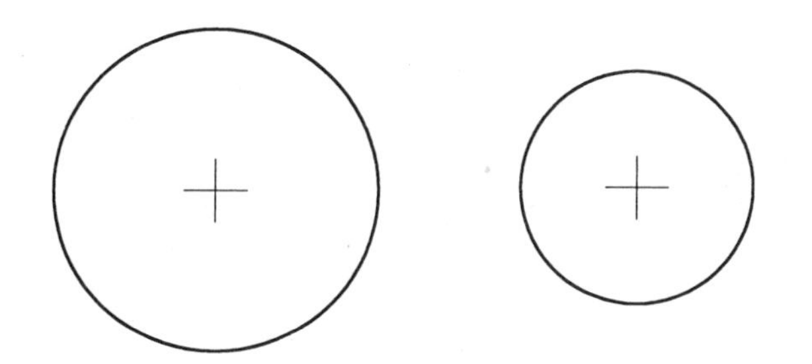

EXERCISE 4–19

Draw a circle of 2.25" (57 mm) diameter tangent to the given circles but containing only the smaller one. (A specific construction is not given that corresponds to this exercise. By referring to Constructions 11 and 12, you should be able to solve this problem.)

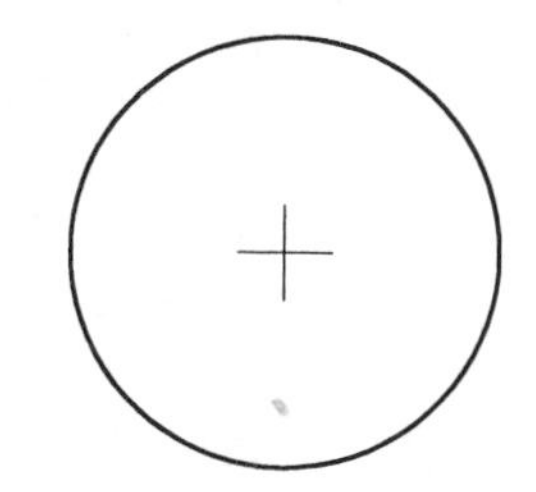

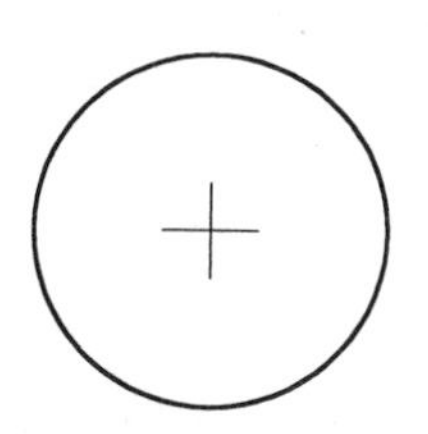

DRAWN BY DATE

EXERCISE 4–20

Draw a hexagon with 1.5" (38 mm) across flats (often written as 1.5" A/F). Refer to Construction 13.

EXERCISE 4–21

Draw a hexagon with 1.5" (38 mm) across corners (often written as 1.5" A/C). Refer to Construction 14.

DRAWN BY DATE

EXERCISE 4–22

Using line AB as the base of a triangle ABC, construct a triangle. Let side AC equal 2.25" (57 mm) and side BC equal 3" (75 mm). This you will have to solve for yourself!

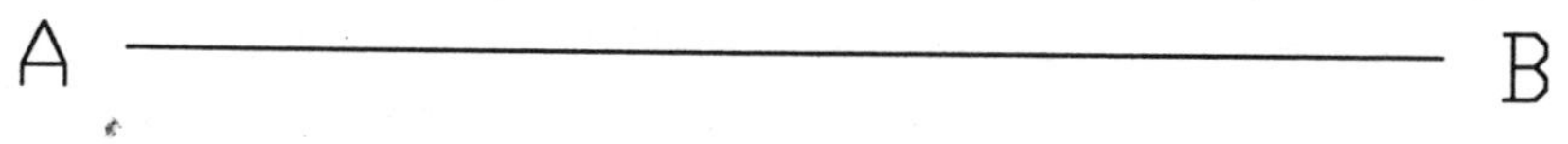

A B

EXERCISE 4–23

At X, circumscribe a square about a circle of 1" (25 mm) radius. At Y, circumscribe a regular hexagon about a circle of 1.25" (32 mm) radius. Refer to Constructions 13 and 15.

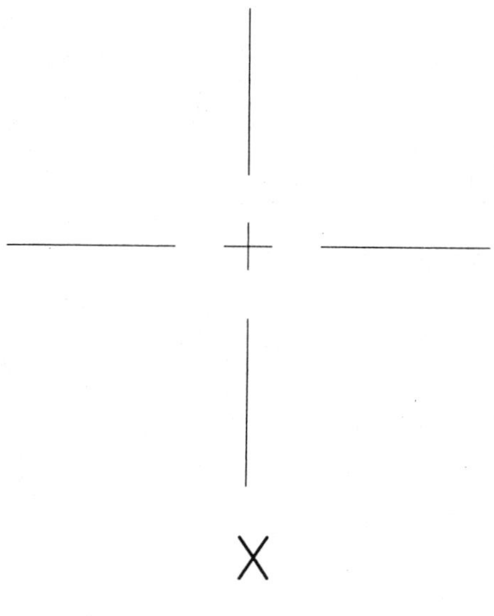

X

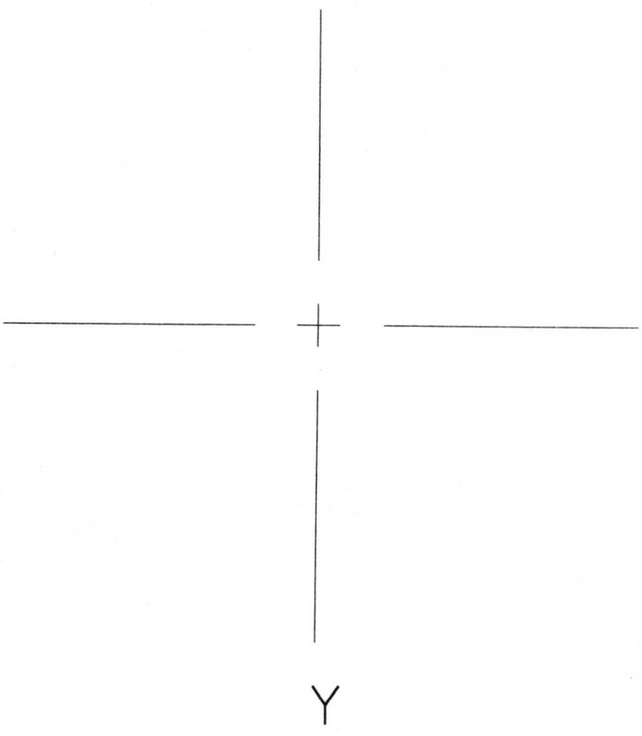

Y

DRAWN BY DATE

EXERCISE 4–24

Draw an octagon with 2" (50 mm) across flats. Refer to Construction 15.

EXERCISE 4–25

Draw an octagon with 2" (50 mm) across corners. Refer to Construction 16.

DRAWN BY DATE

This page and the following ones show some shapes that you should try to draw on blank 8 1/2" by 11" (metric A4) paper. These exercises will give you practice with your instruments. Specifically, you should try to draw all lines with an even line quality of thickness and darkness, make all corners very neat, and measure accurately. The experience from doing the previous problems in this chapter will help you, but you will have to be creative with the use of your triangles and compass to draw some of these figures!

NOTES:

a. The equal signs indicate that a line has been divided into equal parts.

b. Make the size of the drawing according to the dimensions given, *not* according to its size in this book.

c. "TYP" indicates that this is a typical dimension that would apply to another similar feature. In Exercise 4–27, all angular lines are at a 60° angle to the horizontal. In Exercise 4–28, all angular lines are at 45° to the horizontal.

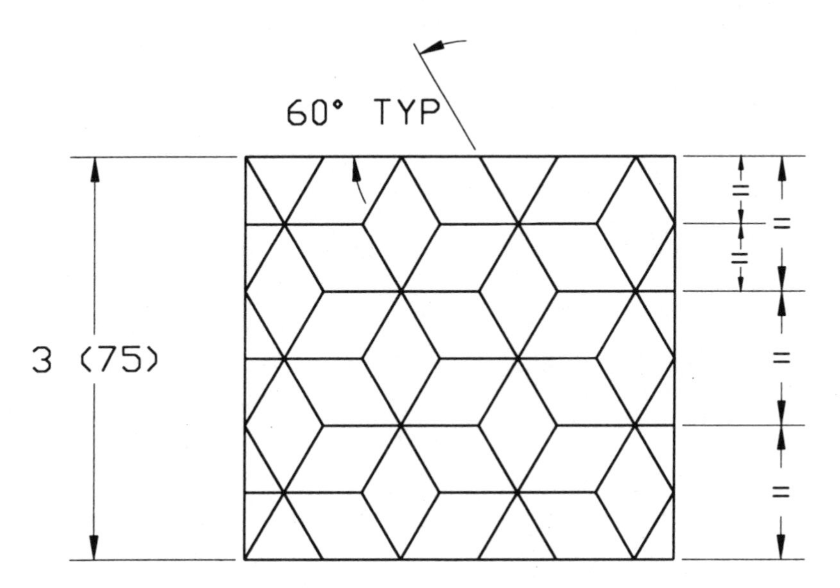

EXERCISE 4–27

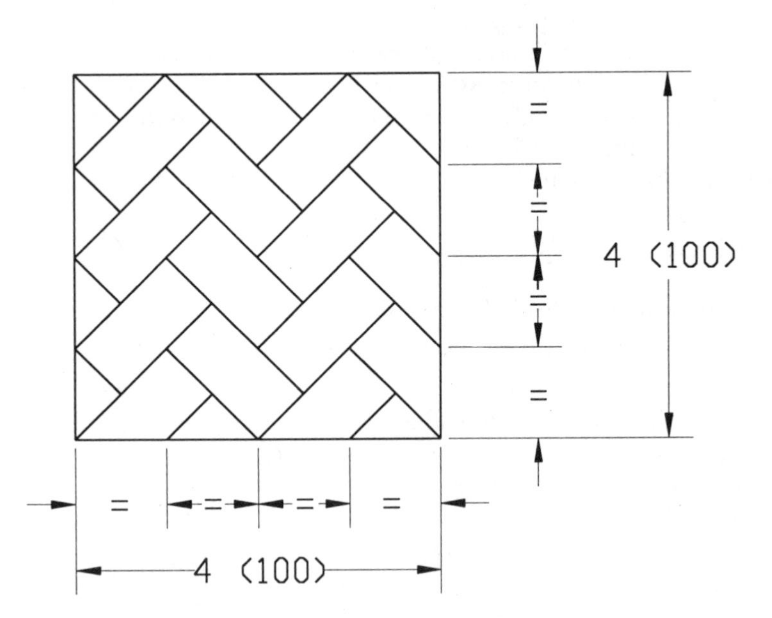

EXERCISE 4–28

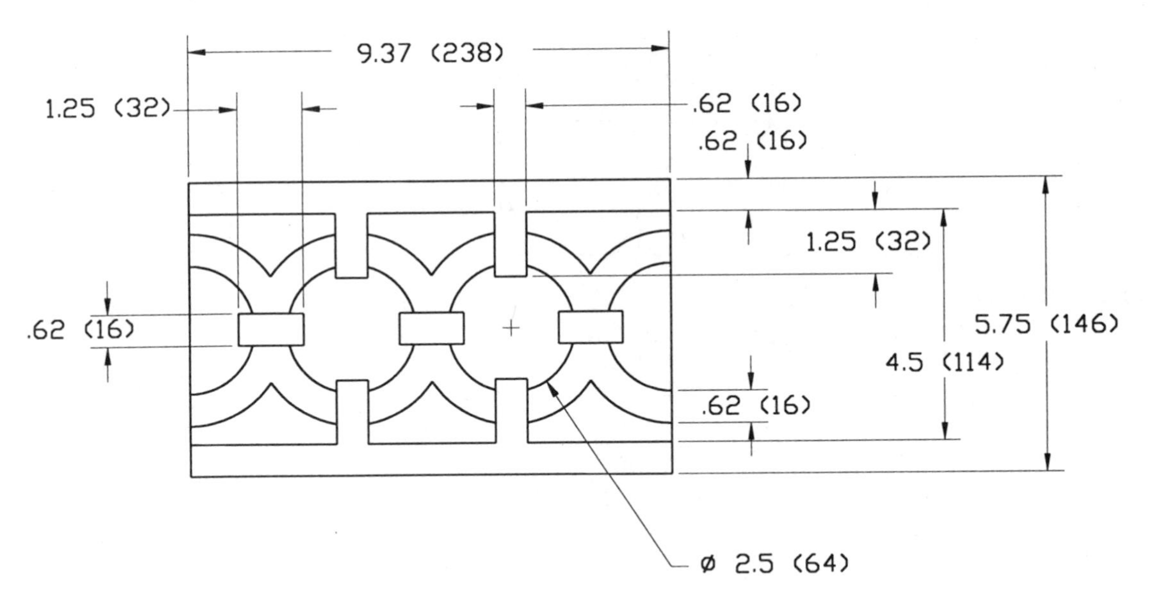

EXERCISE 4–29

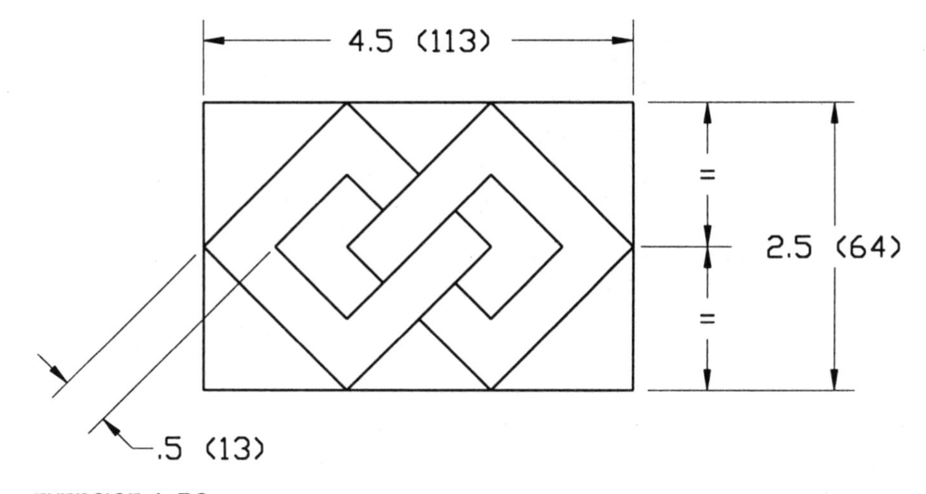

EXERCISE 4–30

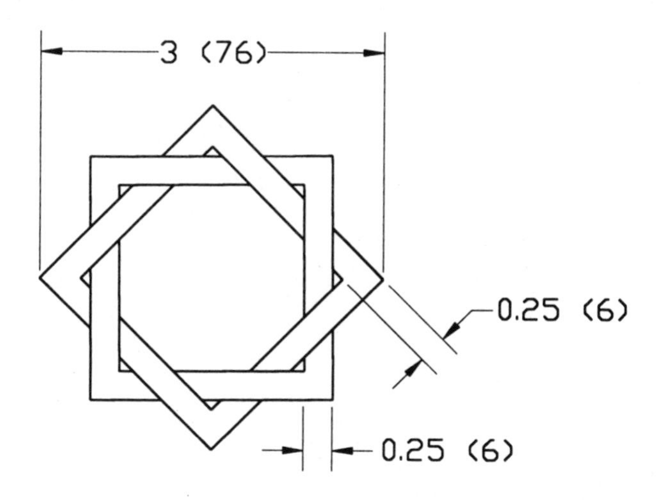

EXERCISE 4–31

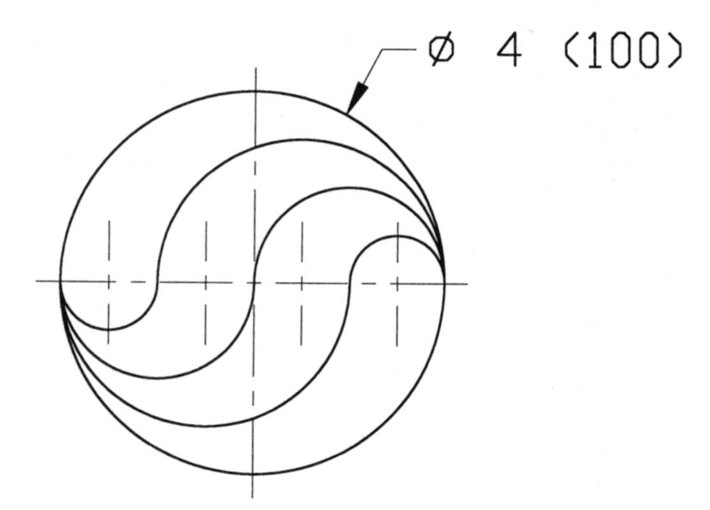

EXERCISE 4–32

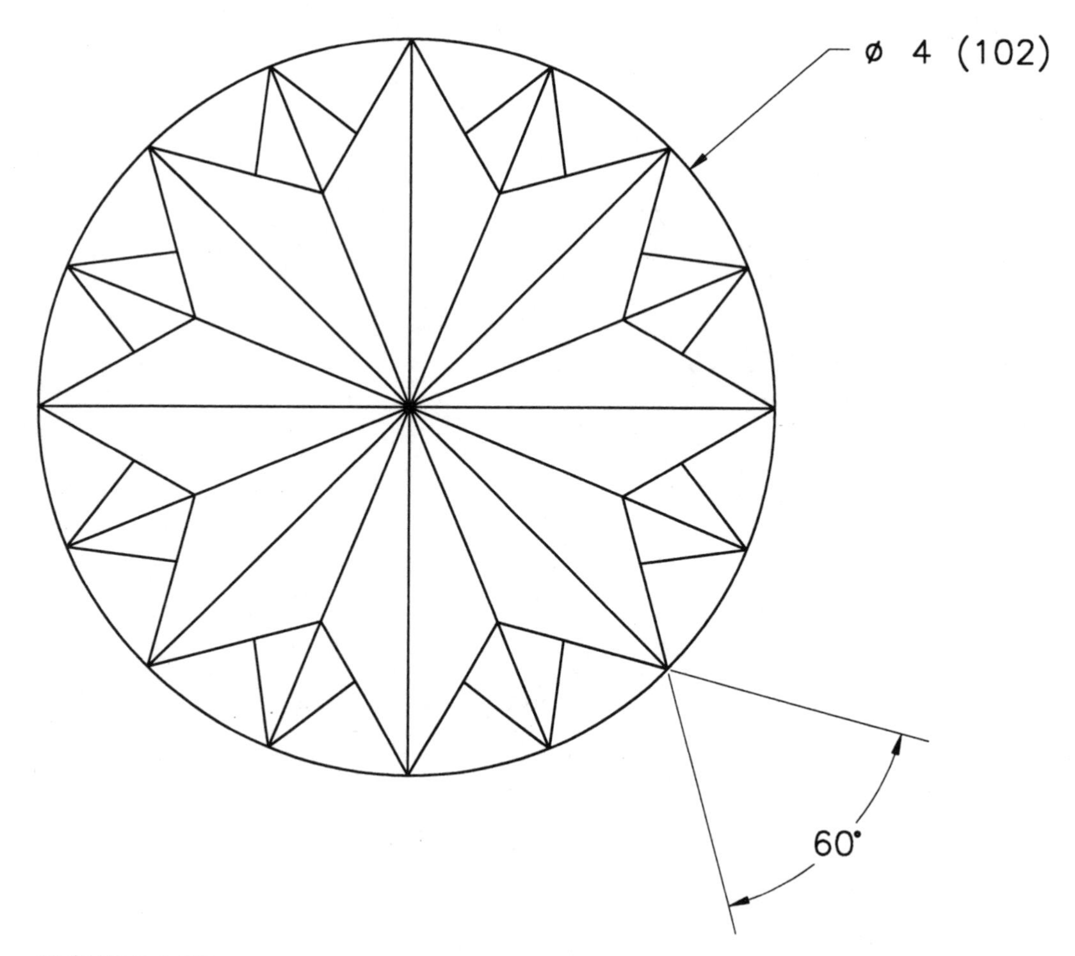

EXERCISE 4–33

THE HELIX

The HELIX is a curve generated by a point that revolves uniformly around and up or down the surface of a cylinder. If you were to trace a thread around a bolt, you would trace a helix. Think also of the spiral around a barber shop pole.

The LEAD (pronounced LEED) is the vertical distance the point rises or drops in one complete revolution.

CONSTRUCTION 18: To draw a helix (Fig. 4–23)

STEP 1 Given the diameter of the cylinder and the lead, draw the top and front views. The top view will only be a circle. The front view at this stage will only be a rectangle.

STEP 2 Draw horizontal and vertical center lines through the center of the circle using your parallel straightedge and 30°–60° triangle. Then, holding this triangle firmly against the parallel straightedge, draw lines through the center of the circle to cut the circumference (edge of the circle) at 30°, 60°, 120°, and 150°. Extend each of these lines through the circle so that each line cuts the circumference twice. You should now have the circumference of the circle divided into 12 equal parts. Number these intersection points as shown in Fig. 4–23. (Actually, you may choose to divide the circumference into any convenient number of parts, not necessarily 12.)

STEP 3 Project these 12 points that are on the circumference down to the bottom of the front view, and number the corresponding new points of intersection according to the corresponding number from the top view, as illustrated in the figure.

STEP 4 Divide the lead into the same number of equal parts, in our case 12 and label them as shown. (Refer to Construction 3.)

STEP 5 The points of intersection of lines with corresponding numbers lie on the helix. Use a French curve to join the points in the front view. There is no scientific way to use a French curve. Simply try using various parts of the curve to join three or more points. Move the French curve so as to use the part that will best join three or more points (the more the better). This is a trial-and-error method!

NOTE: Because points 8 to 12 lie on the back portion of the cylinder, the helix curve starting at point 7 and passing through points 8, 9, 10, 11, 12, and 1 will appear as a hidden line.

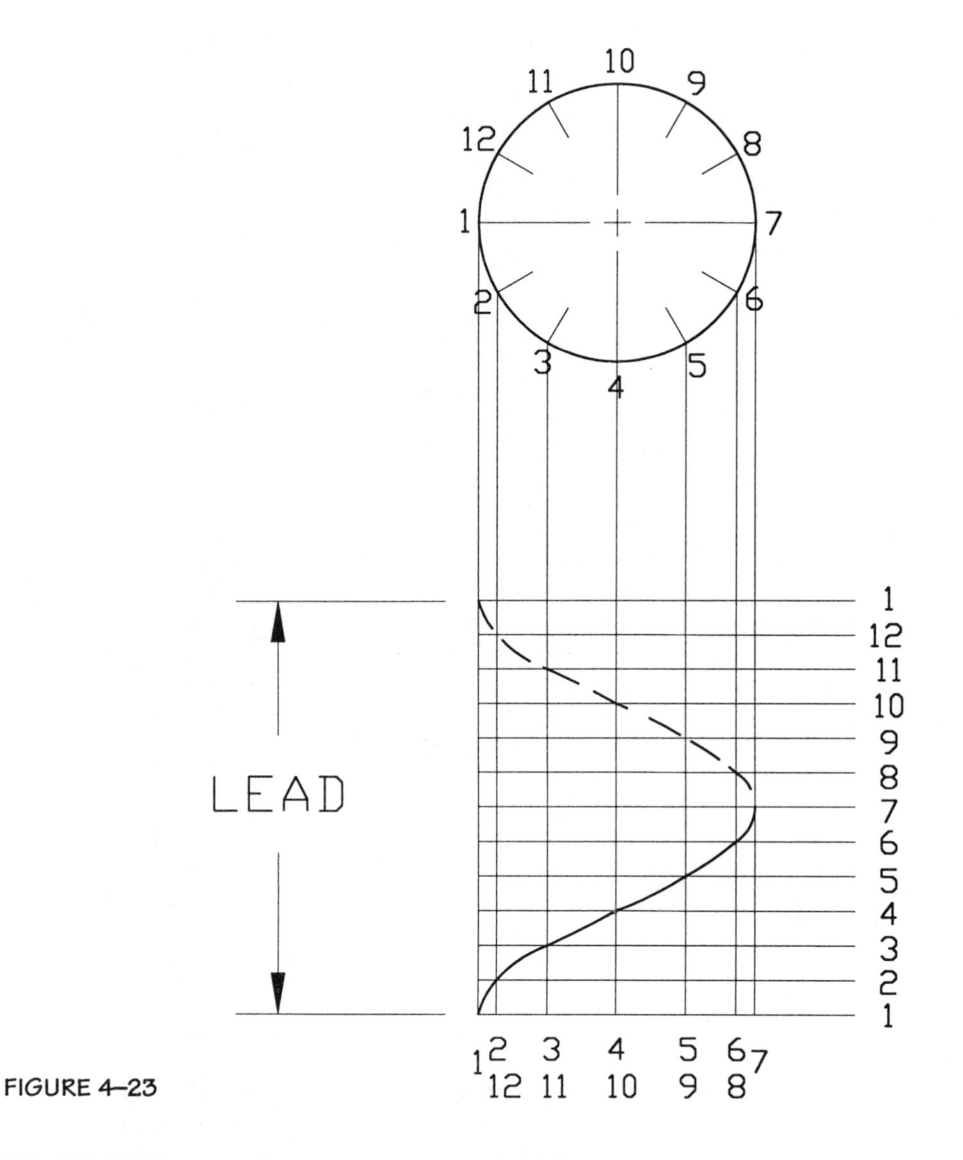

FIGURE 4–23

CONSTRUCTION 19: To draw a conical helix (Fig. 4–24)

The CONICAL HELIX is a curve generated by a point that revolves uniformly around and up or down the surface of a cone. If you were to trace a thread around a wood screw that tapers to a point, you would be tracing a conical helix.

STEP 1 Given the diameter of the base of the cone, the height of the cone, and the lead, draw the top and front views. The top view is only a circle. The front view is only a triangle at this stage.

STEP 2 Divide the circumference of the top view in Fig. 4–24 into a convenient number of parts (in our example, eight) by using the 45° triangle held firmly against the parallel straightedge and drawing lines passing through the center point. Number these points, as shown in the figure.

STEP 3 Project these points of intersection down to the bottom of the front view, and number the corresponding new points of intersection, as shown in the front view of Fig. 4–24.

STEP 4 Draw light lines from the points in step 3 to the apex (top point) of the cone, labeled C in our figure.

STEP 5 Divide the lead into the same number of equal parts and label them as shown.

STEP 6 The points of intersection of lines with corresponding numbers (in the front view) lie on the conical helix. Join these points with a French curve. *Note:* Since points 6, 7, and 8 lie on the back portion of the cone, the conical helix curve starting at 5 and passing through points 6, 7, 8, and 1 will appear as a hidden line in the front view.

STEP 7 Project the point where the curve intersects the horizontal line up to the top view. Note where these lines intersect the radial lines in the top view. Join these points with a French curve in the top view.

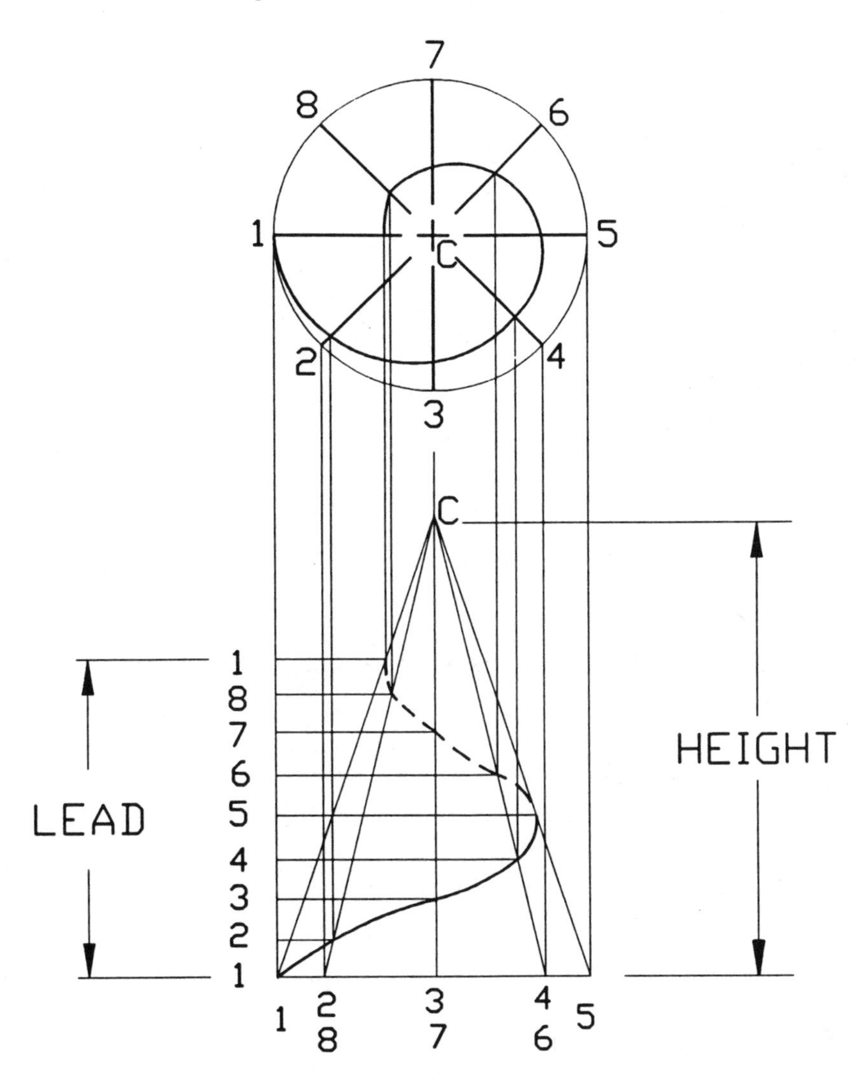

FIGURE 4–24

EXERCISES FOR THE HELIX

EXERCISE 4–34

On an 8 1/2" by 11" (metric A4) sheet of paper, draw a helix with a diameter of 2" (50 mm) and a lead of 3" (75 mm).

EXERCISE 4–35

On an 8 1/2" × 11" (metric A4) sheet of paper, draw a conical helix with a diameter of 2" (50 mm), a lead of 2" (50 mm), and a height of 3" (75 mm).

ANSWERS TO TANGENT QUESTIONS ON PAGE 94

The number of straight line tangents that may be drawn to two circles that are not touching is four, as illustrated in Fig. 4–25(a).

An infinite number of curved line tangents may be drawn to two circles that are not touching. Some of these curved tangents are illustrated in Fig. 4–25(b).

FIGURE 4–25(A)

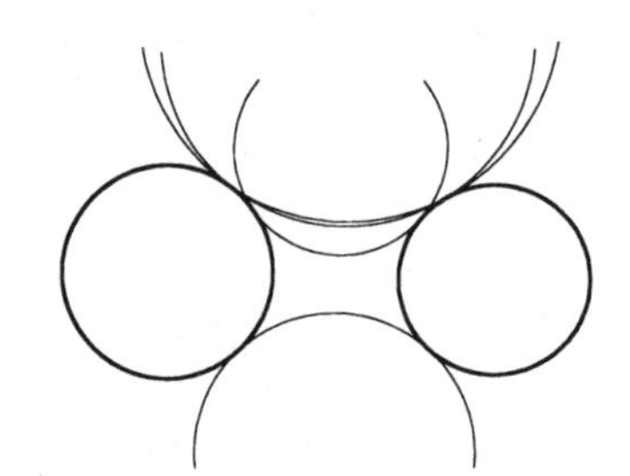

FIGURE 4–25(B)

ISOMETRIC VIEWS USING DRAFTING INSTRUMENTS

5

OBJECTIVE

When you have finished this chapter, you should be able to make isometric views using your drafting instruments.

EQUIPMENT

You are required to have all of your drafting instruments for this chapter.

STEPS FOR DRAWING AN ISOMETRIC VIEW WITH DRAFTING INSTRUMENTS

PROBLEM: We want to draw the isometric view illustrated in Fig. 5–1 using drafting instruments. We will use the dimensions shown and will draw to full scale. Of course, your drawing will be larger than this one, since the size of each illustration has been reduced to fit this book. The dimensions are in inches, with millimeters in parentheses.

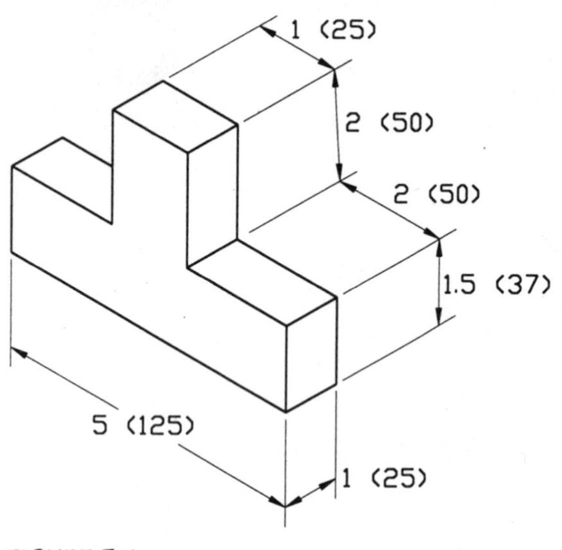

FIGURE 5–1

The steps we follow are similar to those studied in Chapter 1 to sketch an isometric object.

STEP 1 Tape down an 8 1/2" by 11" (metric A4) sheet of paper on your drawing board by lining up the top edge of the parallel straightedge with the top of your paper (the wide side), using about 1 1/2" (about 40 mm) of drafting tape at each of the four corners. Place the drawing sheet near the middle of the drawing board, not at the bottom. (See Fig. 5–2.)

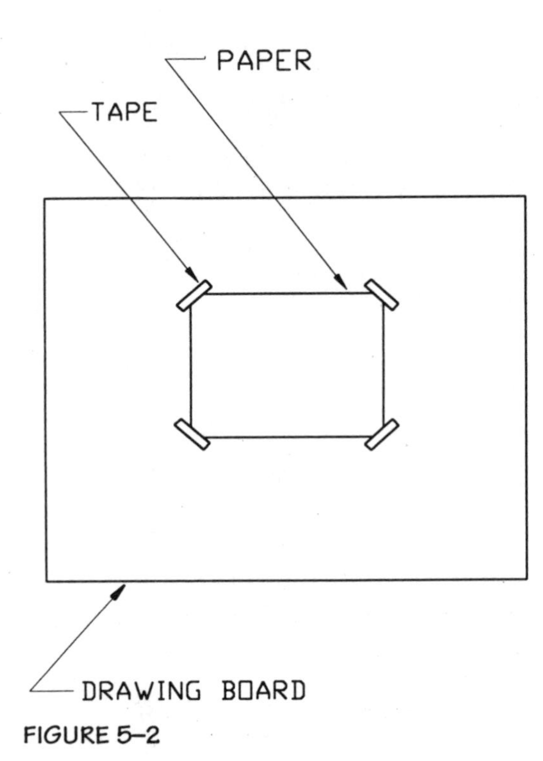

FIGURE 5–2

STEP 2 Most companies and many schools use drafting paper with preprinted border lines and title block. If you are using a blank sheet of paper, draw a border line 1/2" (13 mm) from the edge of the paper and draw a title block. A sample title block is shown in Fig. 5–3. Although the style chosen for the title block is optional, it must include the information listed below. The letter labels correspond to the format suggested in Fig. 5–3.

a.Scale of drawing
b.General specifications and notes
c.Company name
d.Title of drawing
e.Drafter name or initials and, if necessary, approvals
f.Drawing number

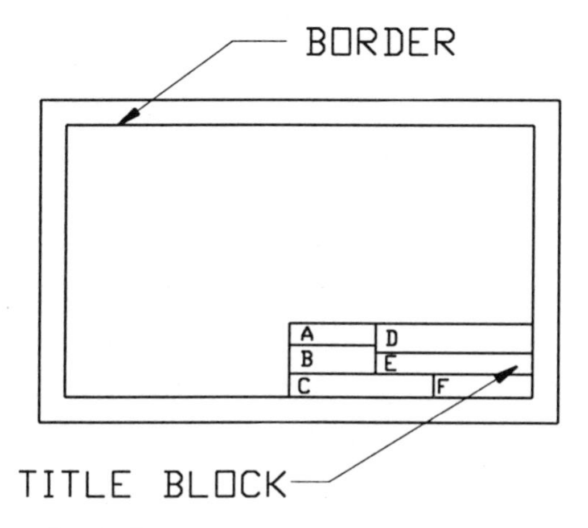

FIGURE 5–3

STEP 3 Center your isometric drawing by approximating the starting point, keeping in mind that the drawing will fit into an isometric box for which you know the overall size.

Lightly draw the isometric axes using your 30°–60° triangle (set square) held firmly against the parallel straightedge. Figure 5–4(a) illustrates how to orient your triangle. Start to draw near the bottom center of your drawing paper using your 0.3 mm technical pencil.

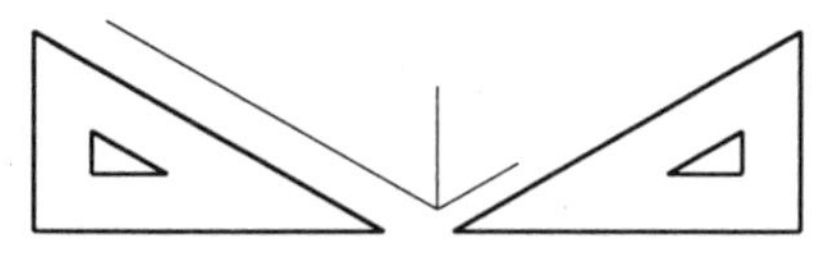

FIGURE 5–4(A)

STEP 4 On the isometric axes, and using the appropriate scale, mark off the height, width, and depth of an isometric box to enclose this object. (When measuring a length on an isometric axis, always hold the scale parallel to the axis.) In this case, we will use dimensions 3.5" (87 mm), 5" (125 mm), and 1" (25 mm), respectively. See Fig. 5–4(b).

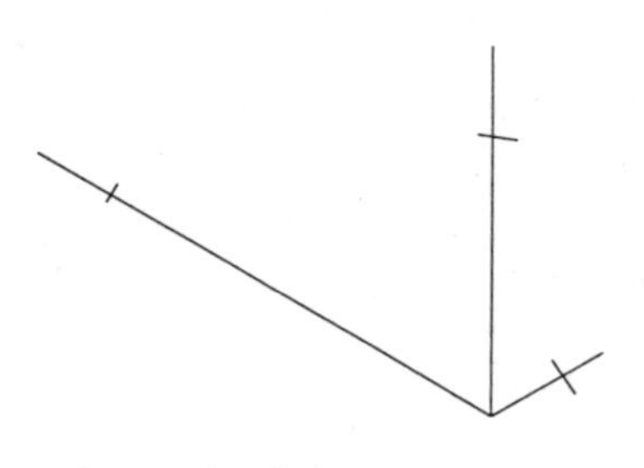

FIGURE 5–4(B)

STEP 5 Slide your 30°–60° triangle along the parallel straightedge to the appropriate location and, holding it firmly against the parallel straightedge, draw vertical lines at the appropriate location. See Fig. 5–4(c).

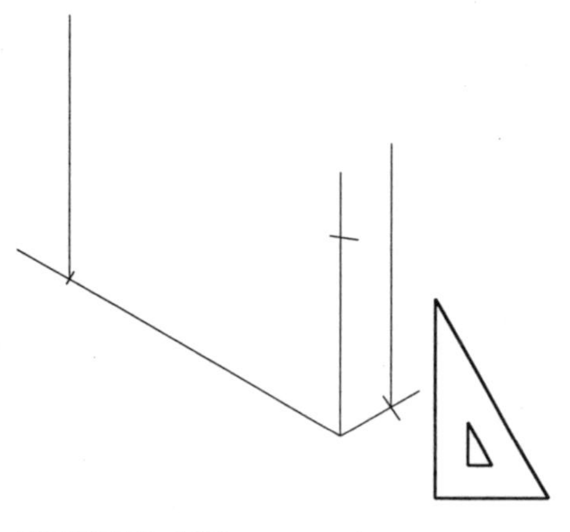

FIGURE 5–4(C)

STEP 6 Turn your 30°–60° triangle so that it is oriented according to the illustration. Make sure that one side is placed firmly against the parallel straightedge. Lightly draw lines as shown in Fig. 5–4(d).

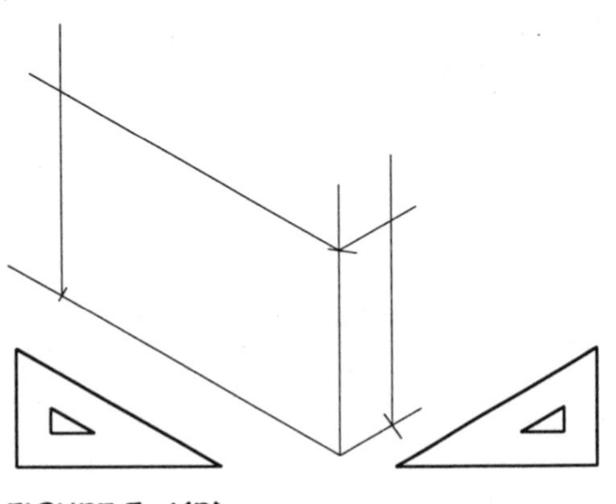

FIGURE 5–4(D)

STEP 7 In a similar fashion, draw lines to finish the box. If the construction lines were drawn lightly, they do not have to be erased. Otherwise, erase them using an erasing shield to protect the good lines. See Fig. 5–4(e).

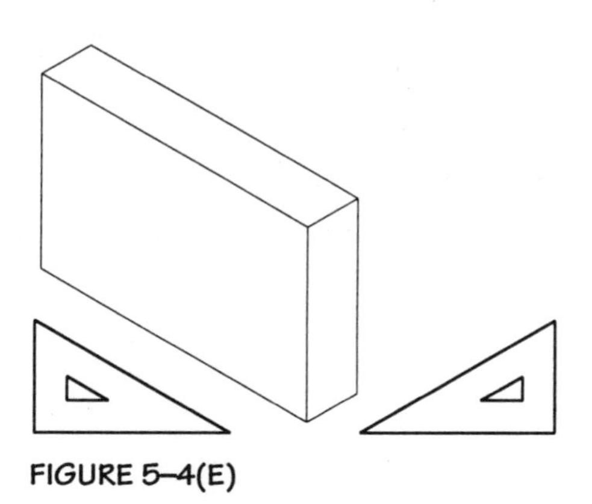

FIGURE 5–4(E)

STEP 8 Using the appropriate scale, mark off some distances to help locate edges on the object. See Fig. 5–4(f). Use measurements from Fig. 5–1.

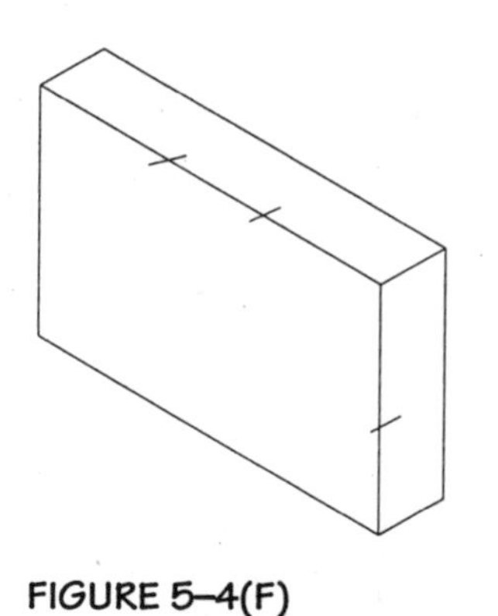

FIGURE 5–4(F)

STEP 9 Lightly draw lines through these points, holding your triangle firmly against the parallel straightedge. See Fig. 5–4(g).

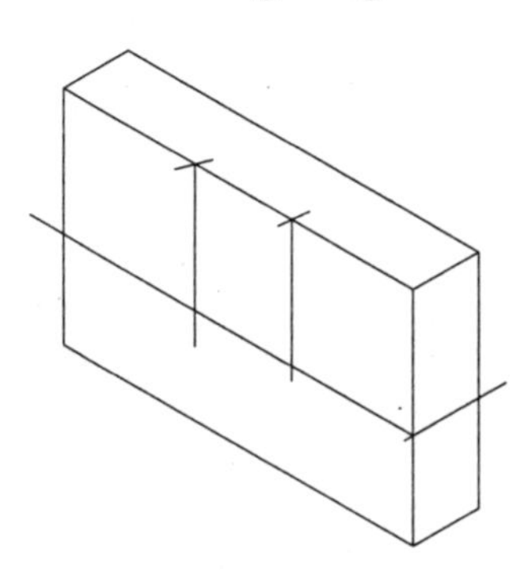

FIGURE 5–4(G)

STEP 10 Add appropriate lines as shown. See Fig. 5–4(h).

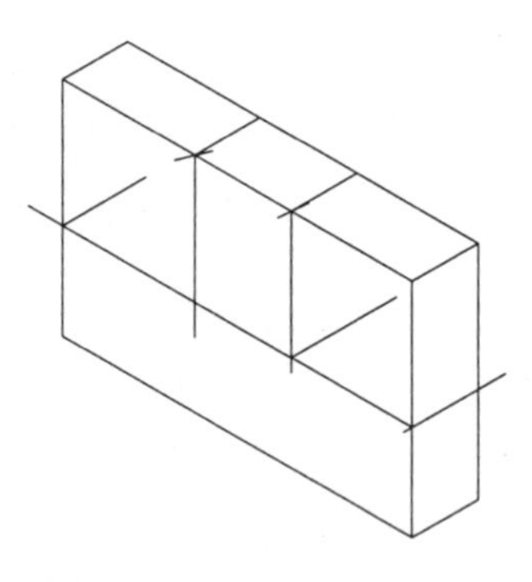

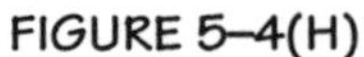
FIGURE 5–4(H)

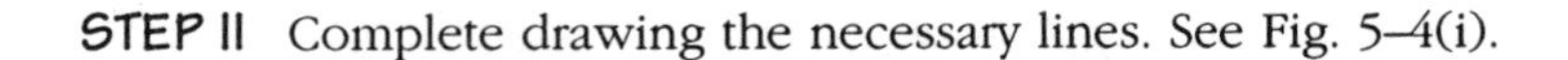
STEP 11 Complete drawing the necessary lines. See Fig. 5–4(i).

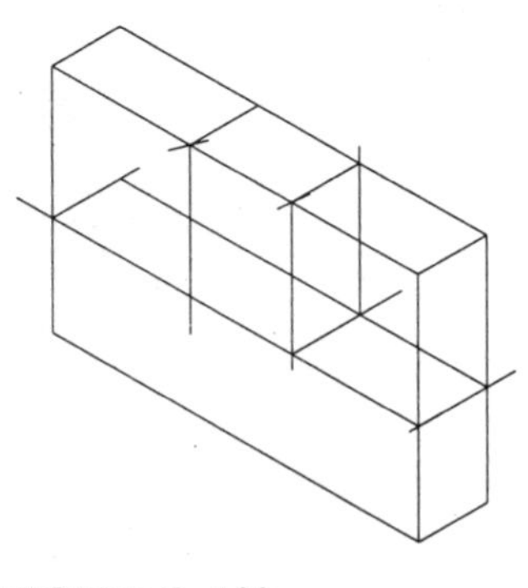

FIGURE 5–4(I)

STEP 12 Using the erasing shield to cover the lines that you wish to keep, erase any unnecessary lines. Darken the remaining lines with the 0.7 mm pencil, using your triangle held firmly against the parallel straightedge. Do not try to darken these lines by sketching, and do not try using the triangle without having it held firmly against the parallel straightedge. Do not place any dimensions on your isometric view. See Fig. 5–4(j).

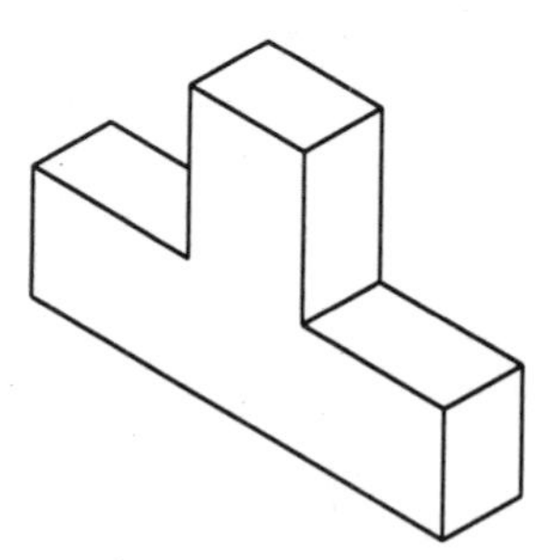

FIGURE 5–4(J)

HINTS ABOUT USING DRAFTING INSTRUMENTS

a. *Reminder:* If you are using a lead holder and *not* a 0.3 or 0.7 mm technical pencil: To achieve an even line quality on an instrument drawing, roll the pencil between your thumb and finger as you draw a line. With this method, the lead will wear evenly and the line will not get wider as you draw. As explained in Chapter 4, this method is not necessary for sketching.

If you are using a 0.3 or 0.7 mm technical pencil: You can simply hold the pencil in a vertical position—the thickness of the lead will maintain the line thickness. You do not have to rotate this type of pencil.

b. Use the scale for *measuring* only, *not* for drawing lines. Use the triangle or parallel straightedge to draw lines. If you run your fingertip along the edge of the scale, you will feel that it has indentations for each marking or graduation. This edge will wear your lead quickly, can dirty your scale, and consequently dirty your drawing.

c. Use only the TOP of your parallel straightedge. Avoid using the bottom edge for drawing, and avoid putting your triangle on the bottom edge. The reason is that parallel straightedges move up and down to form parallel lines, but if you use the top and bottom edges, you are assuming that these two edges are parallel to each other. This assumption may or may not be true, depending upon the quality of the straightedge.

d. After you have used your instruments to make a drawing, do not go over the lines freehand.

e. DECIMAL EQUIVALENTS TABLE
Refer to page 326 in the Appendix. This table converts a fraction of an inch to a decimal and vice versa. It also converts a decimal or a fraction of an inch to the metric equivalent in millimeters.

Example: 1/64" equals .015625" in decimal form; 1/64" also equals 0.39688 millimeters. Similarly, 1/8" equals .125" in decimal form, which is also equal to 3.175 mm.

f. When using triangles, avoid using the tip when drawing a line. Figure 5–5(a) illustrates a poor way of using the triangle. Figure 5–5(b) shows a better way.

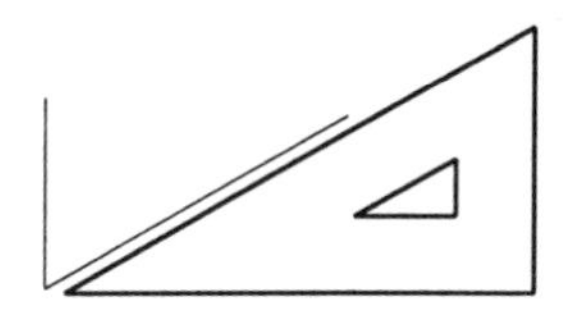

NO

FIGURE 5–5(A)

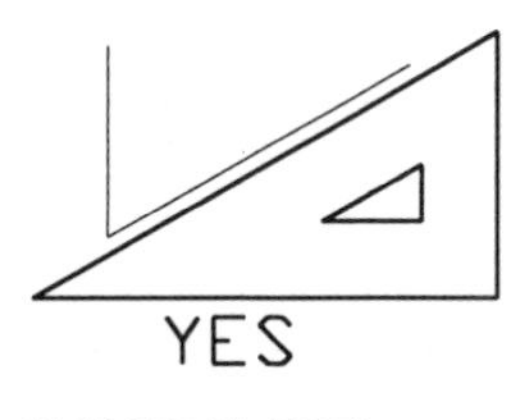

YES

FIGURE 5–5(B)

g. Use your triangles, rather than your protractor, for drawing 30°, 60°, 90°, or 45° angles because they will give you a more accurate angle. See page 323 of the Appendix for information on triangles.

h. Remember to tape your good drafting paper in the middle of the drafting board (not at the bottom of the board), using about 1 1/2" (about 40 mm) of masking or drafting tape at each of the four corners.

SOURCES OF ERROR IN ISOMETRIC DRAWINGS

If your isometric drawing does not look correct, you may have made one of the following errors:

a. You made an inaccurate measurement.

b. You drew a line that was not parallel to one of the isometric axes. One side of the triangle must always be firmly held against the parallel straightedge. Place triangles on the top of the parallel straightedge only, and avoid the bottom edge in case the top and bottom are not parallel to each other.

c. You used the 45° triangle when you should have used the 30°–60° triangle.

d. You used the 60° angle instead of the 30° angle on the 30°–60° triangle.

NOTES TO REMEMBER ABOUT ISOMETRIC DRAWINGS

a. Always start your isometric drawing by making an isometric box to completely enclose the object.

b. Locate the starting point by estimating where the bottom corner should be.

c. Draw lightly at first, using the 0.3 mm technical pencil, and finish with the 0.7 mm pencil.

STEPS FOR DRAWING AN ISOMETRIC CIRCLE

NOTE: The diagrams accompanying the following steps illustrate an isometric circle drawn on the top surface of an isometric cube. The steps apply equally well to circles drawn on the right or left surfaces of a cube.

STEP 1 Start by drawing an isometric square to enclose the circle. Note that there are two *large angles* (larger that 90°, or obtuse angles) and two *small angles* (less than 90°, or acute angles) in each isometric square. These angles are illustrated in Fig. 5–6(a).

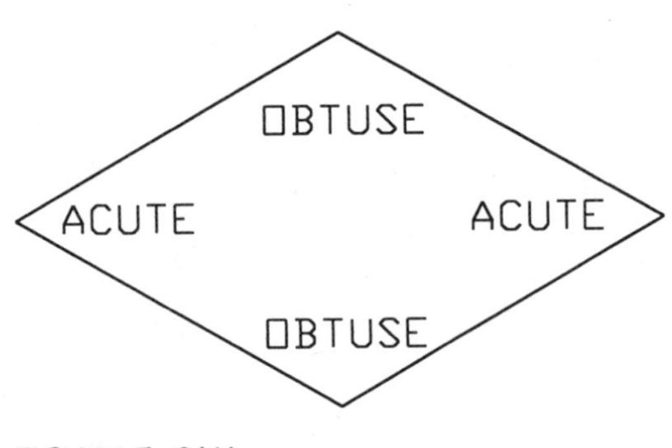

FIGURE 5–6(A)

STEP 2 Draw lines from the vertex of one of the large (obtuse) angles to the midpoint of the opposite sides. See Fig. 5–6(b). You can find the midpoint by using your scale or by referring to Construction 4 in Chapter 4.

A second method applies to isometric views only, as follows. Using one of the large angles, draw perpendiculars to the opposite side. A perpendicular is a line that is 90° to another line. Be sure to draw these lines with your 30°–60° triangle held *firmly* against the parallel straightedge. See Fig. 5–6(c). (When drawing on the right and left surfaces, some perpendiculars are horizontal lines. See Fig. 5–7. These horizontal lines are drawn with the parallel straightedge only.)

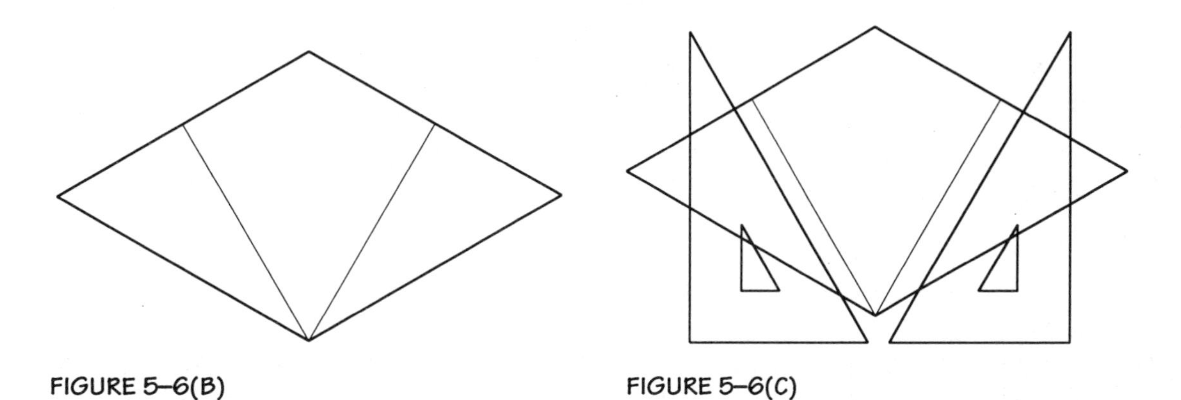

FIGURE 5–6(B)

FIGURE 5–6(C)

STEP 3 As in step 2, draw lines from the vertex of the other large angle to the midpoint of the opposite side. Alternatively, in the case of isometric drawings, draw lines from the vertex of the other large angle perpendicular to the opposite side. See Fig. 5–6(d).

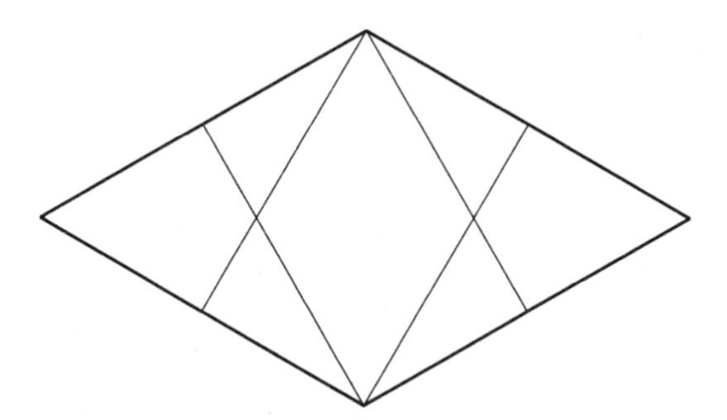

FIGURE 5–6(D)

STEP 4 Where these perpendiculars cross will be the center of an arc. The radius will be from this center to the closest side. Draw two arcs using your compass. See Fig. 5–6(e). See Hint (a) on page 148.

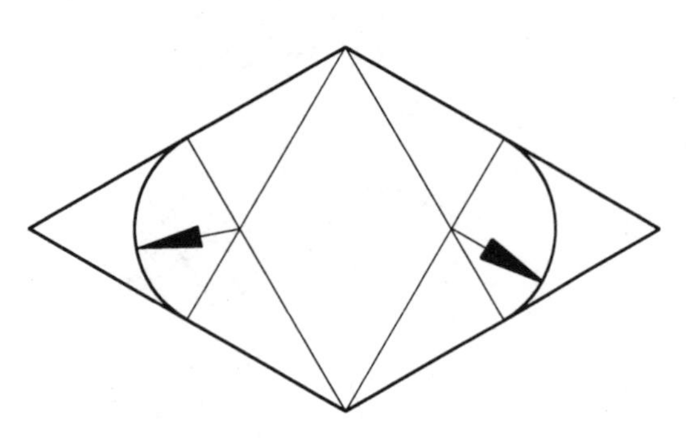

FIGURE 5–6(E)

STEP 5 With the vertex of one of the large angles as center, and radius to the opposite side as measured along the perpendicular lines that were drawn in steps 3 and 4, draw an arc with your compass. Draw the last arc in a similar manner from the other vertex to complete the isometric circle. See Fig. 5–6(f).

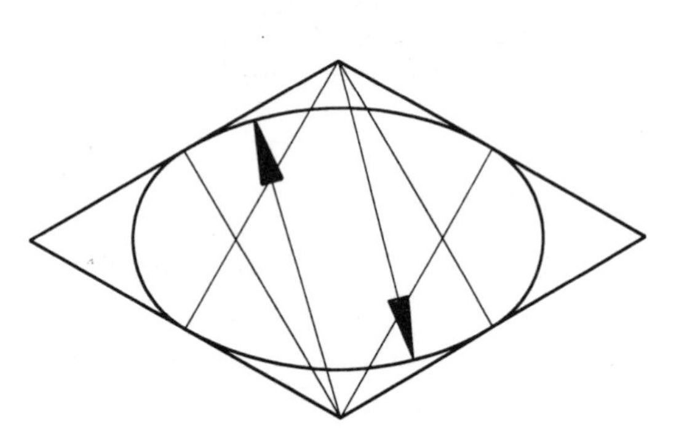

FIGURE 5–6(F)

To draw isometric circles on the two remaining isometric surfaces, follow the steps on the preceding pages and refer to Fig. 5–7.

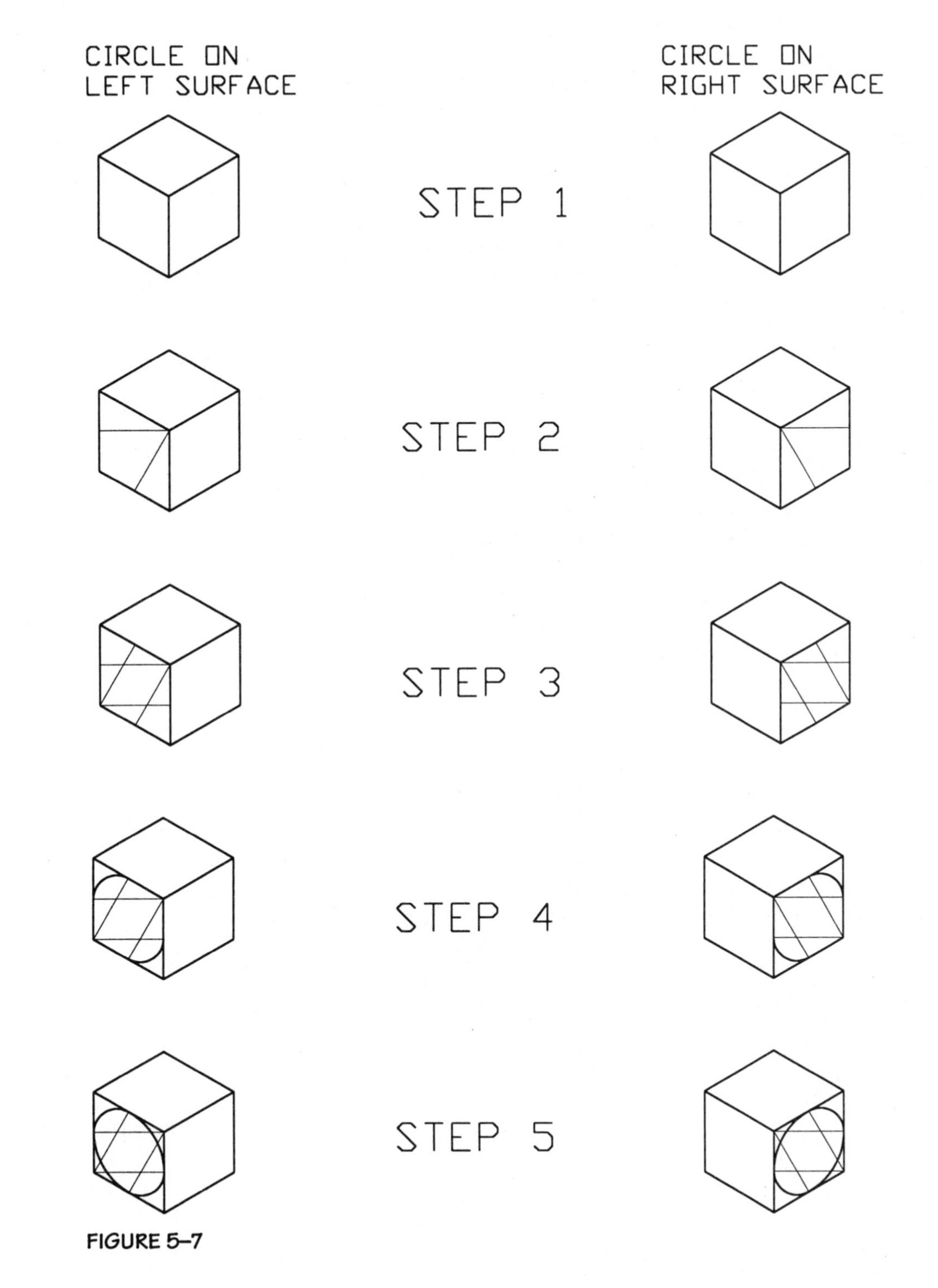

FIGURE 5–7

HINTS ABOUT DRAWING ISOMETRIC CIRCLES USING DRAFTING INSTRUMENTS

a. You will be using the compass in this chapter. The lead in your compass should protrude about 1/2" (13 mm). Over a waste basket, sharpen your compass lead with the sandpaper pad to a flat chisel edge. The metal point in your compass should protrude about 1/16" (2 mm) beyond the lead, since some of the point will penetrate the paper and soft drawing surface. When purchased, your compass included a piece of lead and may have had some extra lead in a small tube. You probably do not know the grade of this lead. If you do not like the line quality of this lead, discard it and buy H lead to fit your compass. If you are using a lead holder, break off a piece of lead from your H lead holder and sharpen it to a flat chisel edge with your sandpaper pad. You should now be able to make a circle or arc with the same thickness and darkness as a straight line drawn with a pencil. To keep your drawing clear and clean, use your compass only once for each arc or circle. Avoid drawing a light arc or circle and darkening it later.

b. Each time that you want to draw an isometric circle, *always* start with an isometric square with the sides equal to the diameter of the circle. Then follow the steps previously described. Also use this principle to draw an isometric arc (part of a circle).

c. To save time when drawing an isometric cube, pick up your scale only once. Start by drawing the isometric axes. Pick up your scale and mark off the required distance on each of the three lines. Now put down your scale and finish the cube.

SOURCES OF ERROR IN DRAWING ISOMETRIC CIRCLES

These errors are similar to those outlined on page 144, but are specific to isometric circles.

a. All sides of the square used to draw the circle may not have been drawn of equal lengths. You cannot draw a circle in a rectangle and have the circle touch all the sides.

b. Your construction lines may not have been drawn accurately. Was one side of the triangle always held *firmly* against the parallel straightedge when drawing the lines?

Figure 5–8 shows an example of an isometric cube with an isometric circle on each side. Notice the THIN construction lines and THICK outlines.

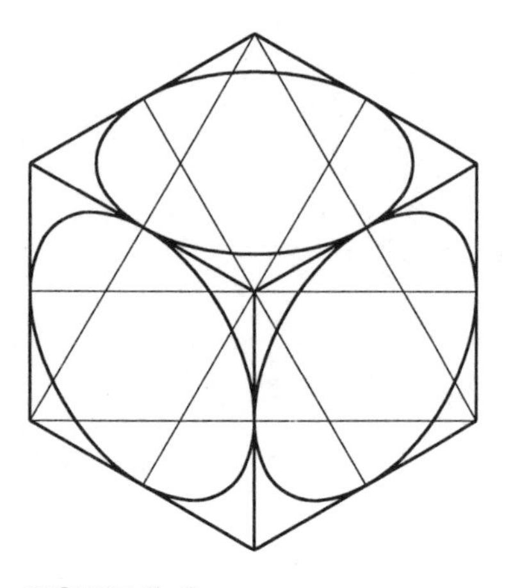

FIGURE 5–8

EXERCISE 5–1

On a blank sheet of paper, draw a 3" (75 mm) isometric cube using your drafting instruments. On each face of this cube, draw an isometric circle of 3" (75 mm) diameter. Work on only one face of the cube at a time. Work carefully. Your answer should look similar to the diagram in Fig. 5–8.

HINT: Tape down the paper by lining up its top edge with the top of the parallel straightedge. Avoid using the bottom of the parallel straightedge. Put a 1 1/2" (about 40 mm) piece of masking or drafting tape at each corner of the paper.

NOTE: Do not go over the lines freehand!

FOR EXERCISES 5–2 TO 5–9

On the following pages are eight objects shown in their isometric views. Dimensions are given in inches, with millimeters in parentheses. The symbol Ø means diameter, and R means radius. Do the following steps for each figure.

a. On an 8 1/2" × 11" sheet of paper (metric A4), make a freehand isometric sketch of the object. You will be copying the view as it is drawn in the figure. Do not use a straightedge or scale.

b. Using drafting instruments, draw the isometric view of the object. Use C size drafting paper (metric A2) or choose another suitable size according to your needs. Draw full scale (or to a reduced scale) using the dimensions shown, but do not put any dimensions on your drawing. Of course, your drawings will be larger than those in the figure. Small isometric circles are usually drawn with an isometric circle template. If you do not have one, omit the small holes isometrically (those under 1", or 25 mm). To simplify the task, draw an isometric square in place of the small isometric circles.

c. On an 8 1/2" × 11" sheet of paper (metric A4), make a freehand sketch of the necessary orthographic views of the object. Choose your own front view by following the theory learned in Chapter 2.

EXERCISE 5–2

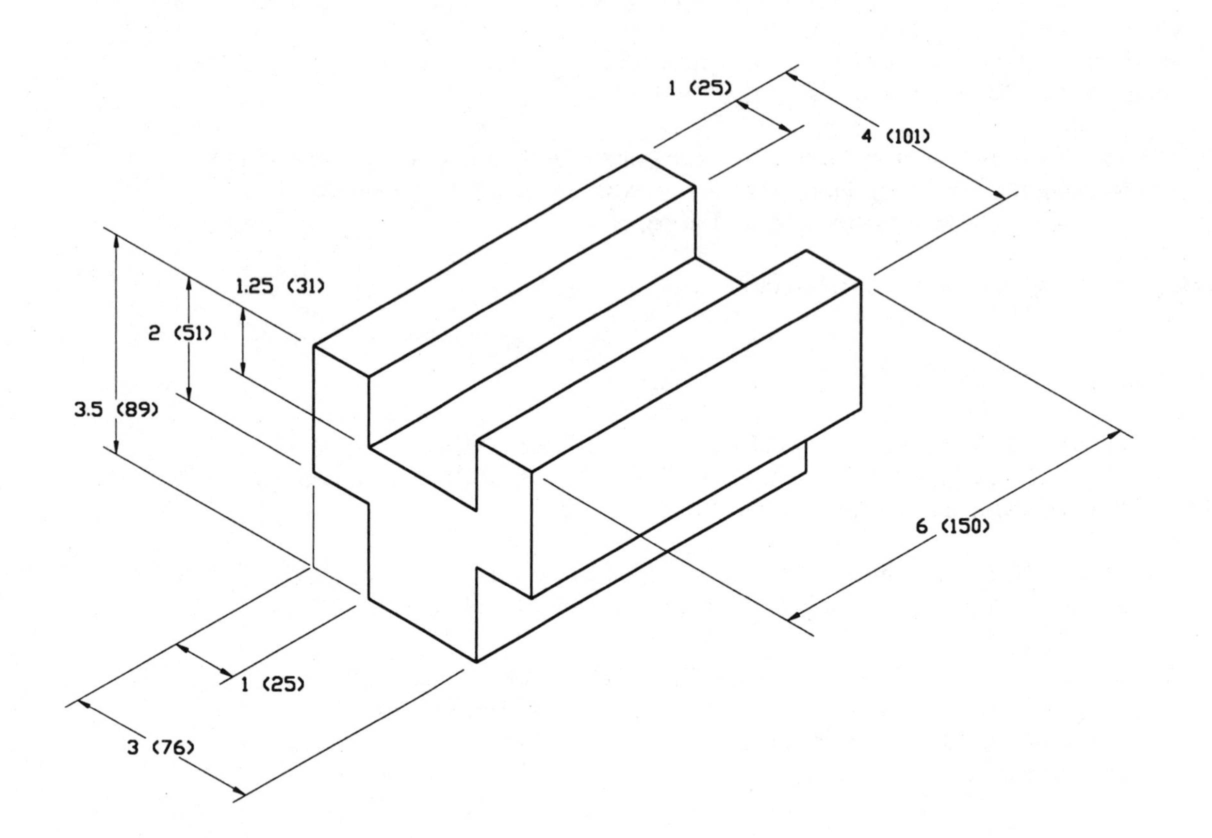

EXERCISE 5–3

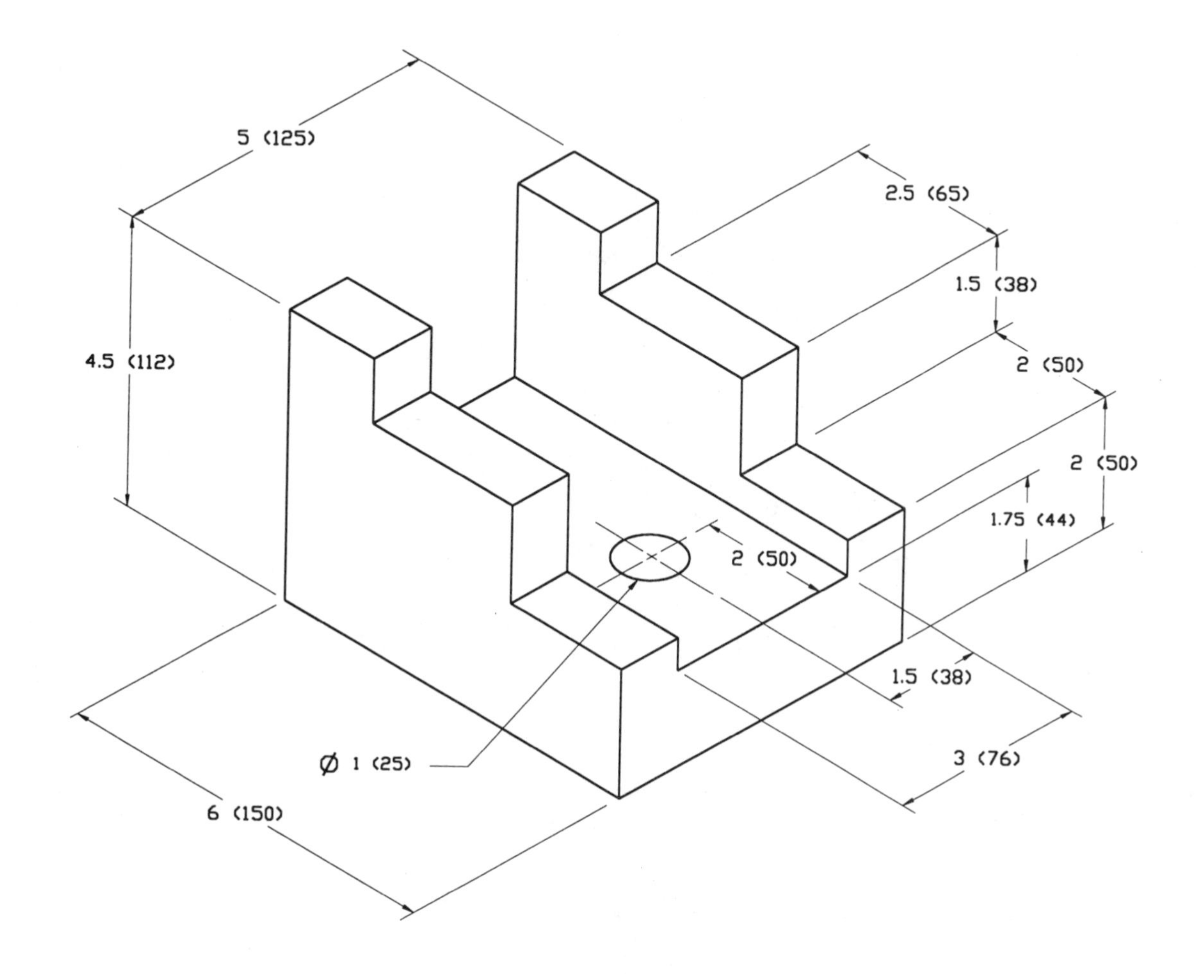

EXERCISE 5–4

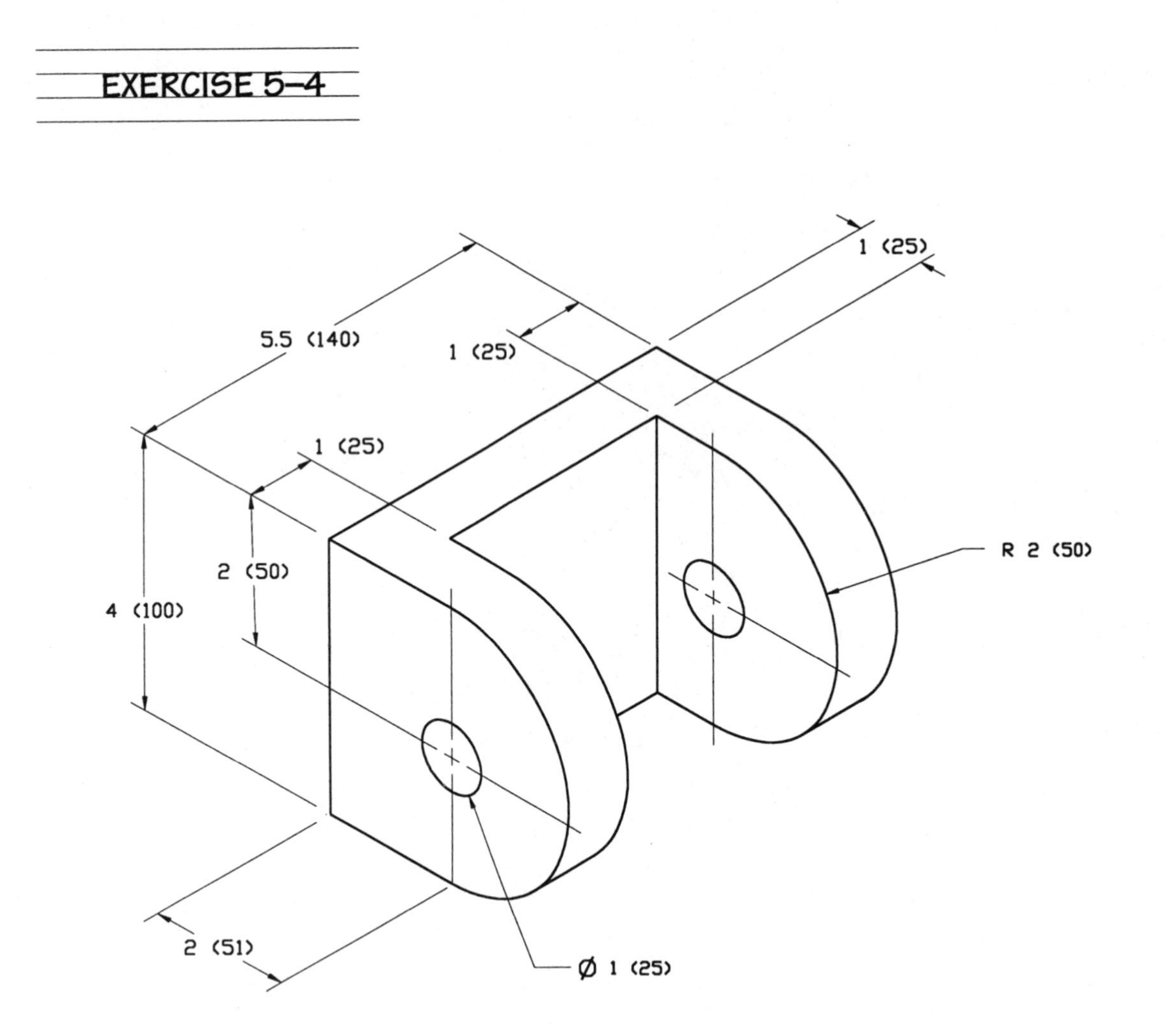

EXERCISE 5–5

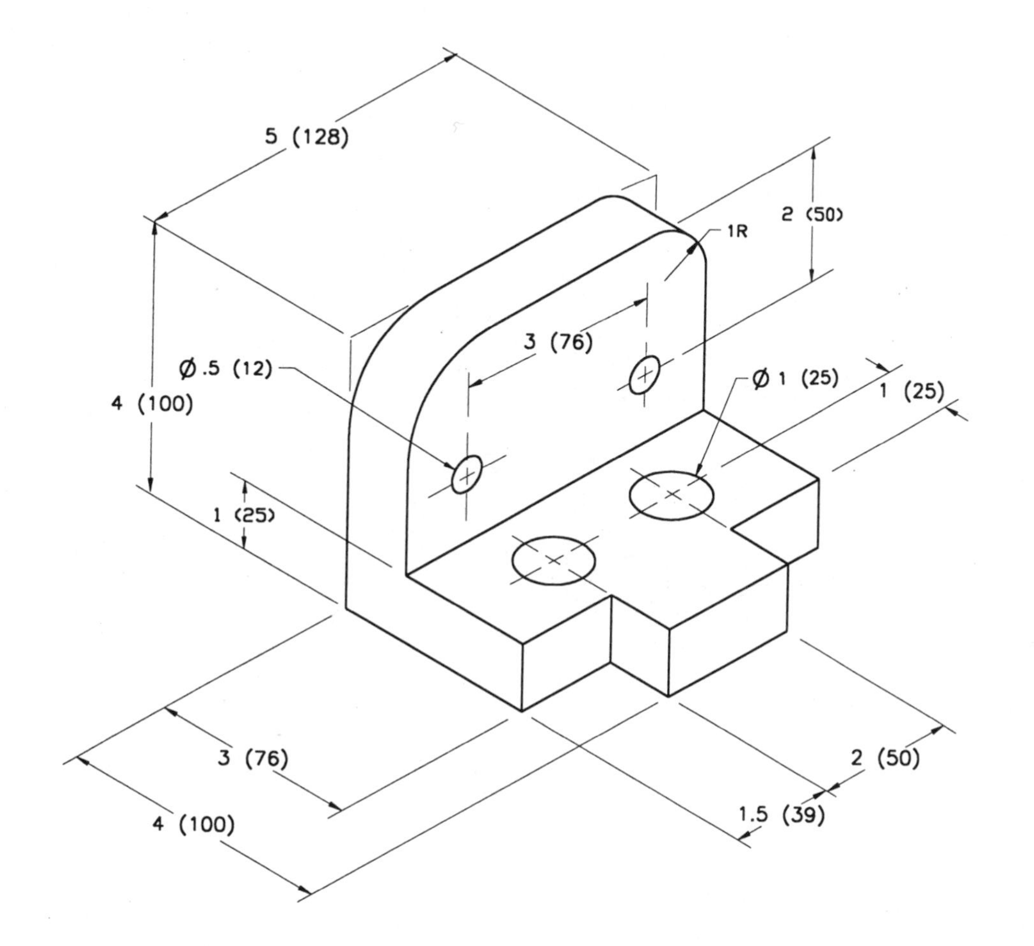

EXERCISE 5–6

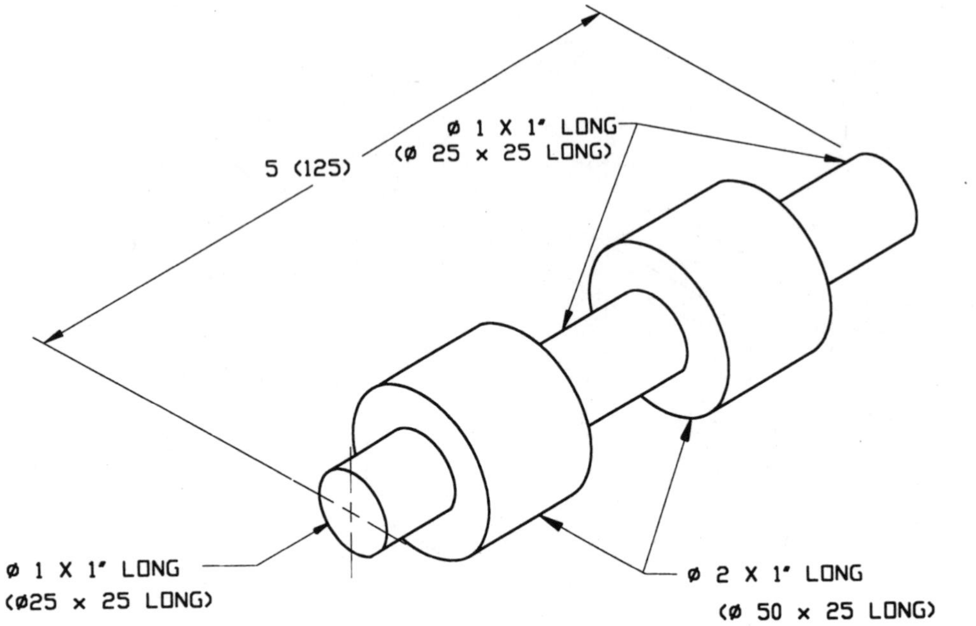

5 (125)

Ø 1 X 1" LONG
(Ø 25 x 25 LONG)

Ø 1 X 1" LONG
(Ø25 x 25 LONG)

Ø 2 X 1" LONG
(Ø 50 x 25 LONG)

EXERCISE 5–7

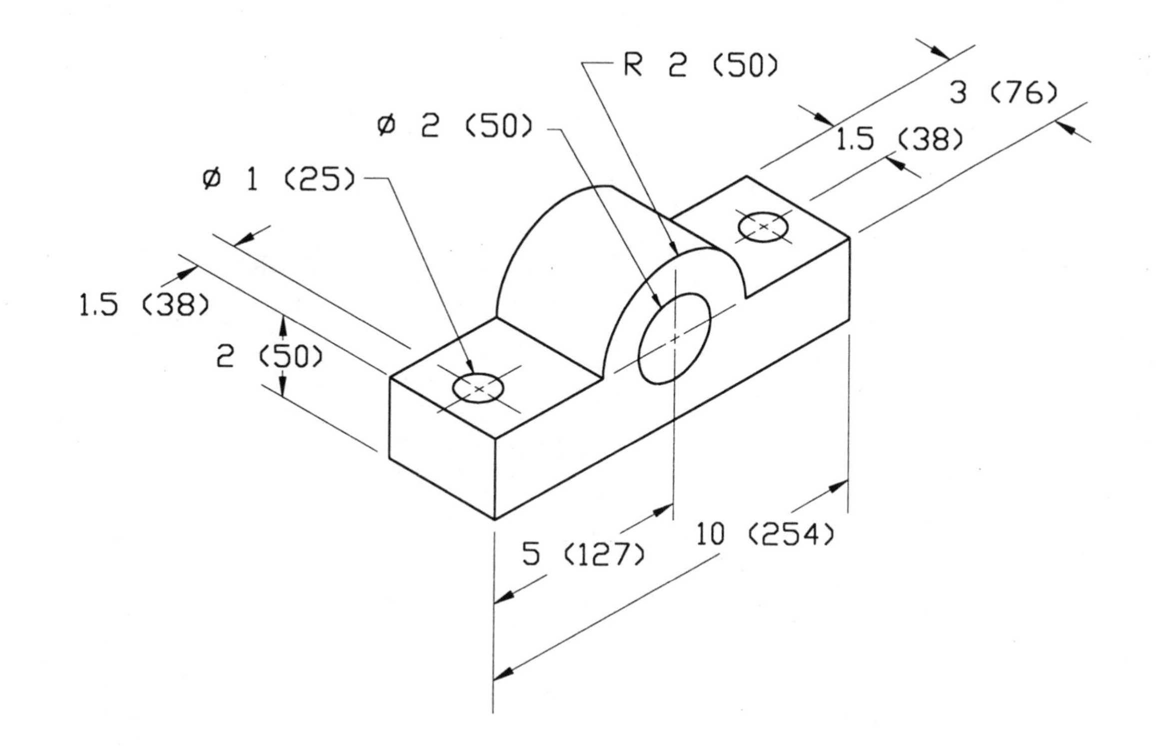

EXERCISE 5–8

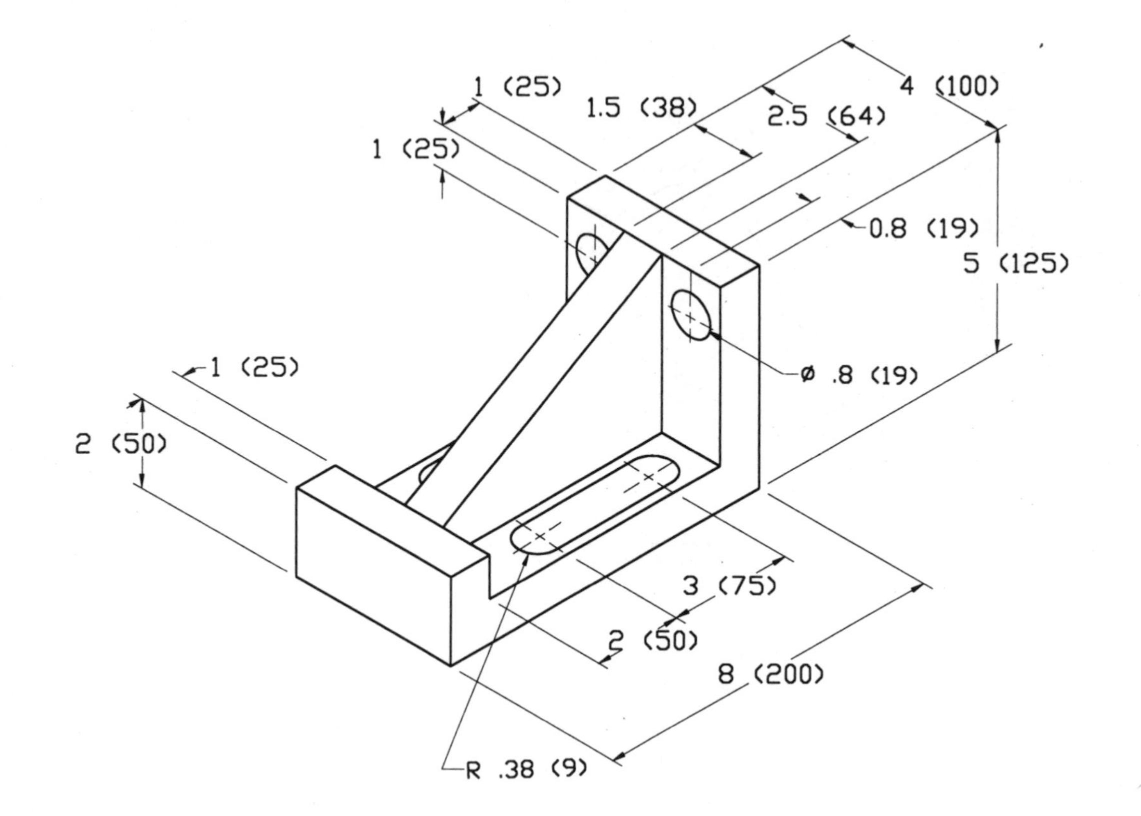
1 (25)
1.5 (38)
4 (100)
2.5 (64)
1 (25)
0.8 (19)
5 (125)
1 (25)
ø .8 (19)
2 (50)
3 (75)
2 (50)
8 (200)
R .38 (9)

EXERCISE 5–9

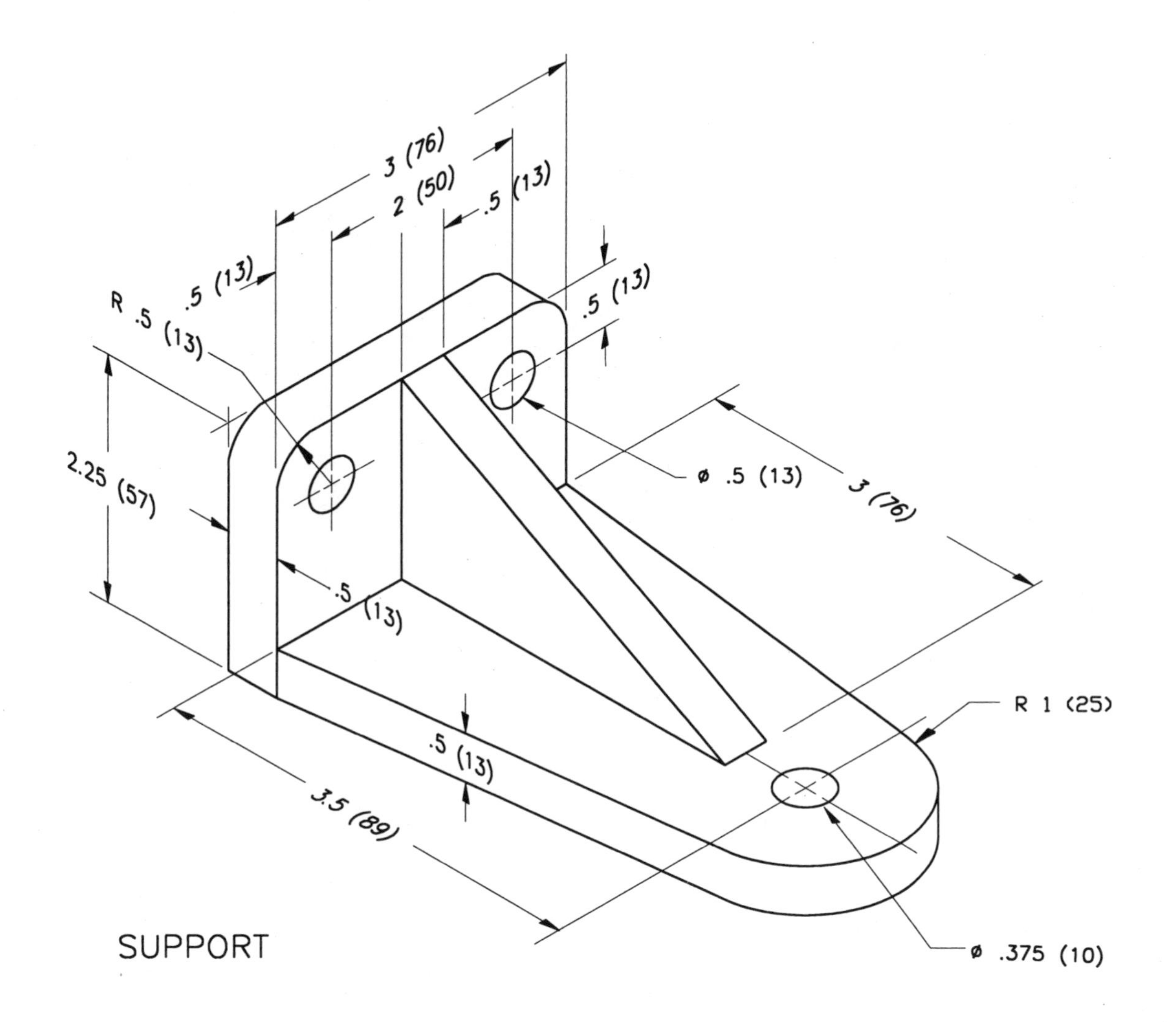

SUPPORT

ORTHOGRAPHIC VIEWS USING DRAFTING INSTRUMENTS

6

OBJECTIVE

When you have finished this chapter, you should be able to draw orthographic views using your drafting instruments.

EQUIPMENT

You are required to have all of your drafting instruments for this chapter.

STEPS FOR MAKING A THREE-VIEW ORTHOGRAPHIC DRAWING WITH INSTRUMENTS

Most drawing done in industry uses orthographic projection. By following the steps below, you will be able to draw orthographic views properly centered on drafting paper.

STEP 1 Clean all drawing instruments, using a soft cloth for the plastic scales, triangles, and templates. If they are very dirty, use mild soap and water. Similarly, clean the drafting table. Cleanliness is very important in drafting. If you are not careful, your drawing will quickly become dirty, and this may lead to errors in reading the drawing. Keep the sandpaper pad in a plastic bag so it does not dirty your instruments. Similarly, keep the pencil pointer, if you have one, in a separate plastic bag.

STEP 2 Sketch freehand on any sheet of paper (graph paper would be helpful) what you plan to draw with instruments. This is a very important step for the beginning drafter. With experience, or when using CAD (computer-aided drafting), you may be able to skip this step to save time.

STEP 3 Refer to Fig. 6–1 on the next page. On your sketch, label the height as H, the width as W, and the depth as D on the appropriate views, and the space between views as S, as shown in the figure. Use 3" (75 mm) for the space between views in the first exercise. (The size of S depends on the size of the views and the number of dimensions placed between them. This topic will be studied further in the next chapter.) Add the appropriate numbers to obtain the total height and width of your drawing. Then tape your sketch to the top of the drawing board to guide you in making a good instrument drawing.

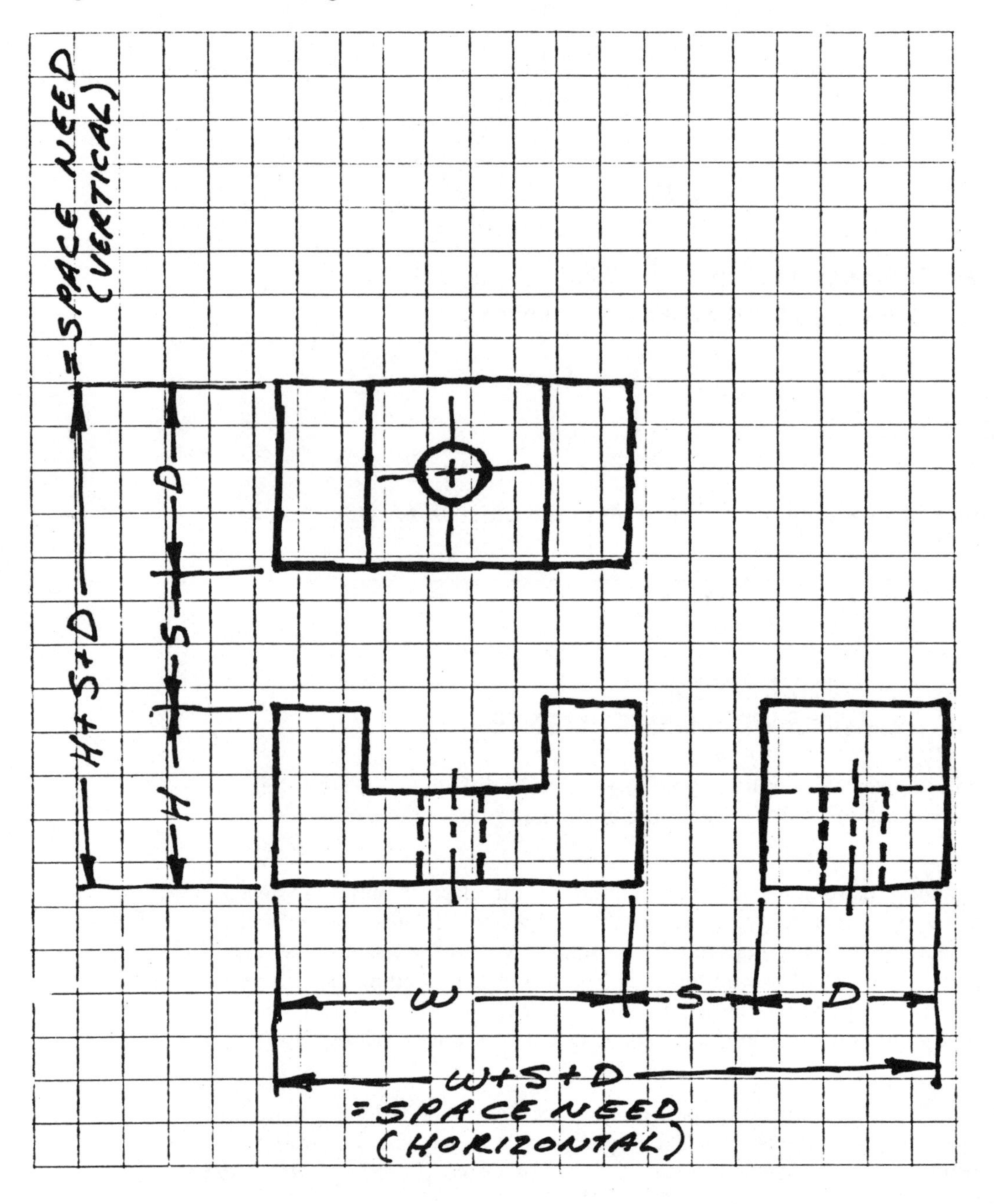

FIGURE 6–1 Example of a freehand sketch used to plan an instrument drawing. (See step 3 for an explanation of the lettered dimensions.)

STEP 4 By referring to the views and dimensions from step 3, choose an appropriate size drafting sheet for your drawing. For this first exercise, use C size (metric A2) drafting paper and full scale, or choose an appropriate scale so that the drawing will fit your paper. To prepare for taping down a clean sheet of this paper on your drawing surface, line it up with a previously drawn line on the paper (if there is one), or with a border line. If the paper is blank, line up the top of the paper with the top edge of your parallel straightedge. Tape the four corners.

STEP 5 Most companies and many schools use drafting paper with preprinted border lines and title block. If you are using a blank sheet of paper, draw a border line 1/2" (13 mm) from the edge of the paper and draw a title block. A sample title block is shown in Fig. 6–2. Although the style chosen for the title block is optional, it must include the information listed below. The letters correspond to the suggested format in Fig. 6–2.

a. Scale of drawing

b. General specifications and notes

c. Company name

d. Title of drawing

e. Drafter name or initials and, if necessary, approvals

f. Drawing number

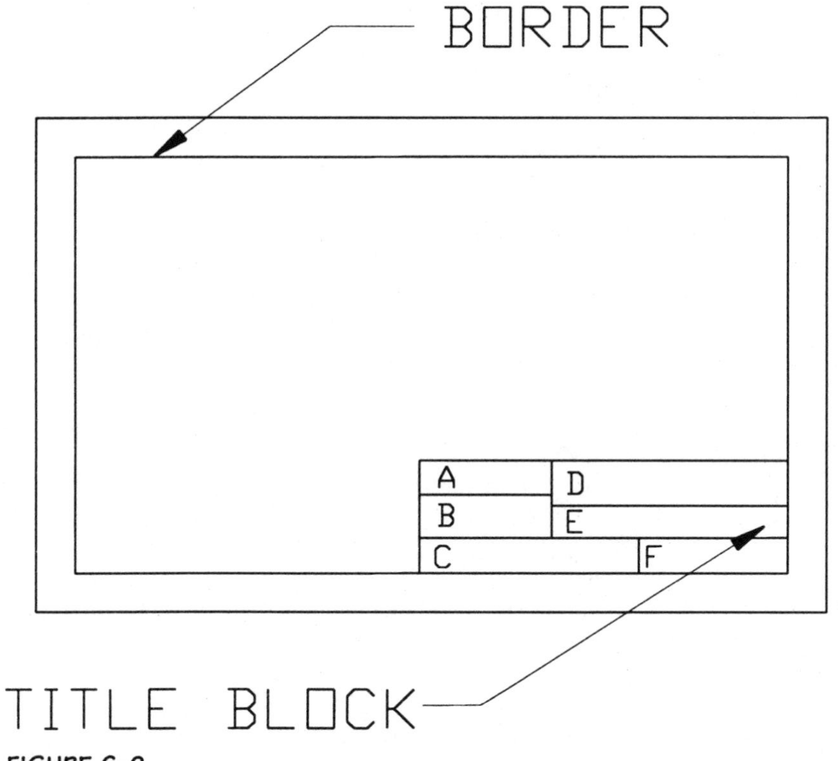

FIGURE 6–2

STEP 6 Center your drawing. You may skip this step and instead center your drawing by estimating the starting point. With experience, you will be able to estimate your starting points. However, the following instructions take only a few minutes.

Measure the horizontal distance available inside the border lines on your blank drafting sheet. This is the "space that you *have* horizontally" in which to draw, as shown in Fig. 6–3. From this, subtract the "space that you *need* horizontally," which is the overall horizontal distance you calculated on your sketch (using Fig. 6–1). Divide the answer by 2.

SPACE THAT YOU
HAVE (HORIZONTAL)

FIGURE 6–3

Expressed as a mathematical formula, we have:

$$\frac{\text{Space that you have} - \text{Space that you need}}{2} = \text{Distance from border to edge of drawing}$$

Measure this distance from the right-hand border and lightly draw a vertical line. Then measure the D, S, and W distances you marked on your sketch and *lightly* draw vertical lines with 0.3 mm lead. Refer to Figs. 6–4(a) through (d).

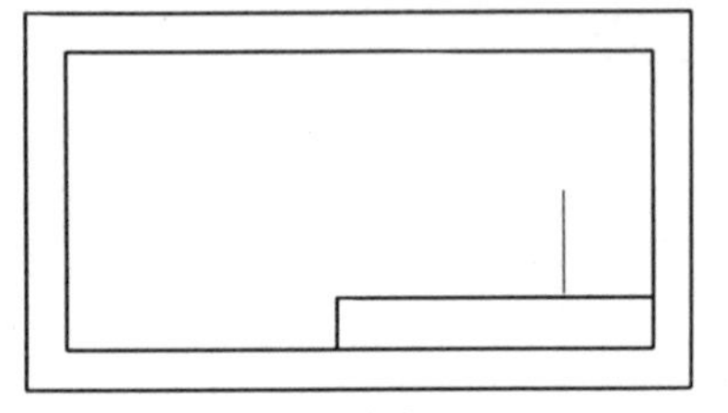

(A)

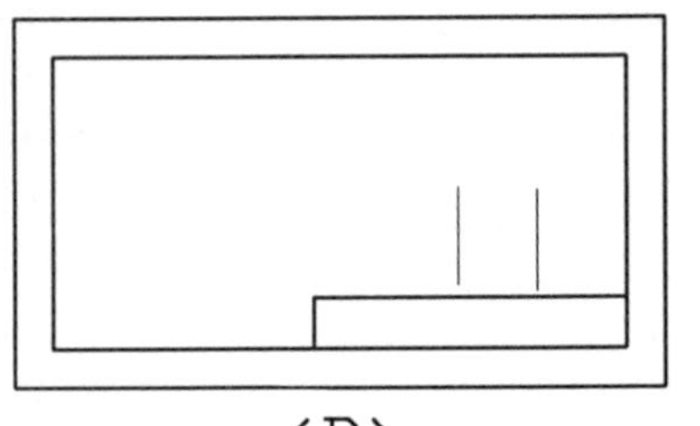

(B)

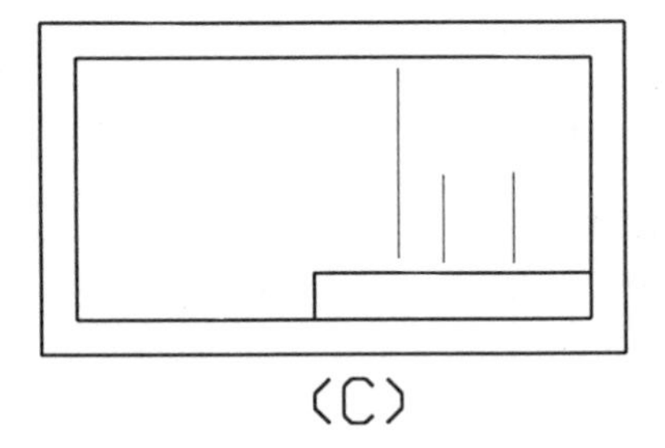

(C)

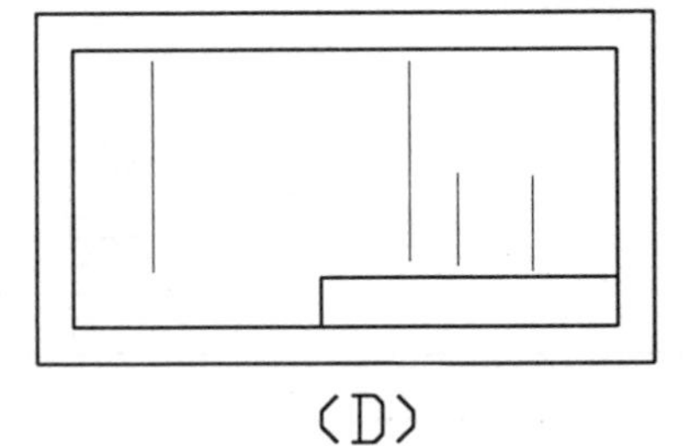

(D)

FIGURE 6–4

Do this same calculation for the vertical distance. See Fig. 6–5.

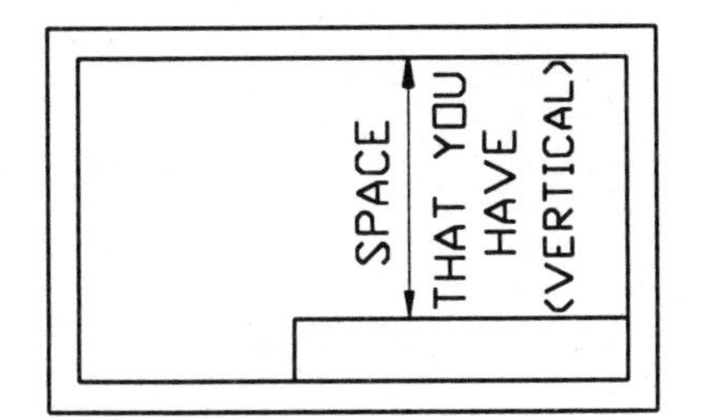

FIGURE 6–5

Draw horizontal lines at the appropriate locations to form boxes for the top, front, and side views. Refer to Figs. 6–6(a) through (d).

You should now have three boxes drawn to represent the top, front, and side views (assuming that this is a three-view drawing). The objective of these steps is to demonstrate that you should proceed by measuring the location of the vertical lines from right to left and the horizontal lines from top to bottom. The right-to-left order allows you to draw and slide your triangle so that it does not smear the lines already drawn. (If you are left-handed, you should work from left to right.)

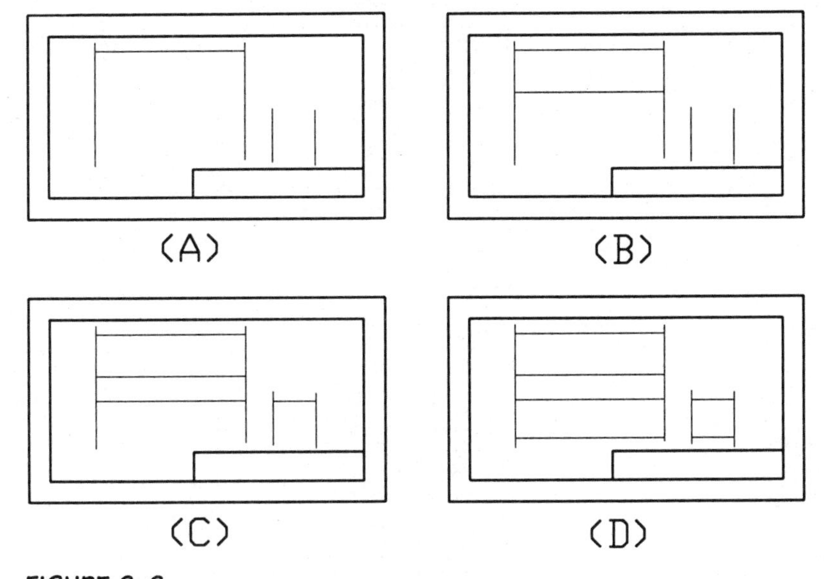

FIGURE 6–6

STEP 7 Erase unnecessary construction lines only if they are too dark. See Fig. 6–7(a). Add center lines to your drawing if needed. *Lightly* add details with your 0.3 mm pencil, as illustrated in Figs. 6–7(b) and (c).

STEP 8 Add circles and curves to your drawing if needed using your 0.7 mm pencil. These lines should be darker than those drawn in step 7. Use the compass for large circles and curves. Use the circle template where possible for small circles and curves, especially for all fillets and rounds, which are illustrated in the Appendix (page 323).

STEP 9 Erase the construction lines if they are too dark. See Fig. 6–7(d).

STEP 10 Darken the required lines. Make sure that the center lines are thinner than the outline. Refer to Fig. 6–7(e).

STEP 11 Fill in the title block (0.7 mm pencil). The choice of a proper title is very important. It should identify the object as briefly and precisely as possible. Refer to Fig. 6–7(f).

STEP 12 Add dimensions (after you have completed Chapter 7).

NOTE: Remember not to erase guidelines used for forming any letters or numbers.

The error some beginners make is to start their drawing as shown in Fig. 6–8 on the next page. *Do not* start your drawing this way.

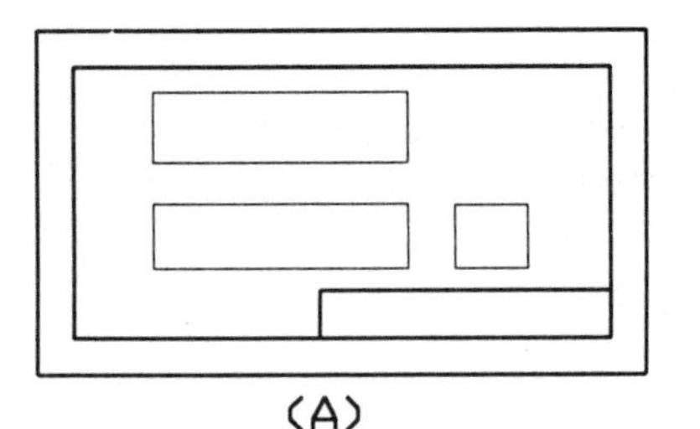

(A)

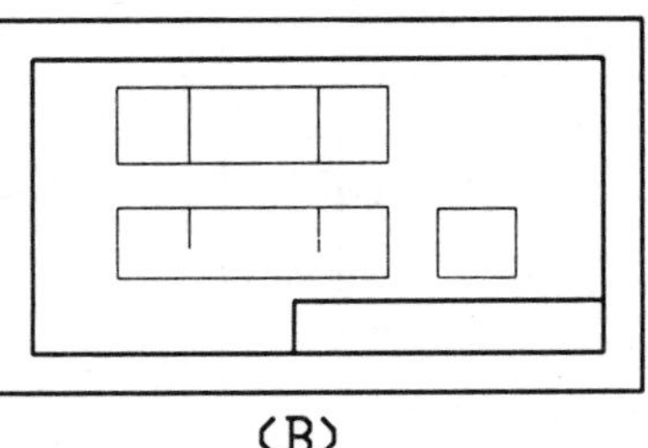

(B)

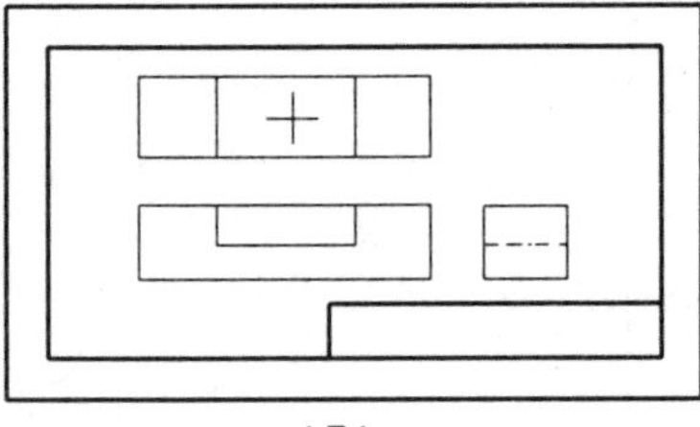

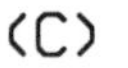

(C)

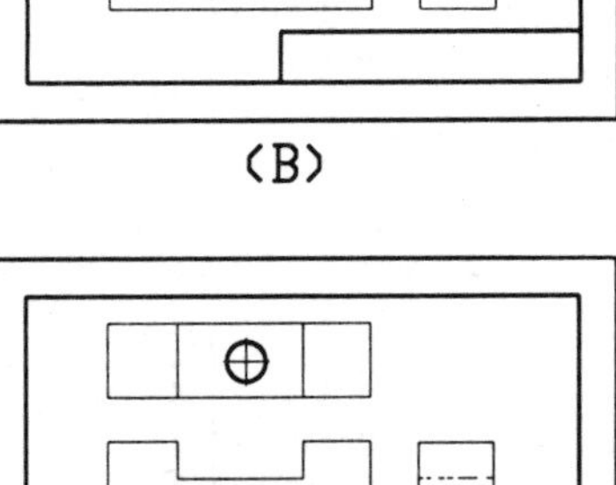

(D)

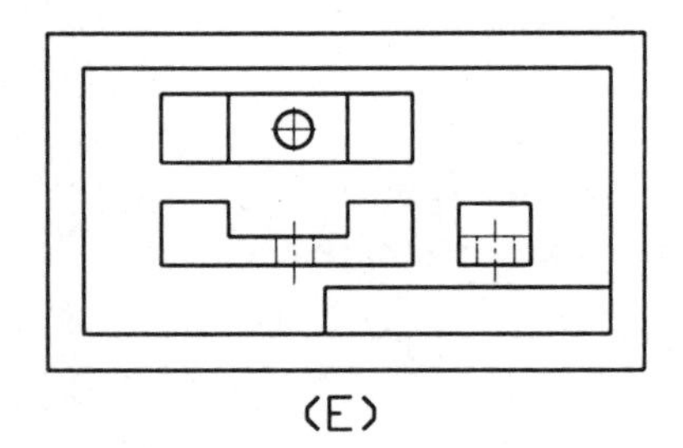

(E)

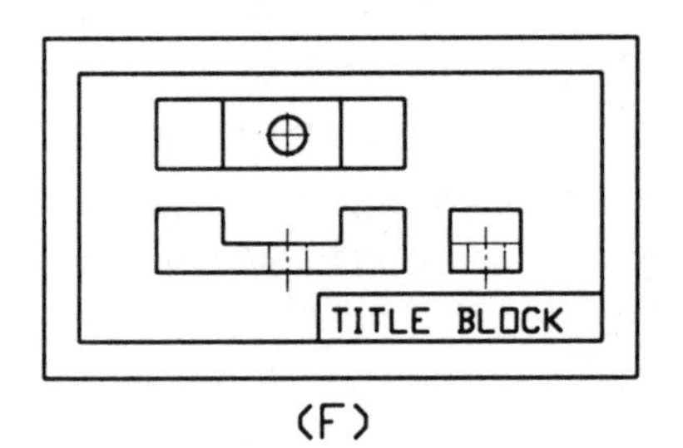

(F)

FIGURE 6–7

INCORRECT WAY TO LAY OUT A DRAWING

Figure 6–8 illustrates poor practice, because the drafter did not work from right to left. Instead, this beginner tried to "box in" the area to draw the three views. Since it is impossible to make exact measurements, slight errors in measurement will accumulate and may result in views too large to fit the big box. This error is illustrated in Fig. 6–8(d). (Notice the two lines on the left.) Always work from right to left and from top to bottom. Any errors that occur when working correctly (from right to left) will be reflected in the distance from the border to the last line drawn, which is an error of minor importance.

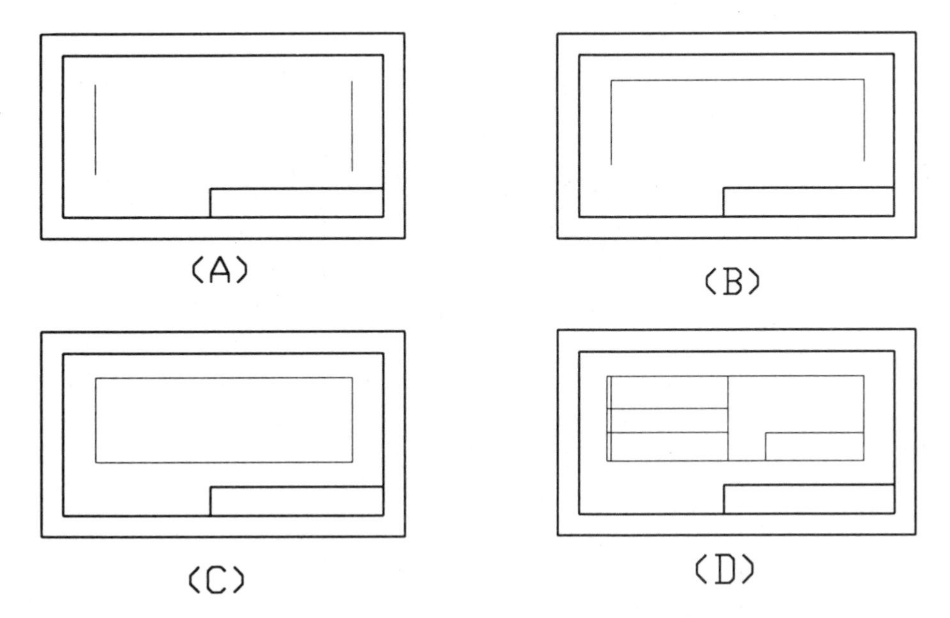

FIGURE 6–8 An example of poor practice in laying out a drawing.

DRAWING EACH VIEW

When drawing *each* view, it is important to make your overall or large measurements first (such as total length, width, or height of each view) and then to work inward making smaller measurements. The reason is that it is impossible to make absolutely accurate measurements. For example, when we measure 1" with our scale, we are relying on the accuracy of the scale, as well as the accuracy of our eye in locating the line with the marking on the scale. In addition, the line representing our measurement has a certain thickness. Although each error may seem insignificant, many consecutive measurements contribute small errors and we can end up with a fairly large error. Illustrations of an incorrect and a correct way of drawing a view appear in Fig. 6–9. This is consistent with what we learned in the steps we have just followed, that is, starting with a box for each view and constructing the view in that box. (Compare with Figs. 6–7(a) through (f).)

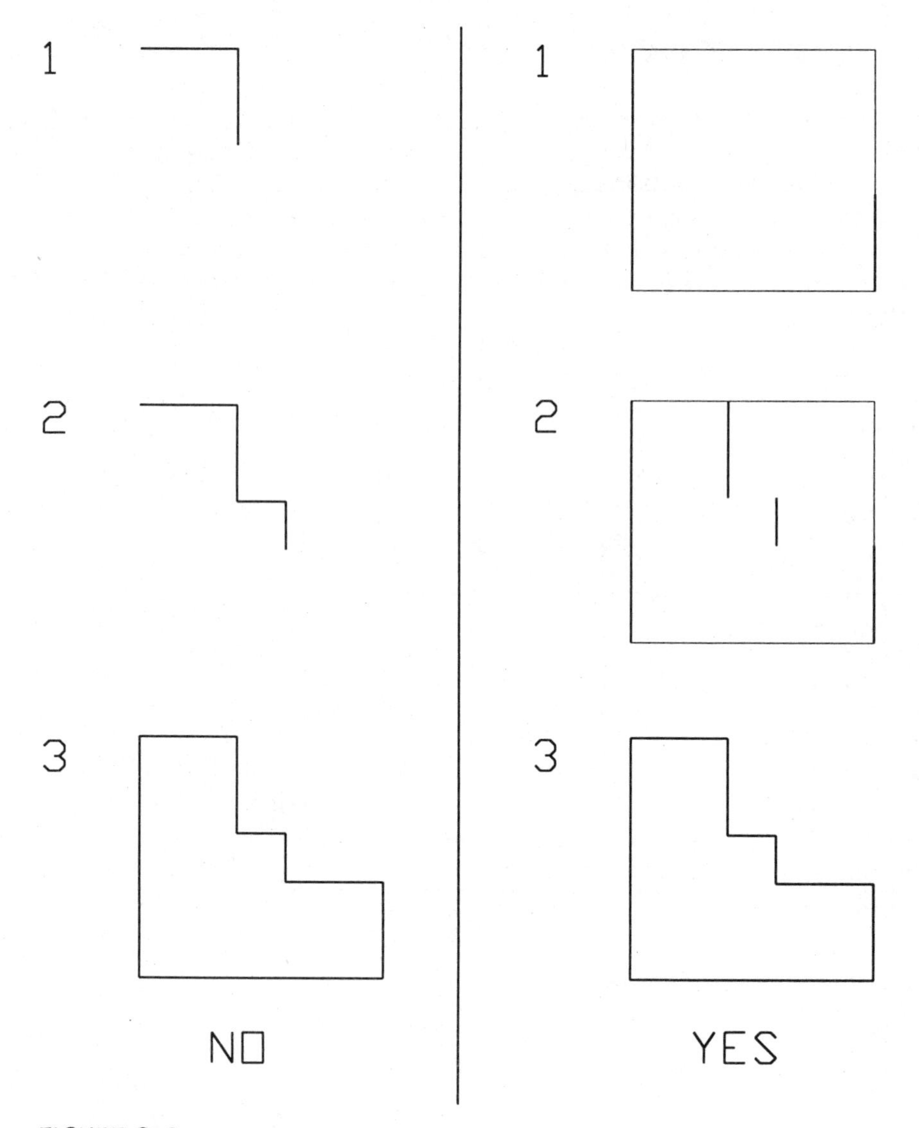

FIGURE 6–9

NOTES ON DRAWING ORTHOGRAPHIC VIEWS

a. Use the parallel straightedge to draw horizontal lines.

b. Use your triangles to draw vertical lines, as well as lines that are at 45°, 30°, and 60° to the horizontal.

c. Always make sure that the triangle is firmly in contact with the parallel straightedge.

d. When using a T-square, make sure that its head is held firmly against the edge of the drawing board.

e. Study page 324 in the Appendix to note common errors that beginning drafters often make. Be sure that you do not make these errors.

f. Study the Appendix for useful information and terminology used in drafting.

g. Note the decimal conversion table on page 326. Use this page to convert fractions to decimals and inches to millimeters.

h. Refer to page 321 in the Appendix for a list of the different types of lines used on a technical drawing, as well as their styles and line thicknesses.

SYSTEMS OF PROJECTION

Two systems of projection are commonly used in orthographic drawings: *third-angle* projection and *first-angle* projection. The method described so far has been third-angle projection, the one used in the United States, Canada, and some other countries.

This book uses only third-angle projection, as do most industries in North America.

To identify which type of projection is used on a drawing, we use the symbol shown in Fig. 6–10 to represent third-angle projection. Think of this object as a solid cone shape with the point cut off.

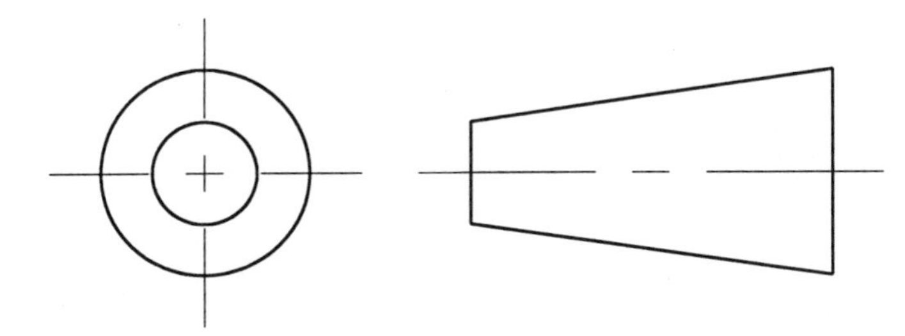

FIGURE 6–10 Symbol for Third-Angle Projection

First-angle projection is used mainly in Europe and Asia. In first-angle projection, we project the view onto a plane (or piece of glass if you think of the object as suspended in a glass box) *behind* the object, rather than onto a plane (or glass) lying between the object and your eyes, as in third-angle projection.

We use the symbol shown in Fig. 6–11 to represent first-angle projection.

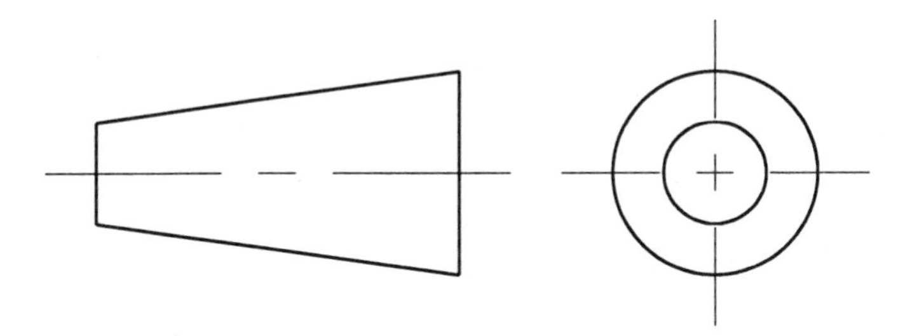

FIGURE 6–11 Symbol for First-Angle Projection

In both cases the views look the same, but their placement relative to the front view has changed. Industrial drawings will usually have one of the above symbols drawn in the title block to tell you which system of projection was used. Illustrations of these two systems appear on the next page.

Figure 6–12(a) illustrates third-angle projection. Figures 6–12(b) and (c) illustrate two examples of first-angle projection of the same object. Since this book teaches only third-angle projection, the one used in North America, no detailed explanation will be given about first-angle projection.

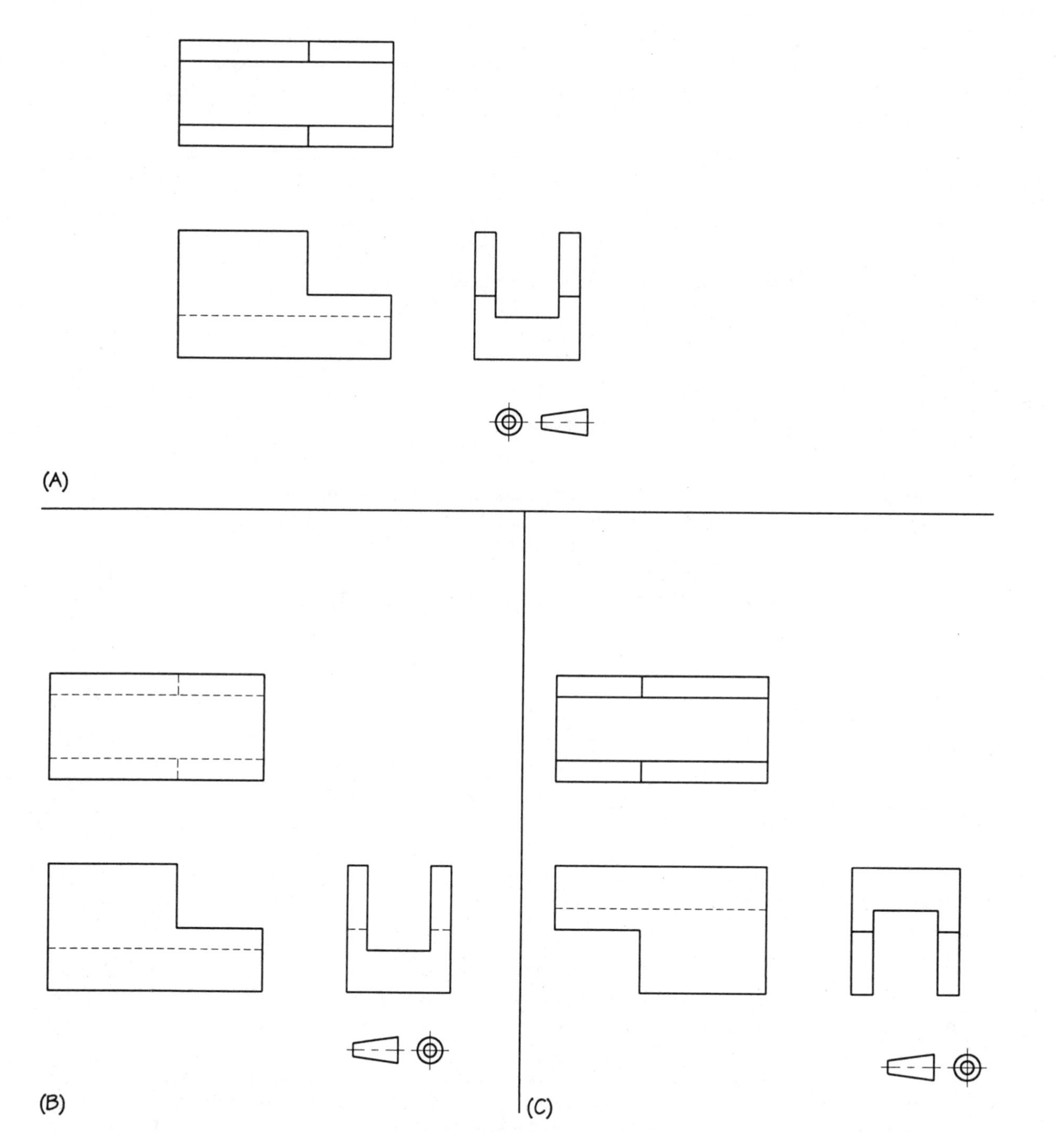

FIGURE 6–12

MITER LINE

One method of drawing a side view, when the top and front views are known, is with a *miter line*. The method becomes cumbersome on complicated drawings, but some beginners may prefer it.

STEP 1 Starting at the top right corner of the front view, and using your 45° triangle, draw a miter line, which is a light line at an angle of 45° to the horizontal (Fig. 6–13a).

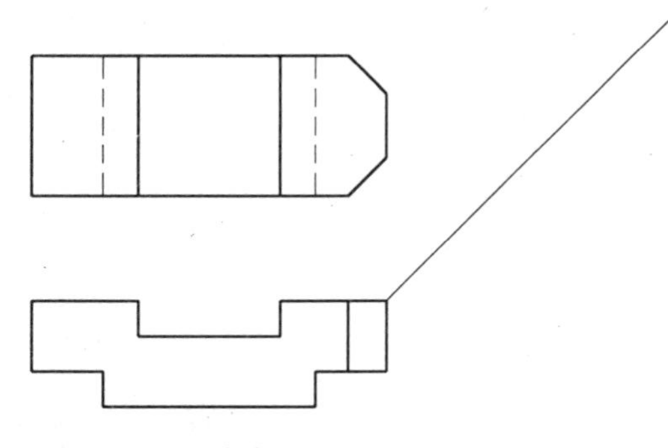

FIGURE 6–13(A)

STEP 2 Draw light construction lines from corners and surfaces in the top view to the miter line (Fig. 6–13b).

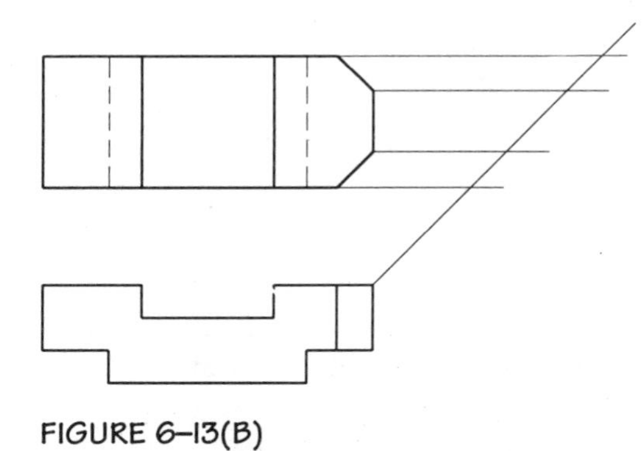

FIGURE 6–13(B)

STEP 3 Where these construction lines intersect, project down light construction lines (Fig. 6–13c).

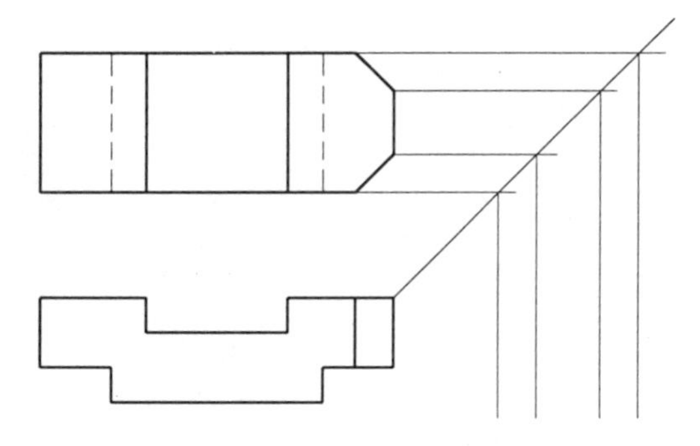

FIGURE 6–13(C)

STEP 4 Project construction lines from the front view to the right to intersect the vertical lines (Fig. 6–13d).

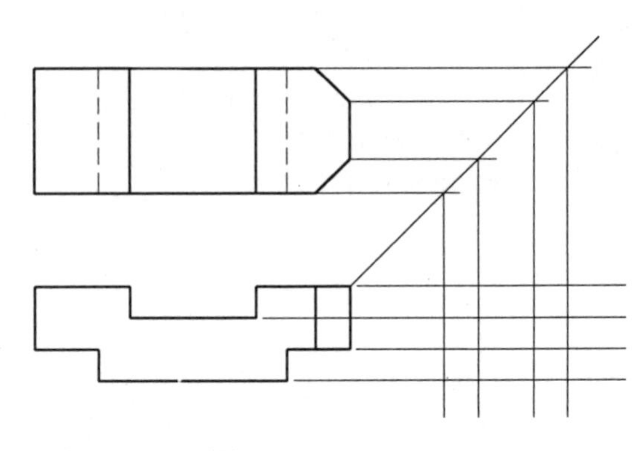

FIGURE 6–13(D)

STEP 5 Complete the side view (Fig. 6–13e).

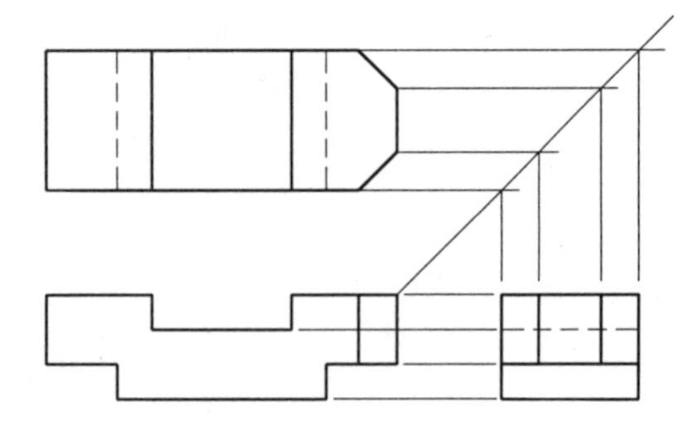

FIGURE 6–13(E)

Notice that in Exercise 6–1 on the next page there is no convenient top right corner in the front view from which you can draw your miter line. You can create a corner for this purpose by drawing a horizontal construction line from the uppermost feature on the front view and a vertical construction line from the extreme right feature on the same view.

EXERCISE 6–1

In the space provided, and following the steps on the previous pages, use a miter line to draw the missing top view of the drawing on this page. The miter line is drawn for you. (If you find the miter line method confusing, then don't use it.) Start your work by taping down this page and drawing construction lines up from the front view. Then draw construction lines up from the side view to intersect the miter line. Finally, from the intersections on the miter line, draw horizontal construction lines toward the left to intersect the vertical construction lines. These intersection points will help you create the top view. Finish the top view.

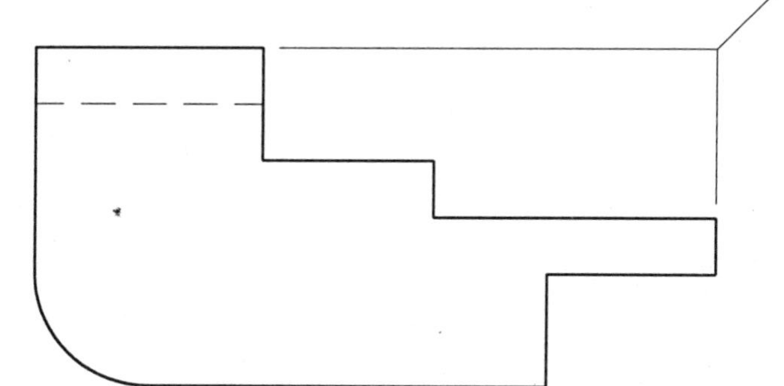

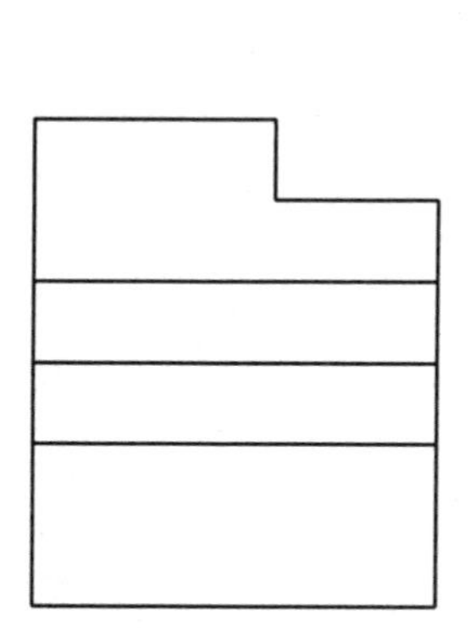

DRAWN BY ______________________________ DATE ____________

EXERCISES 6–2 TO 6–9

On a C size sheet (17" by 22") or the metric A2, use drafting instruments to draw the necessary orthographic views of the objects shown in Exercises 5–2 to 5–9 of the previous chapter. See page 331 in the Appendix for standard drafting sheet sizes.

Use the dimensions shown to make your drawing full scale (1:1), or choose a reduced scale if it will fit your paper better. Do not show any dimensions on your drawing.

Using the decimal equivalent table on page 326 in the Appendix, convert decimals to fractions if you are using the architect's scale.

NOTE: The R symbolizes radius and the Ø symbolizes diameter.

HINT: Allow 3" (75 mm) between views. Use the circle template for small holes and small radii. To use the circle template, first draw centerlines to locate the center of the hole. Then locate the hole in the circle template with the correct (or closest to the correct) diameter. Align the four marks on the template hole with the centerline you have drawn. Draw the circle with your 0.7 mm lead.

DIMENSIONING 7

OBJECTIVE

When you have finished this chapter, you should be able to properly dimension an orthographic drawing.

EQUIPMENT

You are required to have all of your drafting instruments for this chapter.

So far we have learned how to represent an object by drawing orthographic and isometric views. These drawings could not be sent to a tradesperson to make parts or structures, because they did not have any dimensions (sizes) on them. This chapter will show you the proper way to put dimensions on orthographic views.

Study Fig. 7–1 to see an example of a dimensioned view. Note the *extension lines* used to indicate the endpoints of a feature to be dimensioned. Between two extension lines is a *dimension line* with an arrowhead on each end. The *leader line* is used to indicate the size of a hole or to show where a note applies.

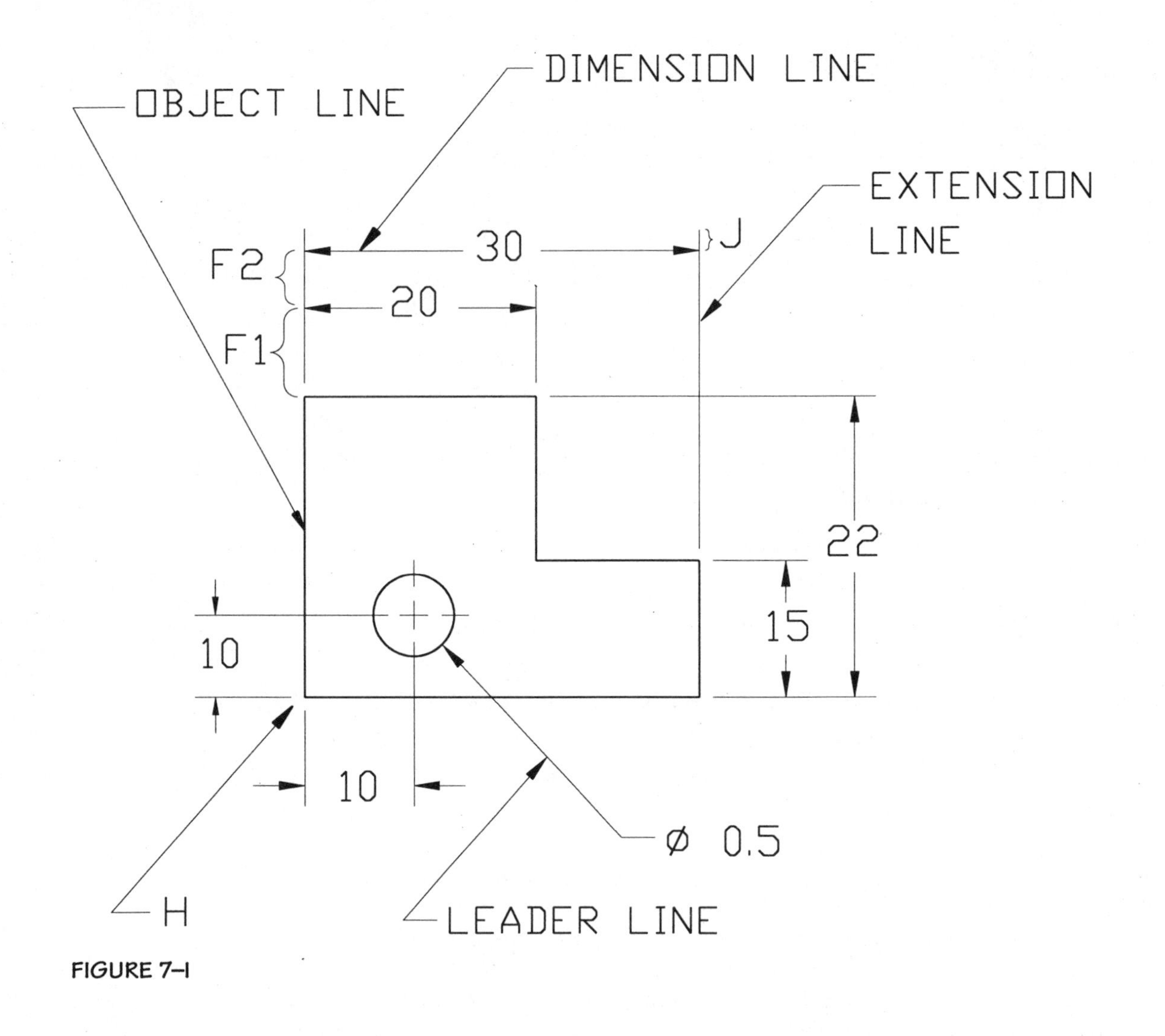

FIGURE 7–1

GENERAL RULES FOR DIMENSIONING AN ORTHOGRAPHIC DRAWING

a. Properly form all numbers and letters. (See Chapter 1.) Use guidelines for all lettering, including all whole numbers, all fractions and all letters. Whole numbers and letters should be 1/8" (3 mm) high, and fractions should be 1/4" (6 mm) high. Use a 0.7 mm technical pencil for all numbers and letters.

b. The numbers appearing on the drawing should describe the size of *all* of the features of the piece. The drawing should not have to be measured to determine the size of a feature. In other words, dimensions must be complete and all size and location dimensions *must* be shown. A *size* dimension gives the size of a particular feature. A *location* dimension locates a center line or some reference line or point. In Fig. 7–1, the two dimensions of 10 used to locate the hole are examples of location dimensions. The dimensions of 15 and 22 are examples of size dimensions.

c. Beginning drafters may choose to plan the dimension placement on the sketch before starting the good instrument drawing.

d. The metric linear unit in general drafting is the millimeter, abbreviated mm. (The international metric system is known as SI or Système International.) The U.S. linear unit in general drafting is the decimal inch. According to ANSI (American) standards, either of these systems may be used, but SI is encouraged. In Canada, SI is the standard. (See page 329 in the Appendix for information on the three standards—ANSI, CSA, and ISO—referred to in this book.)

e. Omit inch (") and millimeter (mm) symbols from individual dimensions. Specify the units on the title block or in a note near the title block, such as, "All dimensions are in millimeters."

f. The standards specify that the first dimension line be a minimum of 3/8" (10 mm) from the object (F1 in Fig. 7–1). All subsequent dimension lines are spaced 1/4" (6 mm) apart (F2 in Fig. 7–1). In practice, it is easier to read the drawing when F1 is 3/4" (20 mm) and F2 is 3/8" (10 mm). Note that the dimension line is broken in the middle to allow space for a dimension. The size of this space is not measured but rather determined by eye. (This is according to ANSI and CSA standards. ISO standards differ, specifying that in most cases horizontal dimension lines not be broken and that dimensions be placed above the line.) The arrowhead on each end of the dimension line touches the extension line.

g. Draw all lines used in dimensioning with a 0.3 mm technical pencil. Make all arrowheads with a 0.7 mm technical pencil.

h. Position extension lines about 1/16" (2 mm) from the object line and perpendicular to the feature to which they apply. Note the small space at H in Fig. 7–1. Extension lines are usually horizontal or vertical, except in the case of a sloping feature.

i. When two extension lines cross, or a leader line crosses a dimension or extension line, break neither line.

j. Extend the extension line beyond the arrowhead about 1/8" (3 mm). See J in Fig. 7–1.

k. Use leader lines to dimension holes, outside diameters, and radii. Try to use a 45° angle, but if necessary use 30°, 60°, or any other angle. Note that the arrowhead on the leader line touches the circle or feature to which it applies, and the other end has a horizontal portion about 1/4" (6 mm) long. Use the symbol Ø to represent a diameter.

l. Draw object lines with a 0.7 mm pencil; they are therefore thicker than extension and dimension lines. (ANSI standards specify a minimum of 0.6 mm lead.)

m. To keep your drawing clean and to avoid overerasing, proceed in the following order to dimension the drawing. This assumes that you have already finished drawing the necessary views. Refer to Fig. 7–1.

STEP 1 Draw ALL extension lines with your 0.3 mm technical pencil.

STEP 2 Draw ALL dimension lines (the ones with the arrowheads) with your 0.3 mm technical pencil.

STEP 3 Add ALL dimensions (numbers) with your 0.7 mm pencil (within guidelines, of course).

STEP 4 Draw ALL arrowheads with your 0.7 mm pencil.

Some beginners erroneously draw two extension lines, followed by one dimension line, two arrows, and a dimension. They then keep repeating this procedure. This method will produce an untidy and dirty drawing. It is also inefficient and therefore wastes time. Try to draw all extension and dimension lines first (steps 1 and 2) before drawing the numbers and arrowheads (steps 3 and 4).

n. Rules for dimensioning orthographic drawings differ from those for isometric drawings. Do not look at the placement of dimensions on an isometric drawing to determine their placement on an orthographic drawing.

o. The rules described in this chapter apply to mechanical engineering drawings. Other fields of technology, such as architecture or civil engineering, have specific rules pertaining to their own drawings.

NOTES:

a. Some of the views in this chapter may not be completely dimensioned. In addition, all necessary views of a piece may not be shown. In this chapter, these partial dimensionings and partial views have been used to illustrate a certain aspect of dimensioning. For example, Fig. 7–1 requires two views.

b. Remember: Different companies may have different drafting standards. You must follow the standards of your company.

c. Abiding by the rules of dimensioning presented in this chapter will ensure that your drawings are clear and easy to read. As with everything in life, however, circumstances may require that you modify a rule. When to do so can be learned only from experience.

ILLUSTRATED RULES OF DIMENSIONING

a. Always locate center lines with a dimension. Never locate a hole by dimensioning to the edge of the hole. See Fig. 7–2.

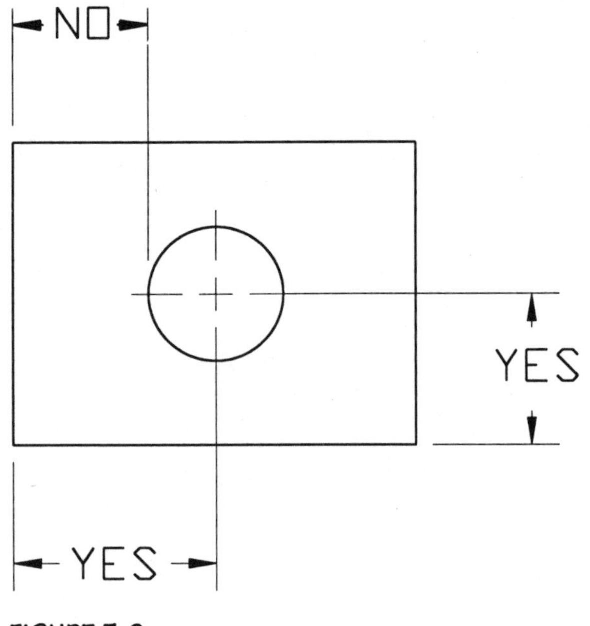

FIGURE 7–2

b. Wherever possible, place the dimension outside of the object. The arrowheads should touch only an extension line, never the object line, except in the case of a leader line where the arrowhead may touch an object line. See Fig. 7–3.

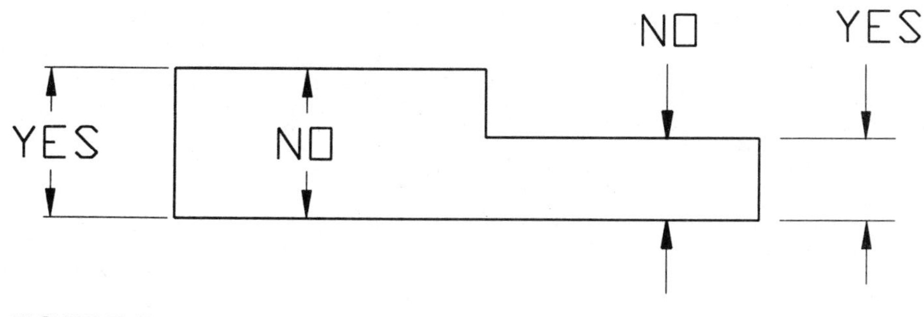

FIGURE 7–3

c. Wherever possible, put the dimensions on the view where you see the profile (shape) of the feature. In Fig. 7–4, the shape of the groove is shown in the front view, so the front view is the preferred placement for the size and location dimensions for the groove. The cut on the corner is seen as a profile in the top view, so that is the preferred view in which to dimension it. The depth of the groove in the middle of the object is seen in profile in the front view, so that is where you would show that dimension. According to CSA standards, a dot may be used to replace two arrowheads in limited space situations.

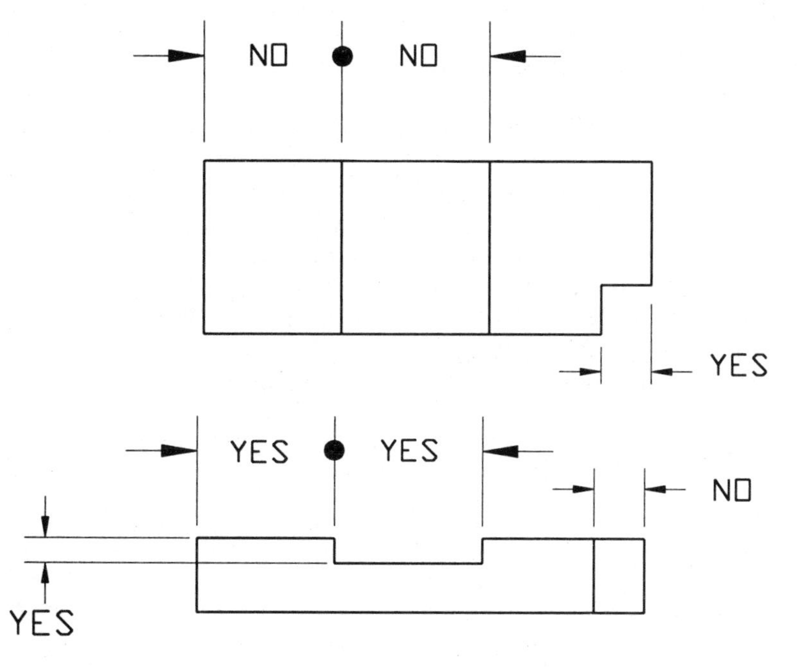

FIGURE 7–4

d. Dimension and locate holes in the view where they are seen as round, rather than where they are seen as hidden lines. Always give a diameter dimension rather than a radius dimension for a hole. See Fig. 7–5.

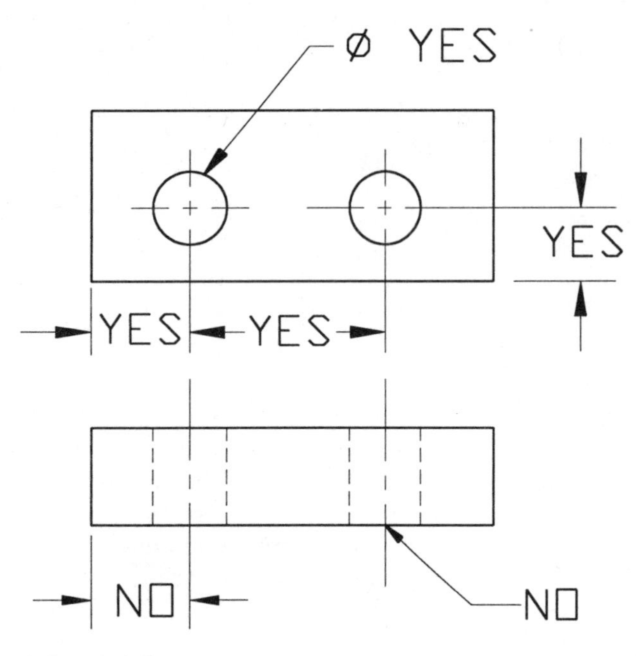

FIGURE 7–5

e. Never repeat a dimension. In Fig. 7–6, use only one of the A dimensions, one of the B dimensions, and one of the C dimensions. Dimensions can be on any view, as long as they follow the other rules of dimensioning. One reason that dimensions are never repeated is to avoid unnecessary errors, such as the one described in the following example. Suppose that a drawing needs to be changed (revised) after it has been issued from the drawing office, and the same dimension on the original drawing has been repeated on two views (such as the height in the front view and the same height in the side view). In making the revision, the drafter might inadvertently change only one of these dimensions and leave the other unchanged. The tradesperson, upon receiving the revised drawing, might work to the original dimension instead of the new one, thus creating a product of the wrong size.

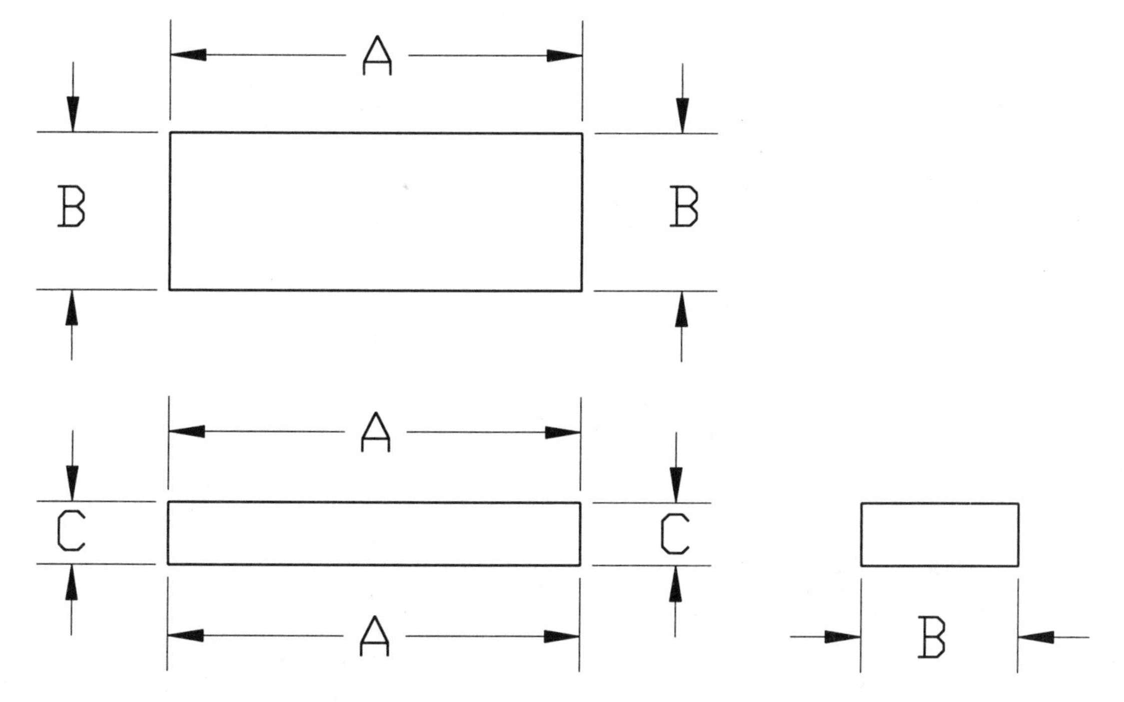

FIGURE 7–6

f. Avoid dimensioning to hidden lines wherever possible. This is illustrated in Fig. 7–7.

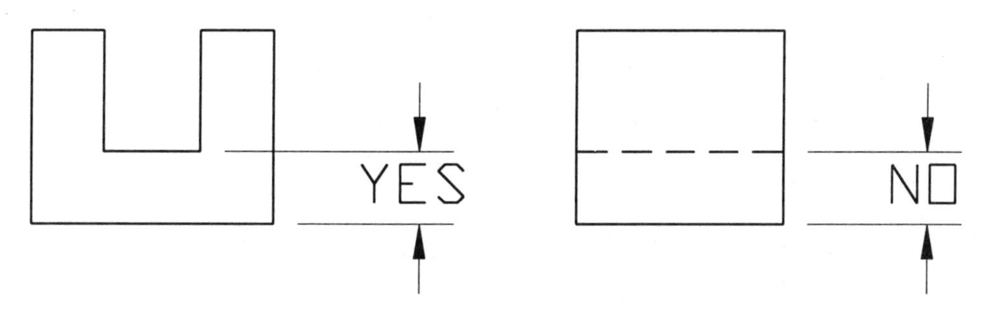

FIGURE 7–7

g. Form arrowheads according to ANSI and CSA standards, so that they are 3 times longer than wide. That is to say, the length to width ratio should be 3 to 1. Use the 0.7 mm lead to draw arrowheads. In Fig. 7–8 are several examples of arrowheads and the standards by which they are acceptable. The important thing to remember is that all your arrowheads should be the same. CSA standards specify that the length of an arrowhead should be equal to the height of the numerals used for dimensions. Refer to page 329 in the Appendix for information on standards.

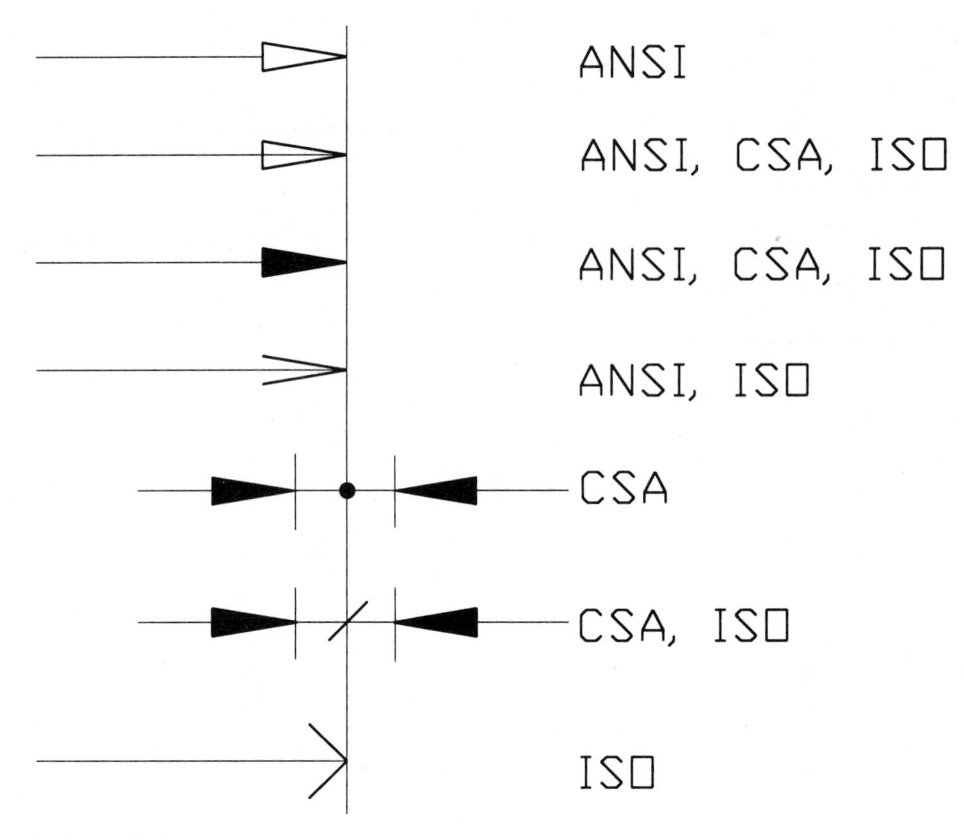

FIGURE 7–8

h. Never crowd the dimensions. If they are crowded, as in Fig. 7–9(a), then place the arrowheads outside the extension lines, as shown in part (b). For very small dimensions, place both the arrowheads and the number outside the extension lines, as illustrated in part (c).

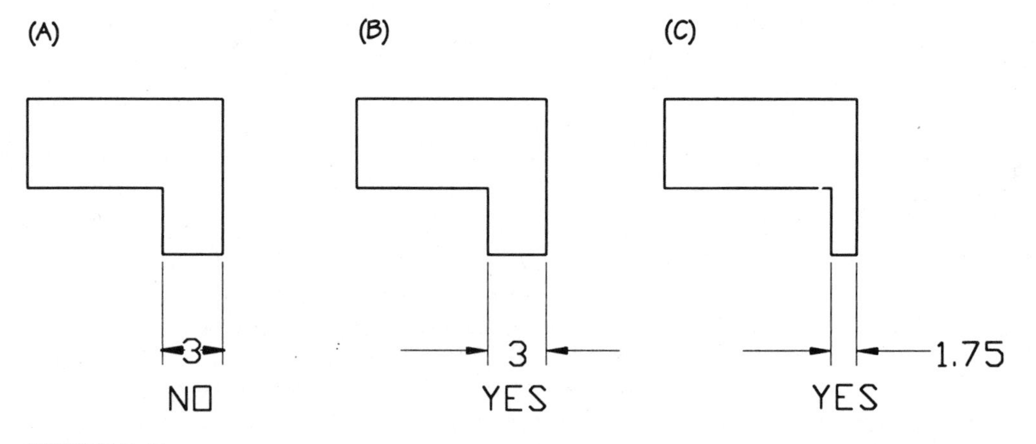

FIGURE 7–9

i. Line up the dimensions wherever possible. In Fig. 7–10(a), the horizontal dimensions are not lined up but staggered. This can lead to errors in reading the drawing. Figure 7–10(b) illustrates horizontal dimensions that are lined up. Note the use of the dot between the 2 and the 3, which may be used to replace two arrowheads in tight spaces. Note also that in part (b), space does not permit the vertical dimensions to be lined up.

When an extension line crosses an arrowhead, break the extension line.

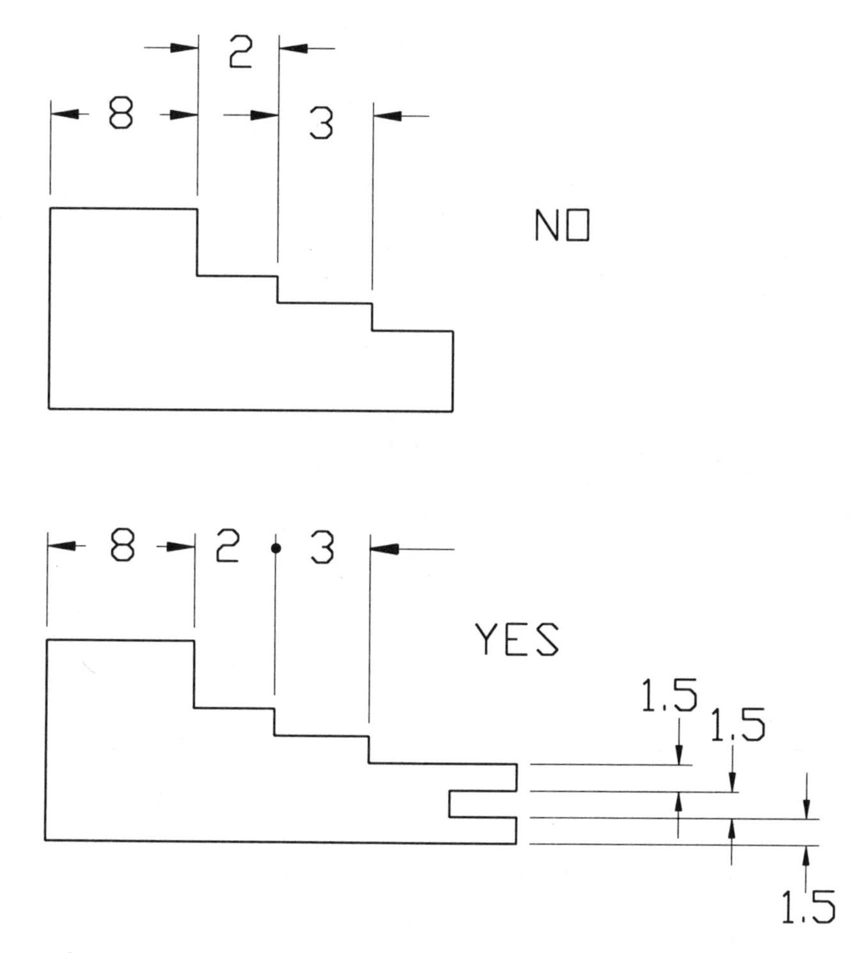

FIGURE 7–10

j. Place shorter dimensions between the object and the larger ones (Fig. 7–11b) to avoid the crossing of extension and dimension lines. To conform to this rule, you may find it helpful to plan your dimensions on the sketch before actually putting them on your good drawing. In this way, you will avoid much erasing.

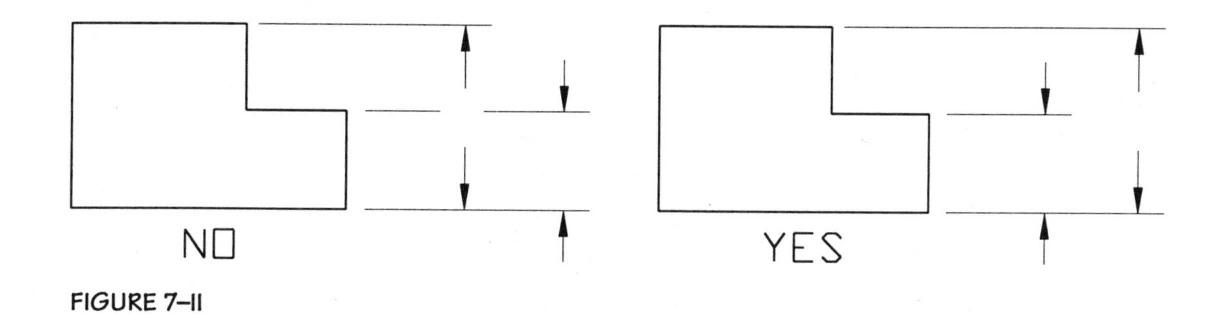

FIGURE 7–11

k. When you use *leader lines* to dimension a diameter, place them so that, if extended, they would intersect the center of the circle (Fig. 7–12).

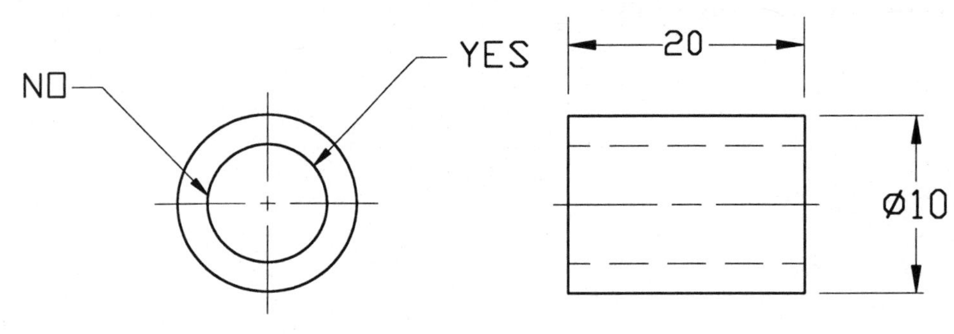

FIGURE 7–12

l. Use the symbol Ø to specify diameter, and place it before the dimension. The symbol Ø is used only with diameters that specify holes or outside diameters, not parts of a circle, such as arcs. An arc, or part of a circle, is dimensioned with the letter R to represent radius. Several examples of the dimensioning of diameters and radii are shown in Fig. 7–13. Notice that the arc with the R 78 dimension has a zigzag in its dimension line. It indicates that the center of the arc is located too far to be conveniently located.

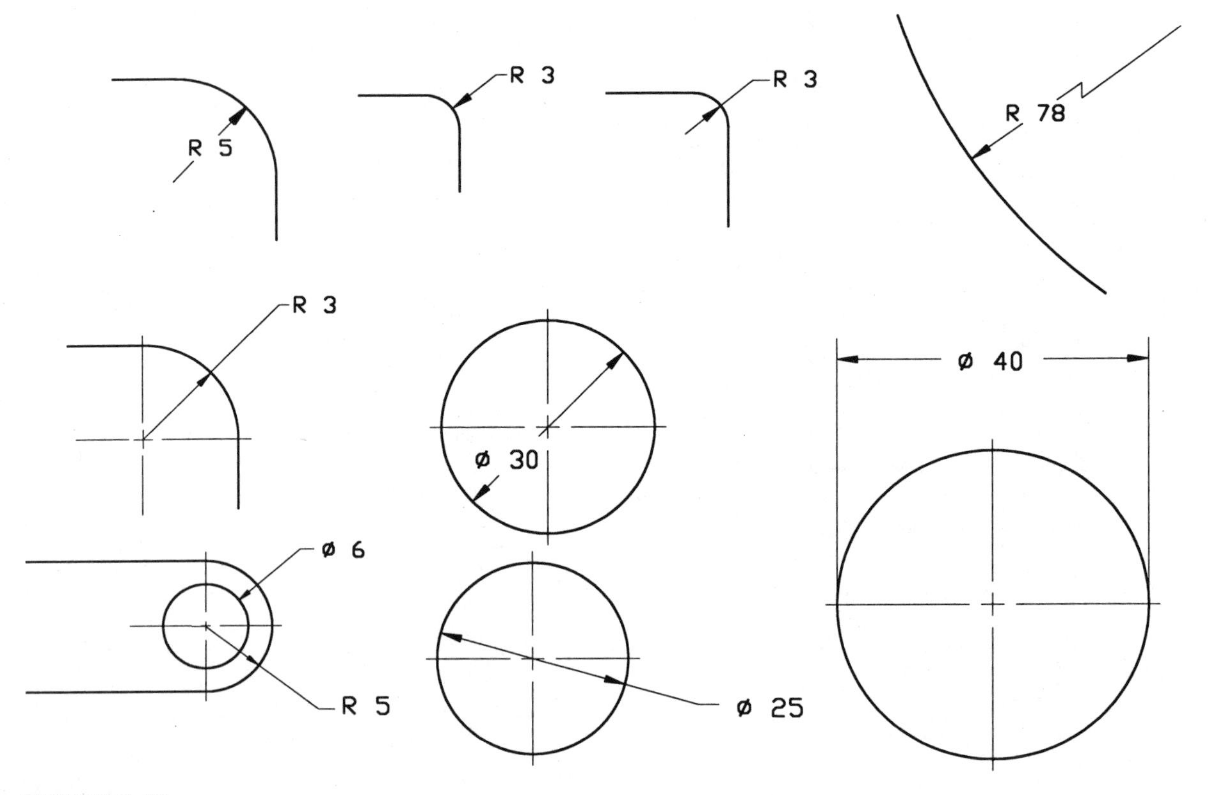

FIGURE 7–13

SYSTEMS OF DIMENSIONING

Figure 7–14 illustrates two systems of dimensioning.

In the *unidirectional system* (Fig. 7–14a), the numbers are read from the bottom of the drawing.

In the *aligned system* (Fig. 7–14b), the numbers are read either from the bottom of the drawing or from the right-hand side.

Both systems are correct, but some company standards prefer one to the other. ANSI standards favor the unidirectional system, whereas CSA and ISO standards do not specify a preference.

Remember that inch (") and millimeter (mm) symbols are omitted. The units should be specified on the title block or in a note near the title block, such as, "All dimensions are in millimeters."

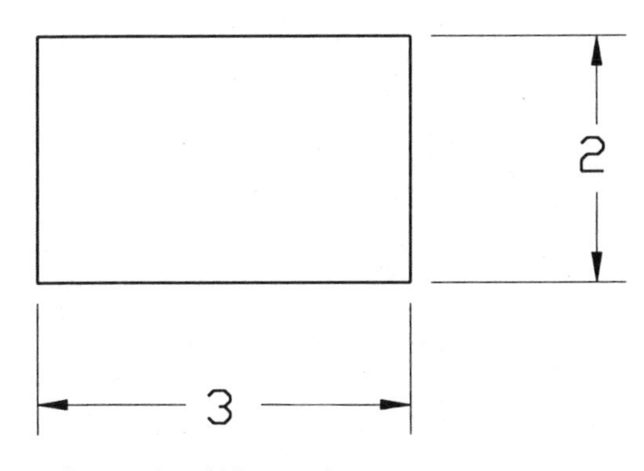

FIGURE 7–14(A) Unidirectional System

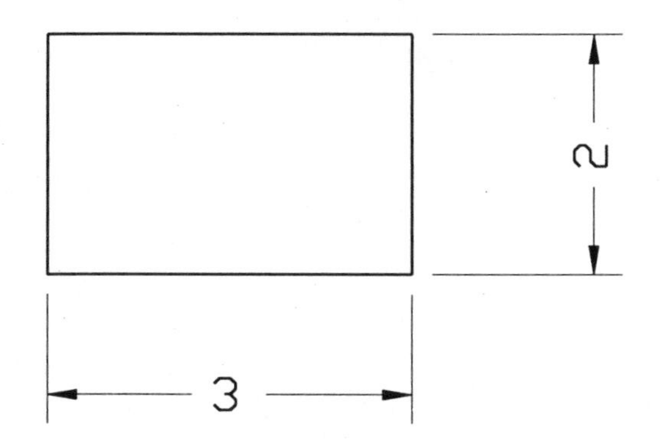

FIGURE 7–14(B) Aligned System

When using the *aligned system* of dimensioning, dimensions are read by looking at the drawing with its top facing up or rotated 90° clockwise. Figures 7–15(a) and (c) illustrate correct dimensioning, since the dimension 4 is read normally and the 3 is read by rotating the top of the drawing 90° clockwise. In parts (b) and (d), we must rotate the top of the drawing 90° counterclockwise to read the 3; this is wrong. In part (b), the 4 is read upside down. This is also wrong.

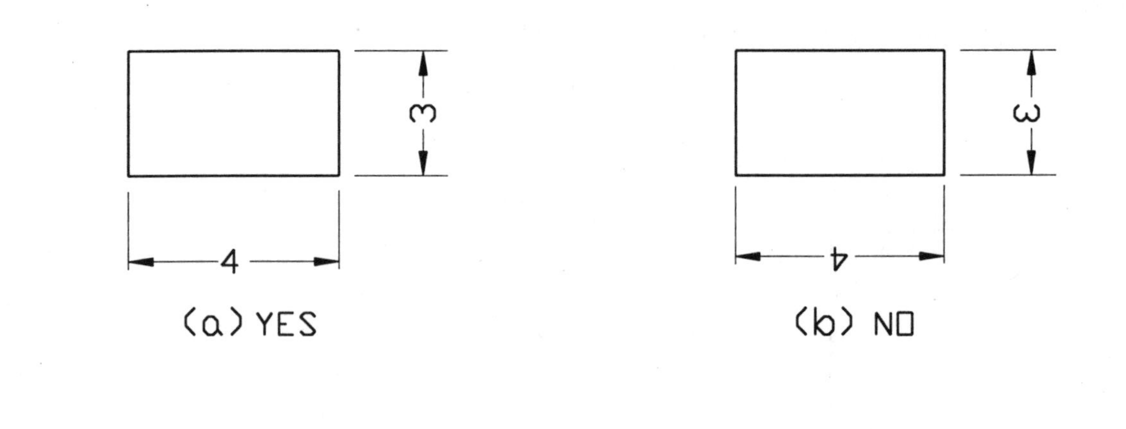

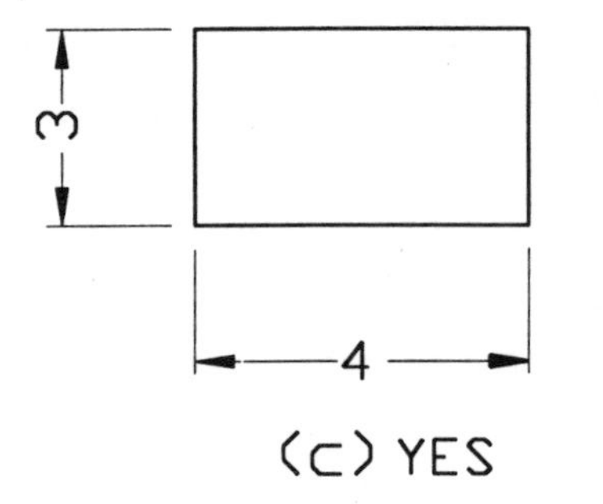

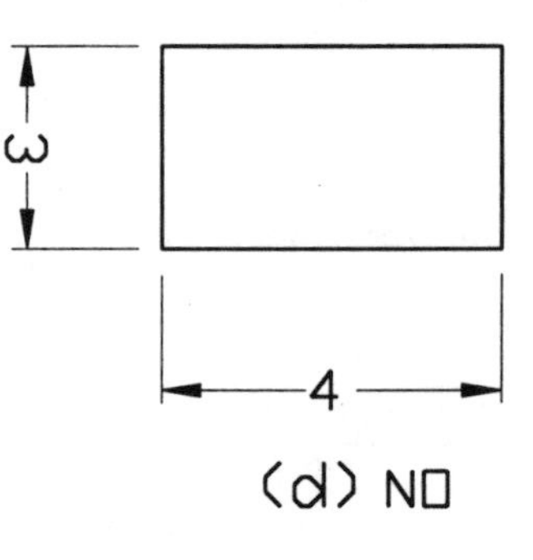

FIGURE 7–15

METHODS OF DIMENSIONING ANGLES

Figure 7–16 illustrates several different ways of dimensioning angles and slopes.

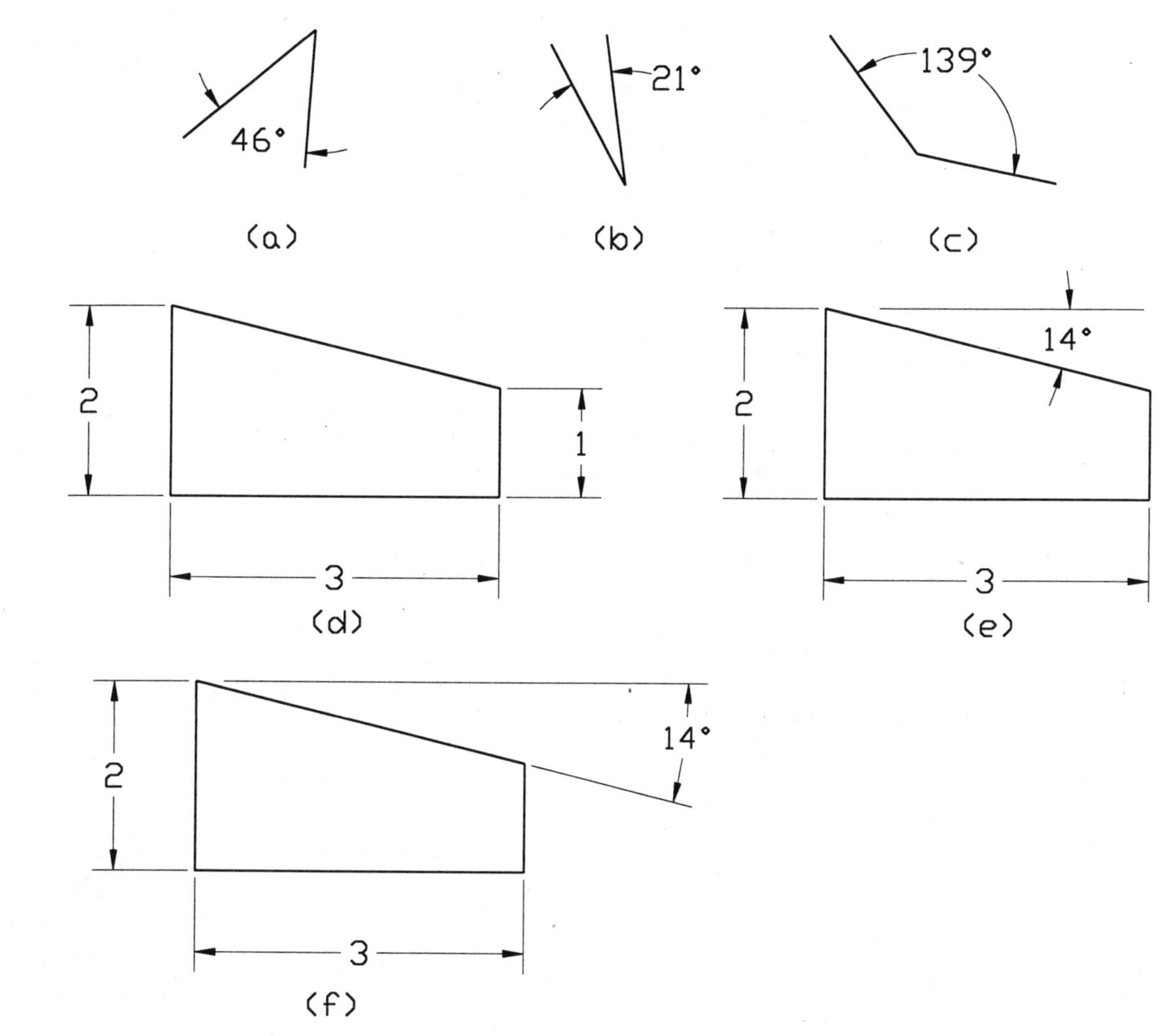

FIGURE 7–16

TOLERANCES

In the manufacturing or building process, all tradespeople work to a certain tolerance for every dimension. *Tolerance* is the total range by which the size of the actual part may vary. If we say, for example, that a part is to be 50 mm long, we must realize that this is an ideal dimension. In actual practice, the finished piece may measure 51 mm. Is this acceptable? What if it were 50.5 mm, or 50.1 mm, or 49.9 mm? Which of these actual sizes would be acceptable? The actual range over (or under) 50 mm that is acceptable depends on many considerations. It may be determined by considering such factors as how the part will be interacting with other pieces, the speed and temperature at which it will be moving, and the materials used in its manufacture. After analyzing these factors, the drafter would specify a certain tolerance to which the tradesperson would then be required to work.

A tolerance note appearing in one location of the drawing would apply to all dimensions on the drawing, unless otherwise specified on certain dimensions. For example, a note that specifies that all dimensions are ±0.1 mm means that a piece dimensioned as 75 mm on the drawing will be acceptable if the finished product measures between 74.9 mm and 75.1 mm. Its total tolerance is 0.2 mm, which is the difference between the larger and smaller values.

In this book, we are not covering the topic of tolerances in detail because it is relatively advanced and beyond the scope of an introductory course. Tolerancing is mentioned at this point to inform you that tolerances are an essential part of all technical drawings and that dimensions are incomplete without them. This brief description of tolerances may help you to better understand some of the dimensioning practices that follow.

TOLERANCE CONSIDERATIONS IN DIMENSIONING

Consider the examples shown in Figure 7–17.

The dimensioning in part (a) is not correct because the overall dimension is not shown. The overall dimension is important for several reasons. It permits the tradesperson to know how big a piece of rough material to start with, without having to add up several dimensions. It also gives the reader of the drawing a sense of how large the part is for shipping it to the customer, what method of shipping should be used, how the part may be packed for shipping, and how large a machine is required to construct the part.

The dimensioning in part (b) is not correct because no allowance is left for an accumulation of tolerances. One of the short dimensions should be left out. Advanced courses in drafting will deal with the concept of tolerancing in more detail. Briefly, if the lengths dimensioned as 2, 2.5, and 2.6 are each manufactured 0.1 unit larger than specified, then the total dimension (7.1) becomes 7.4 on the actual part. This may not be acceptable, and you may have to decide which dimension is the least important of the four dimensions, and leave it out, as was done in parts (c) and (d).

The dimensioning in part (c) is correct because the overall dimension is shown and allowance is left for accumulation of tolerances. Note that any one of the three short dimensions could have been left out, although leaving out the right-hand dimension as in part (c) allows good control of the size of the slot. Compare it to part (b).

Figure 7–17(d) illustrates *datum* line dimensioning, in which one surface or feature is used as a geometric reference for all dimensions in a certain direction. The overall dimension is shown and allowance is left for accumulation of tolerances.

To review: Always show the overall dimensions and allow for an accumulation of tolerances.

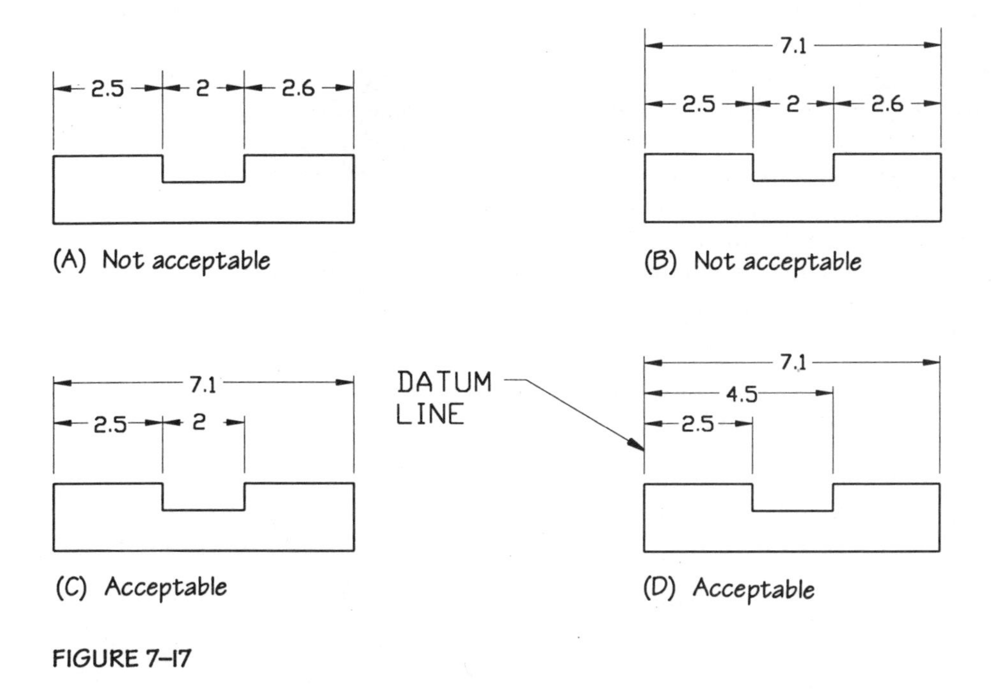

FIGURE 7–17

LOCATION OF HOLES

Figures 7–18(a) and (b) illustrate two examples of how holes may be dimensioned. The dimensioning in part (b) is better practice than the dimensioning in (a), because in (b) we are dimensioning from the center line to the center line of the holes. In the manufacturing process, the dimensioning in part (b) allows a specific tolerance to be applied between the holes. In part (a), the holes are dimensioned from different edges, and we are not specifying a center-to-center distance between the holes. Because a part of this design will probably be attached to another piece with holes having the same center-to-center distance, it is very important that we dimension the center-to-center distance and thus provide a closer control for this dimension.

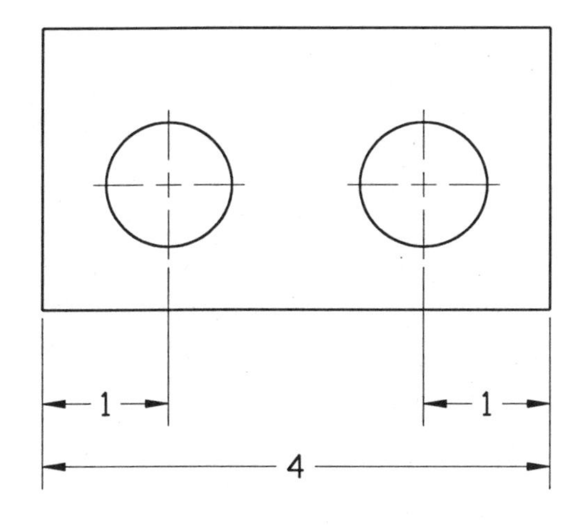

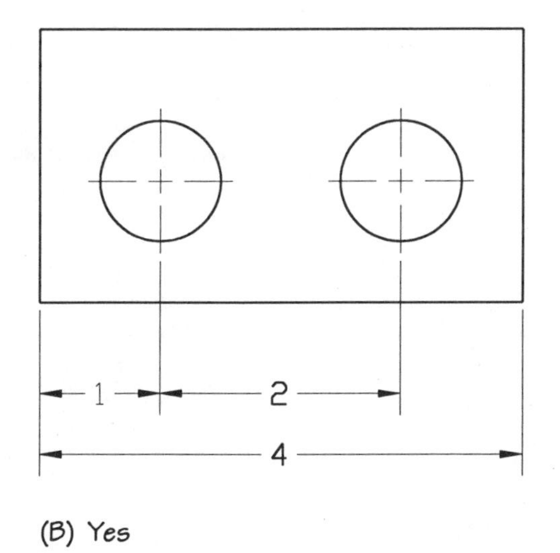

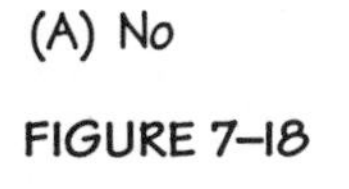

(A) No

(B) Yes

FIGURE 7–18

NOTE CONCERNING THE EXERCISES

You should never scale a feature on a completely dimensioned drawing to determine the size of that feature. The dimension alone that appears on the drawing determines its size. However, in these exercises, and in certain other parts of this book, dimensions are not given on the drawings. In these cases, you are required to measure the views to determine sizes. The dimensions have been deliberately omitted so that you will be required to make your own decisions about where to place the dimensions.

For each of Exercises 7–1 through 7–8, remove the page from this book and tape it down on your drawing board before adding dimensions.

EXERCISE 7–1

Dimension the drawing below (top and front views) by adding extension lines, dimension lines, arrowheads, and dimensions (numbers) in your choice of millimeters or inches. Use the full scale. If possible, you might have an experienced drafter correct this page before attempting the exercises that follow. This way you will not repeat your errors.

QUESTION: How many dimensions do you think you will need to fully dimension the drawing? The answer is 5.

HINT: Draw a box around one of your arrowheads. Is it 3 times longer than it is wide?

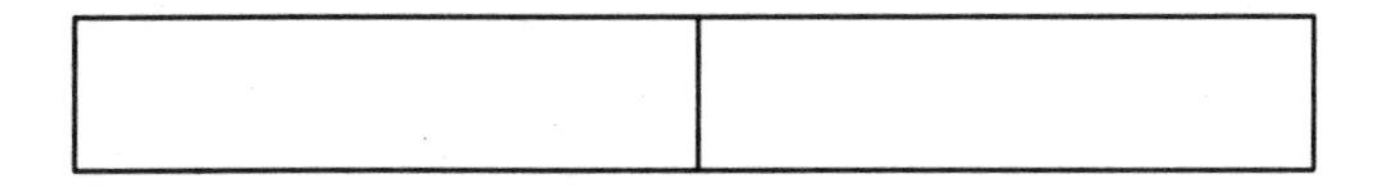

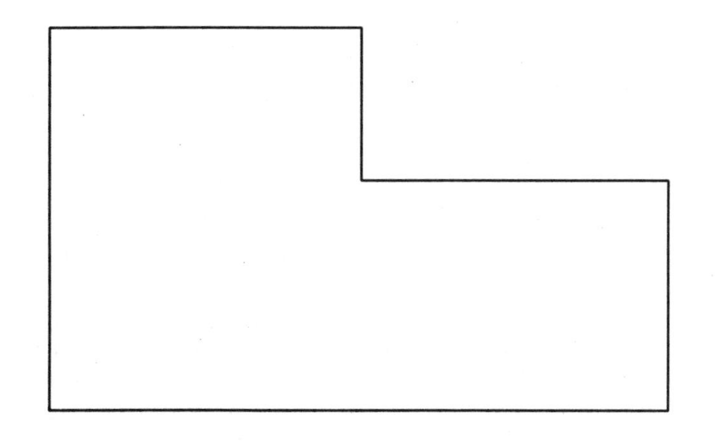

DRAWN BY ____________ DATE ____________

EXERCISE 7–2

Dimension the drawing below by adding extension lines, dimension lines, arrowheads, leader lines, and dimensions (numbers) in millimeters or inches.

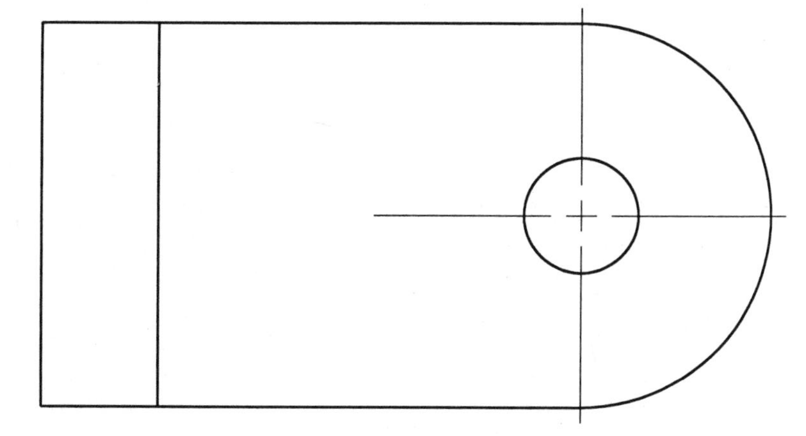

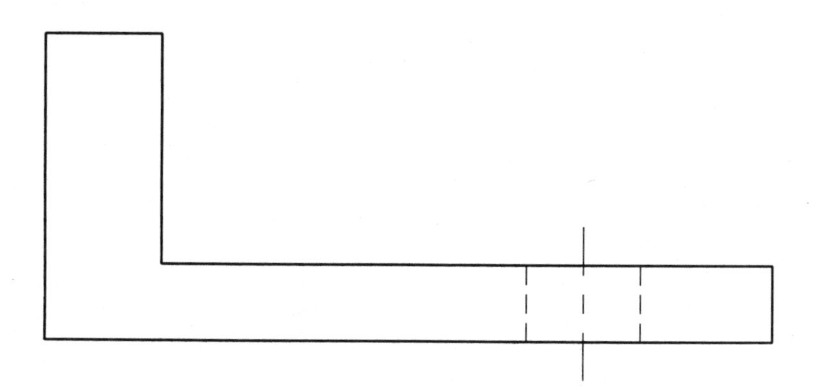

DRAWN BY DATE

EXERCISE 7–3

Dimension the drawing below by adding extension lines, dimension lines, arrowheads, leader lines, and dimensions (numbers) in millimeters or inches.

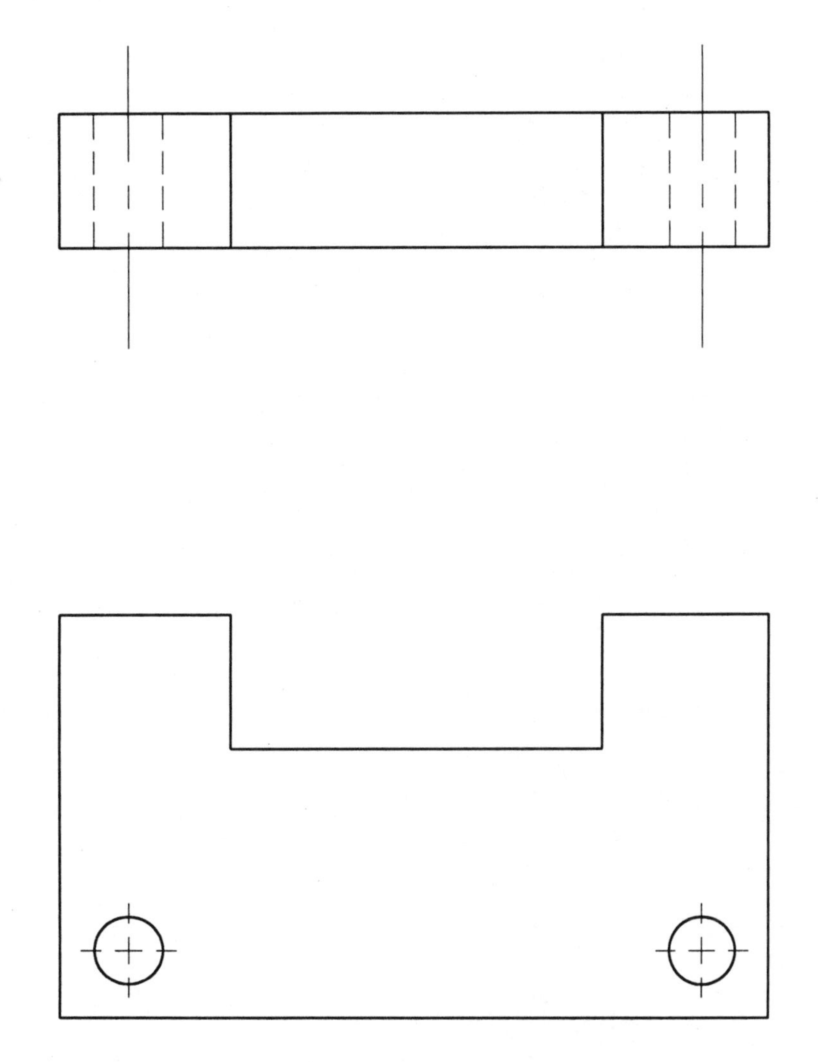

DRAWN BY ______________________ DATE ______________

EXERCISE 7–4

Dimension the drawing below by adding extension lines, dimension lines, arrowheads, leader lines, and dimensions (numbers) in millimeters or inches.

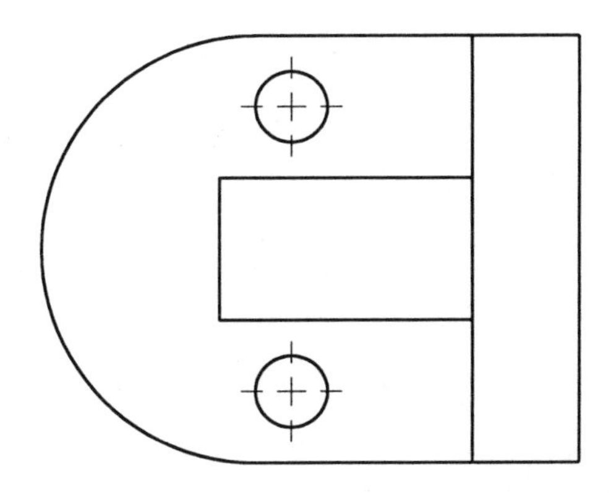

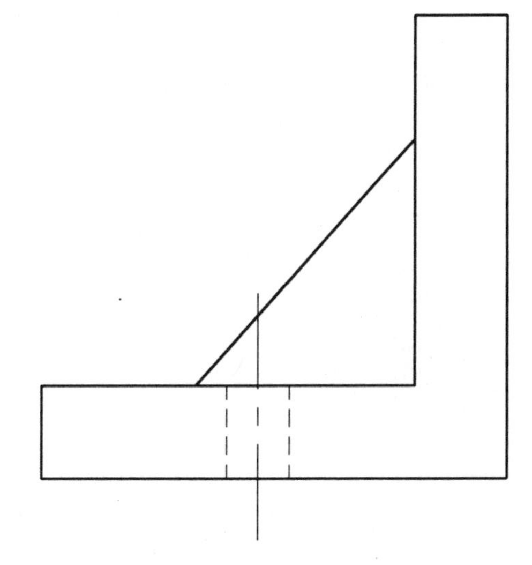

DRAWN BY DATE

EXERCISE 7–5

Dimension the drawing below by adding extension lines, dimension lines, arrowheads, leader lines, and dimensions (numbers) in millimeters or inches.

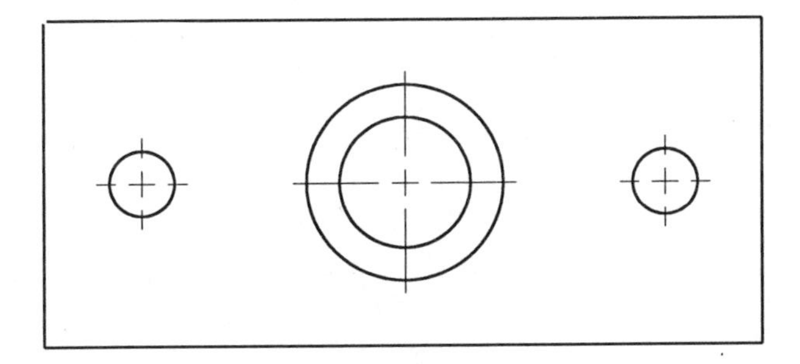

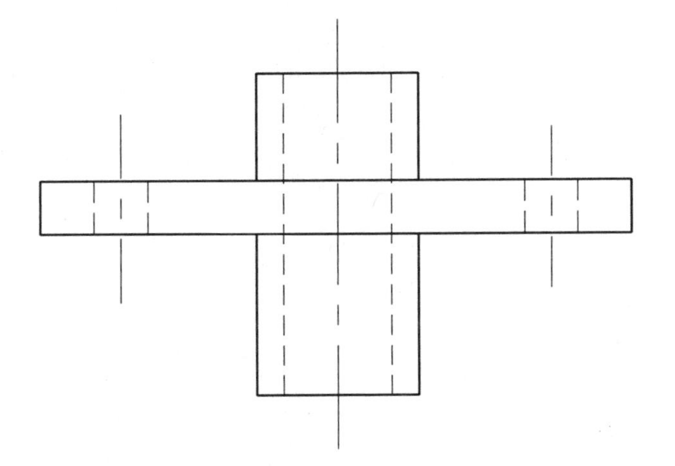

DRAWN BY DATE

EXERCISE 7–6

Dimension the drawing below by adding extension lines, dimension lines, arrowheads, and dimensions (numbers) in millimeters or inches.

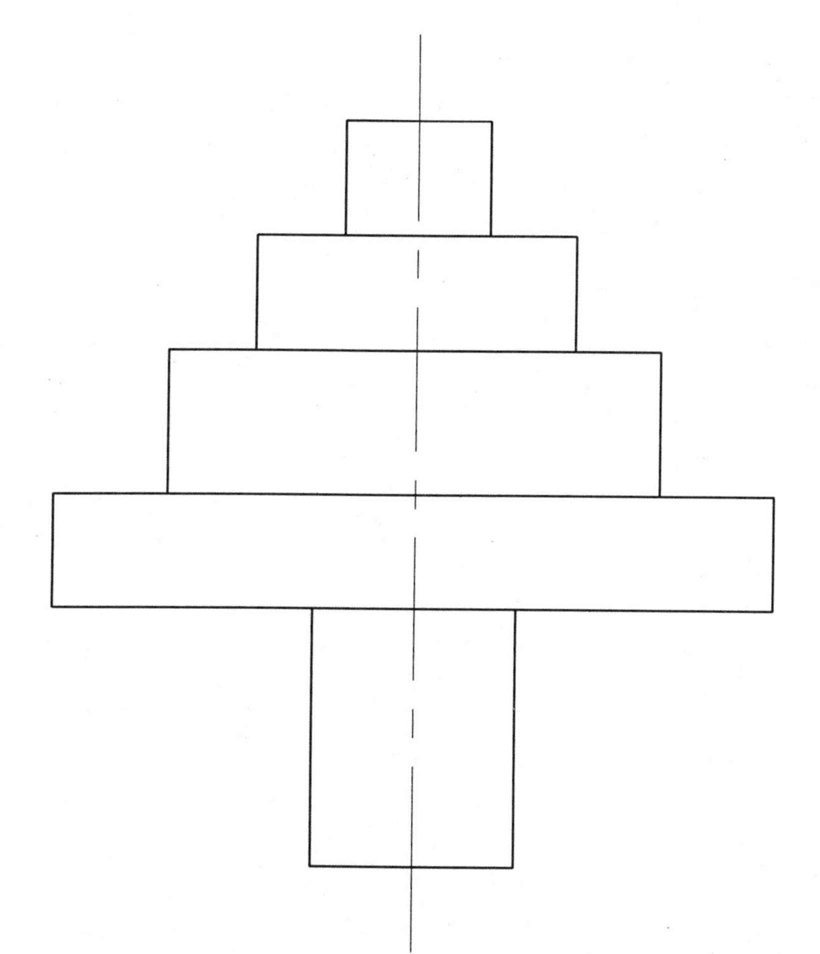

DRAWN BY ______________________________ DATE ______________

EXERCISE 7–7

Dimension the drawing below by adding extension lines, dimension lines, arrowheads, leader lines, and dimensions (numbers) in millimeters or inches.

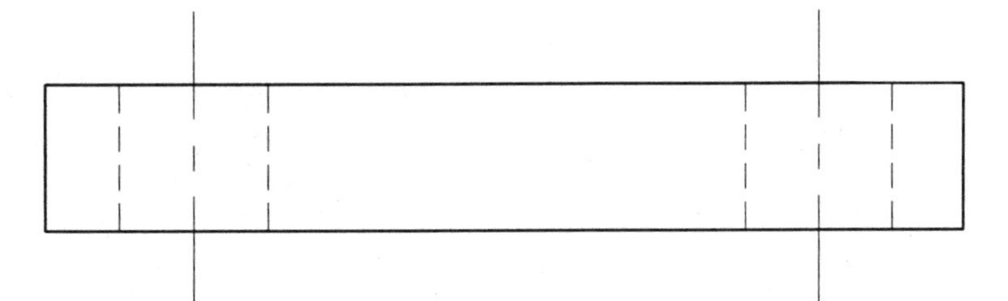

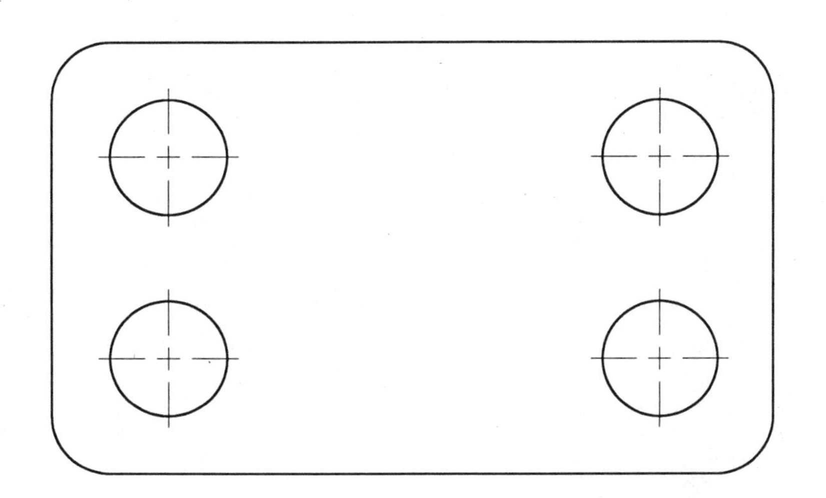

DRAWN BY DATE

EXERCISE 7–8

Dimension the drawing below by adding extension lines, dimension lines, arrowheads, leader lines, and dimensions (numbers) in millimeters or inches.

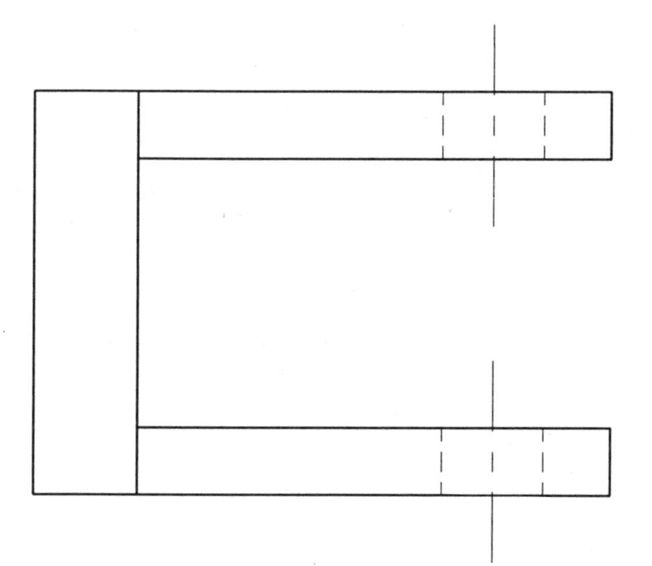

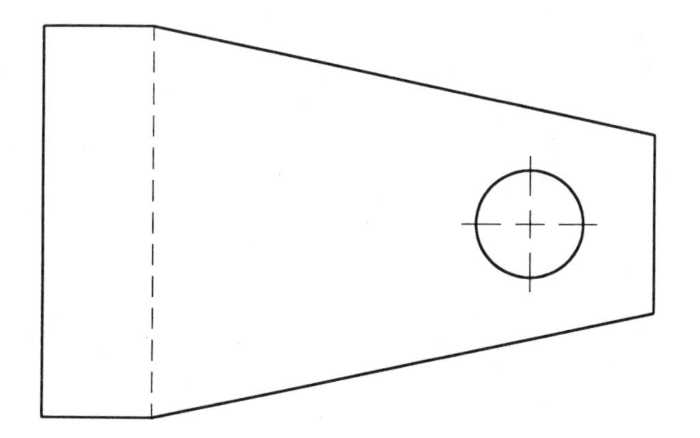

DRAWN BY ______ DATE ______

EXERCISES 7–9 TO 7–16

Add dimensions to the orthographic views drawn at the end of the last chapter in Exercises 6–2 to 6–9, using the dimensions given in Chapter 5 (Exercises 5–2 to 5–9). Do not forget that outlines are *thick*, and that center lines, extension lines, and dimension lines are *thin*.

ACTIVITY: In preparation for studying Chapter 8, cut out Fig. 8–13, and have it with you when you work through the next chapter.

DRAWN BY DATE

AUXILIARY VIEWS: PART I

8

OBJECTIVE

When you have finished this chapter, you should be able to draw auxiliary views.

EQUIPMENT

You are required to have all your drafting instruments for this chapter.

AUXILIARY VIEWS

An *auxiliary view* is an extra view, not one of the six principal orthographic views described in Chapter 2. It is used to show the true shape, and therefore the true size, of a sloping (or oblique) surface, which would appear distorted in any principal view. A sloping surface is not parallel to any of the usual planes of projection. That is, it is neither horizontal nor vertical. The auxiliary view looks directly at the inclined surface, in a direction perpendicular to it, and in this way provides a true size view of the surface.

You can think about auxiliary views as follows. If you wanted to cut a piece of paper to fit over the whole sloping surface in Fig. 8–1, could you cut it the size of the top view? Try it. You would get a similar wrong answer with the side view. The paper would have to be cut the size of the *auxiliary* view in order to cover the whole sloping surface.

An auxiliary view shows only the sloping surface, not the rest of the object. For this reason, it is often referred to as a *partial auxiliary view* or an *auxiliary surface.*

In summary, an auxiliary view is a projection on a plane that is parallel to an inclined (sloping) surface. It looks perpendicularly to the sloping surface. An illustration of a simple auxiliary view appears in Fig. 8–1.

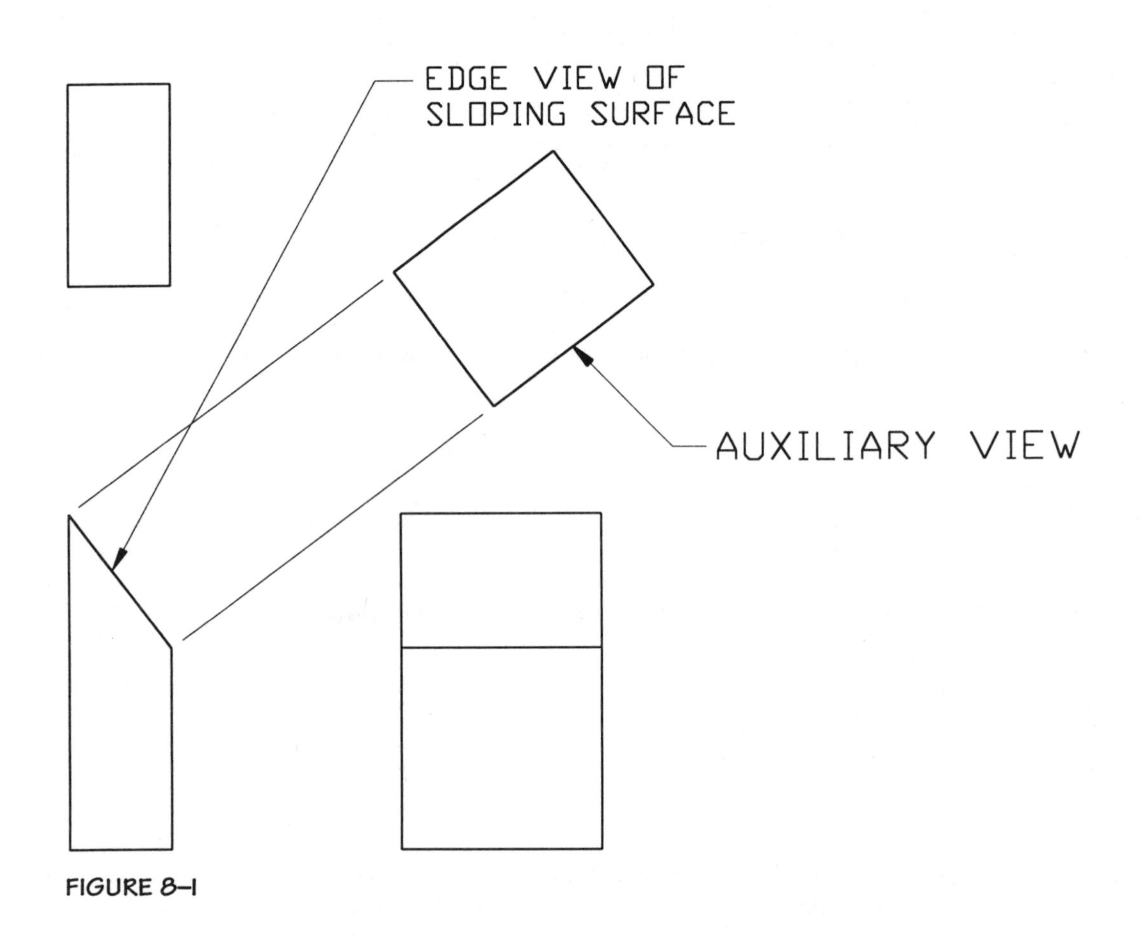

FIGURE 8–1

STEPS TO DRAWING AN AUXILIARY VIEW

STEP 1 Lightly number the corners of *only* the sloping surface as it appears in each view. Note that in the front view of Fig. 8–2, the top corner is labeled 1,2 because it represents the edge from 1 to 2 as seen in the top view or right side view. You must be consistent in your numbering of each view. One method is always to number clockwise around the view. Numbering may be omitted if you have experience with auxiliary views or if you can visualize the auxiliary view easily. *However*, if you do not have experience with auxiliary views and choose to skip this step, you can expect to encounter difficulties with this chapter. The isometric view is shown in Fig. 8–2 to help you visualize the problem. It normally would not appear on a working drawing.

If you are having difficulty placing the numbers, you may shade in the sloping surface.

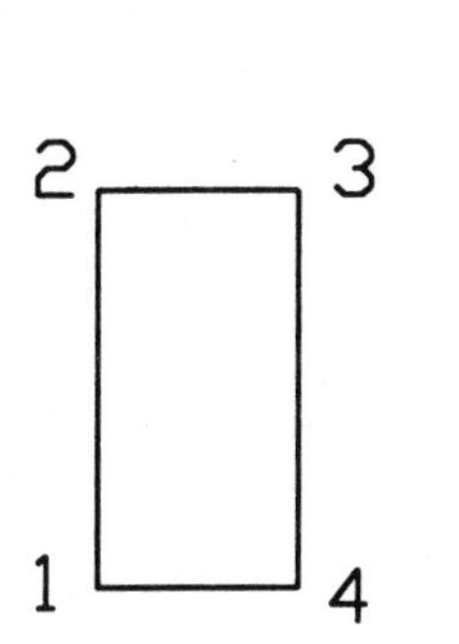

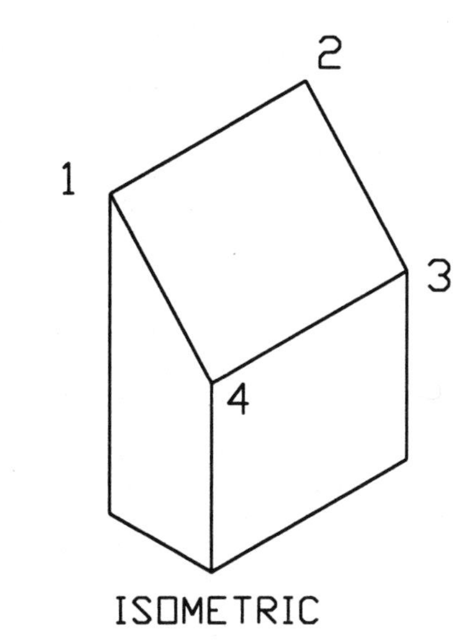

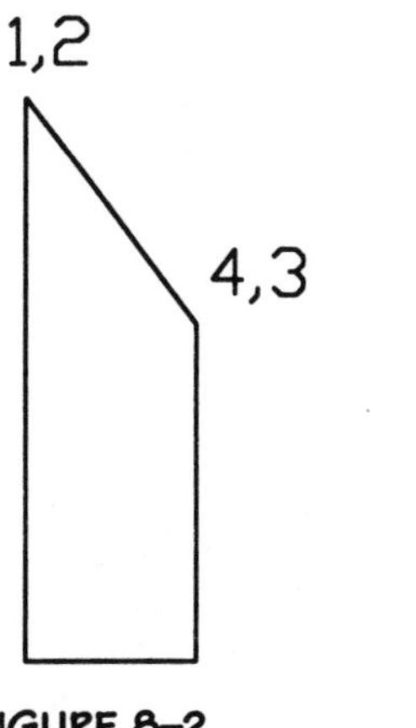

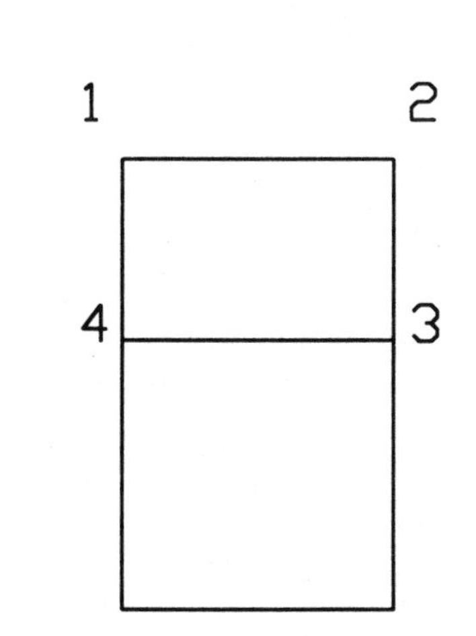

FIGURE 8–2

STEP 2 *Lightly* draw projection lines perpendicular to the slanting-edge view of the sloping plane, as shown in Fig. 8–3. These are called *lines of sight* because they are drawn in the direction in which your eyes would look at the sloping surface if you were looking perpendicularly at it. Lines of sight are *always* at a 90° angle to the sloping surface. (Some students wrongly think that these lines are always drawn at a 45° angle. This is often not the case.)

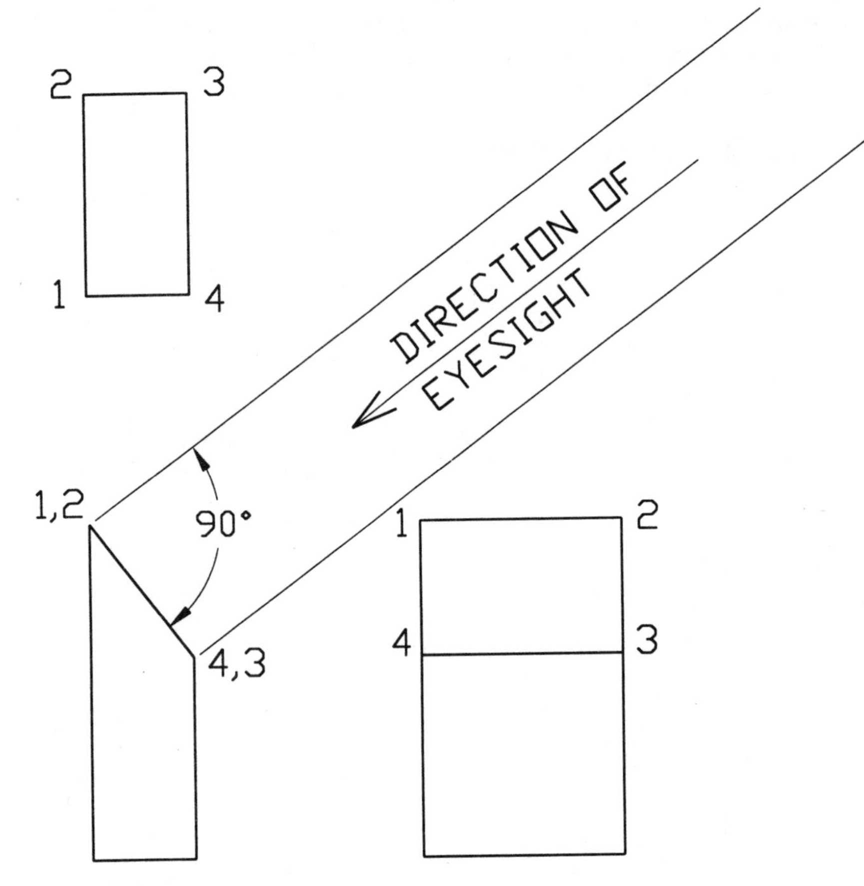

FIGURE 8–3

STEP 3 Draw a line perpendicular to the lines of sight (projection lines) at any convenient distance from the sloping surface (see Fig. 8–4). This line represents an edge of the auxiliary view. The line may also be a reference line, such as a center line, used to construct the auxiliary view. Number the points where this line intersects the projection lines so that the numbers correspond to the numbering on the other views. This line is drawn true length.

NOTE: A true length line means that if you measure that line in a particular view, it will be the actual length of the part. In Fig. 8–4, the distance from 2 to 3 is true length in the front view. In the top view, distance 2 to 3 is not true length. This means that if you wanted to "walk" from 2 to 3, the distance traveled would be that shown in the front view. The distance from 1 to 2 is true length in the top view and right side view. Distance 2 to 3 is not shown true length in the right side view.

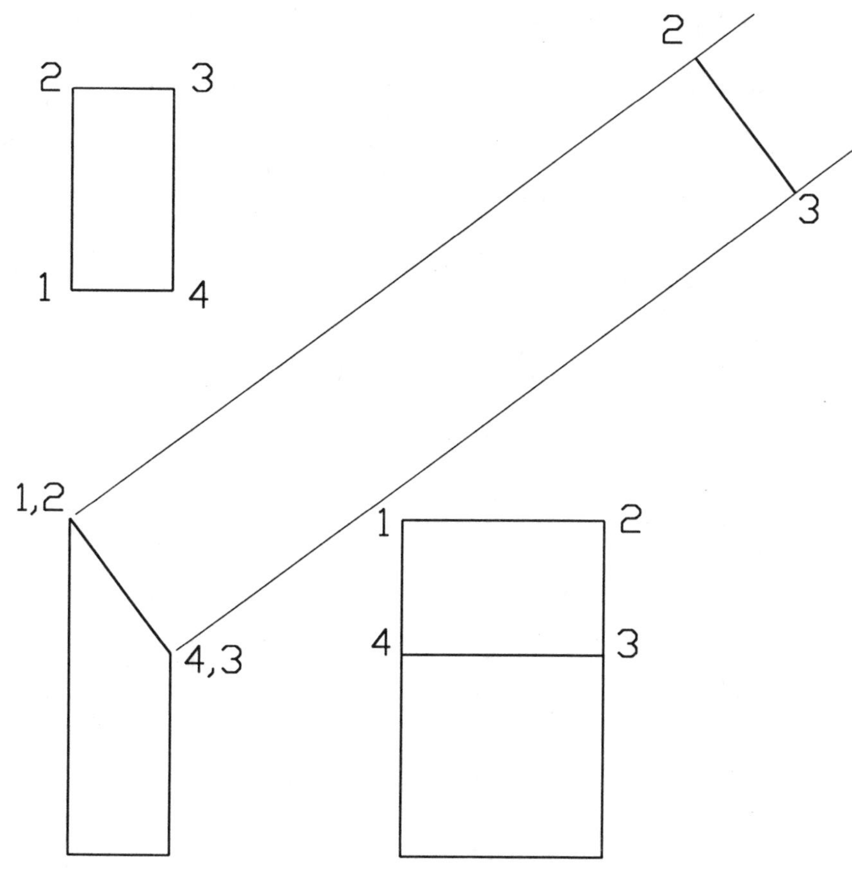

FIGURE 8–4

STEP 4 Locate the remaining corners of the auxiliary view. You must use *true length* distances. Note that the numbers on the corners of the sloping surface are only an aid to drawing the auxiliary view. Erase them when you have finished the view. The use of dividers is helpful in transferring dimensions. The dividers look like a compass with two metal points and no lead. (You may use a compass if you do not have dividers.) Open the dividers to the true length from 1 to 2. In Fig. 8–5, you may choose the top view or the right side view. Put one point at 2 in the auxiliary view and locate point 1. If you do not have dividers or a compass, you may use a scale to transfer the distances, although this method will take longer. Notice the labeling of distance X in three places in Fig. 8–5. Each of these lengths is true length.

NOTES:

a. In our example, no lines are projected directly from the top or side view to the auxiliary view. In general, the auxiliary view is projected only from the one view that shows the sloping surface as an edge.

b. An auxiliary view can be drawn from *any* of the six main orthographic views. An auxiliary view can even be drawn from another auxiliary view.

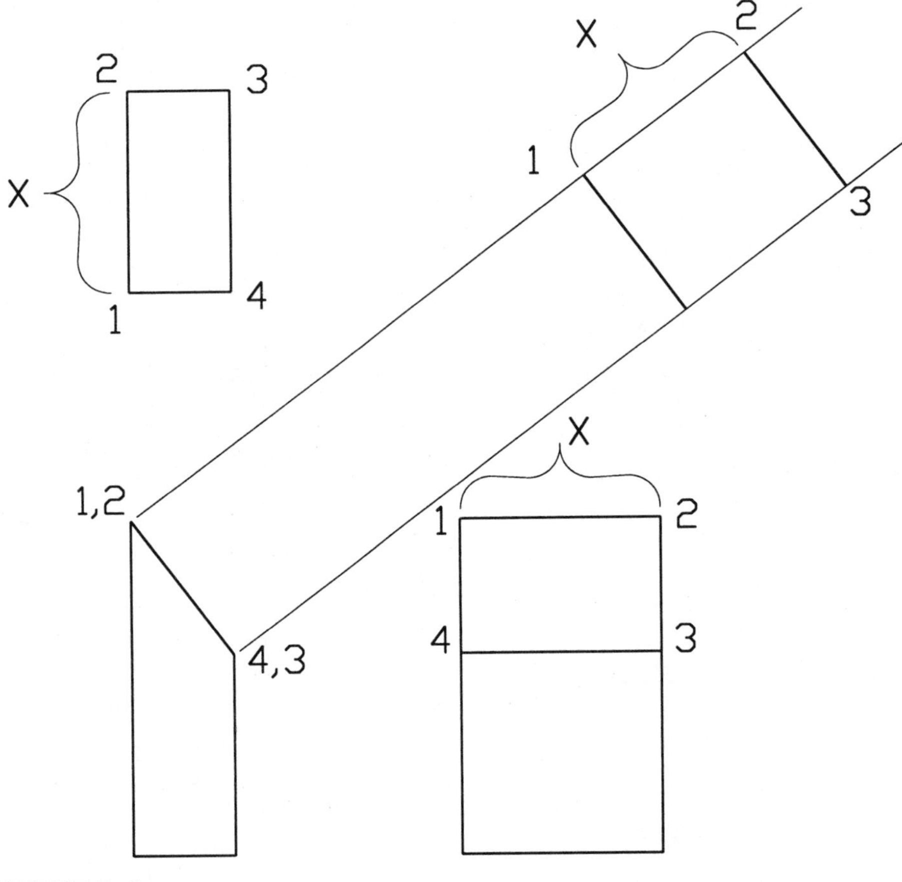

FIGURE 8–5

STEP 5 Locate the centers of holes or any other remaining features.

NOTES:

a. To draw the auxiliary view, you start by projecting in one direction (perpendicular to the sloping surface) and measuring in the other direction (parallel to the lines of sight).

b. Draw only a *partial* auxiliary view, which will show only the sloping surface as in Fig. 8–6(a), *not* the complete auxiliary view shown in part (b).

c. Usually, you do not draw hidden lines on an auxiliary view. You draw only what you see on the sloping surface.

d. Do not use a protractor to draw parallel and perpendicular lines in your auxiliary views. You can slide the hypotenuses of your two triangles together for this purpose, holding one triangle firmly on your drawing board. (This technique is illustrated in Constructions 1 and 2 of Chapter 4.)

e. When you have finished drawing your auxiliary view, you may wish to erase some of the projection lines. However, to help interpret the drawing, it is acceptable by current standards to leave some projection lines between the auxiliary view and the sloping surface.

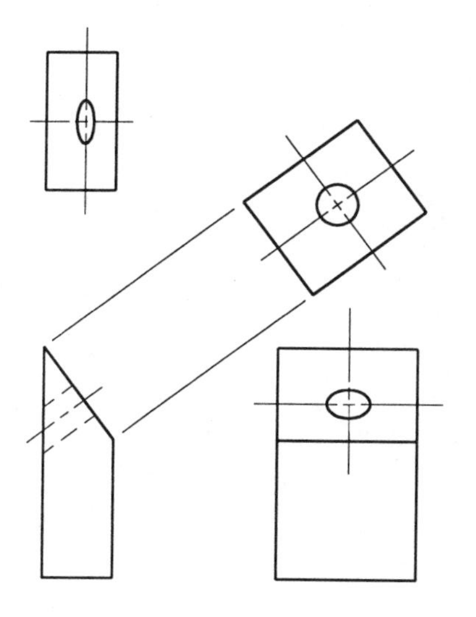

YES

FIGURE 8–6(A)

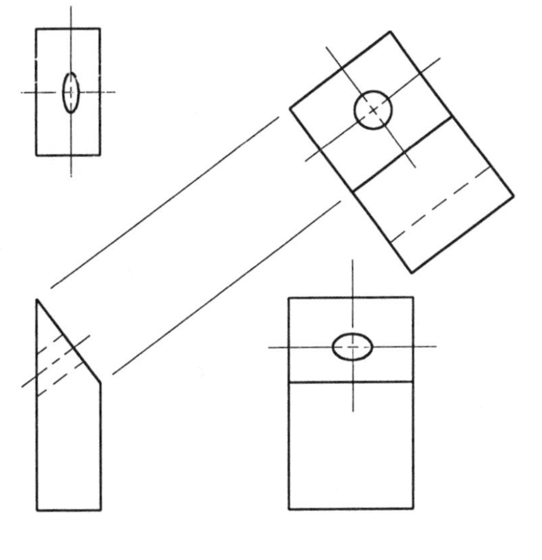

NO

FIGURE 8–6(B)

EXAMPLE

Figure 8–7 shows an example of an auxiliary view. Notice that distance 1 to 6 in the auxiliary view corresponds to distance 1 to 6 in the front view. (Distance 1 to 6 in the top view is not true length.) Notice also that distances 5 to 6 (labeled A in the figure) and 1 to 2 (B in the figure) are true length in the auxiliary view and in the top view.

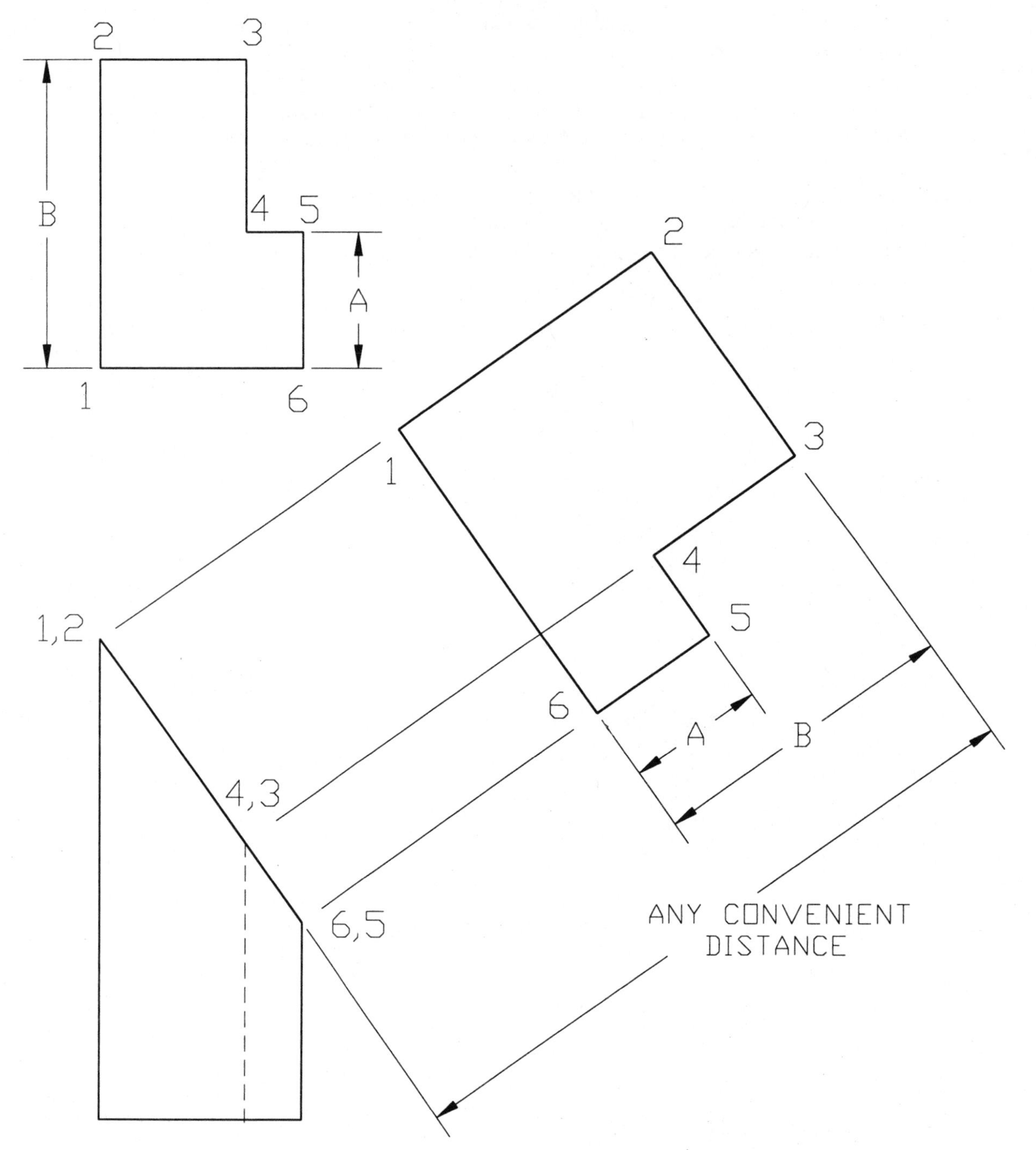

FIGURE 8–7

FOLDING LINES

An alternative way of drawing an auxiliary view is with *folding lines.* These lines represent the edge of each viewing plane. Remember that in Chapter 2 we cut out a figure that, when folded, was a box. Each side of the box represented a plane of projection (or a piece of glass), such as the front plane, top plane, and right side plane. The edges of these planes are represented by the folding lines.

STEP 1 Draw a line parallel to the sloping edge any convenient distance away. See Fig. 8–8. Label as A the side of the line where you will be drawing the auxiliary view, and as S the other side to represent, in this case, the right side plane. Then draw a vertical line to represent the edge of the front and right side planes, labeling its sides F for front and S for side.

Note that we want to look perpendicularly at the sloping surface in the direction of V.

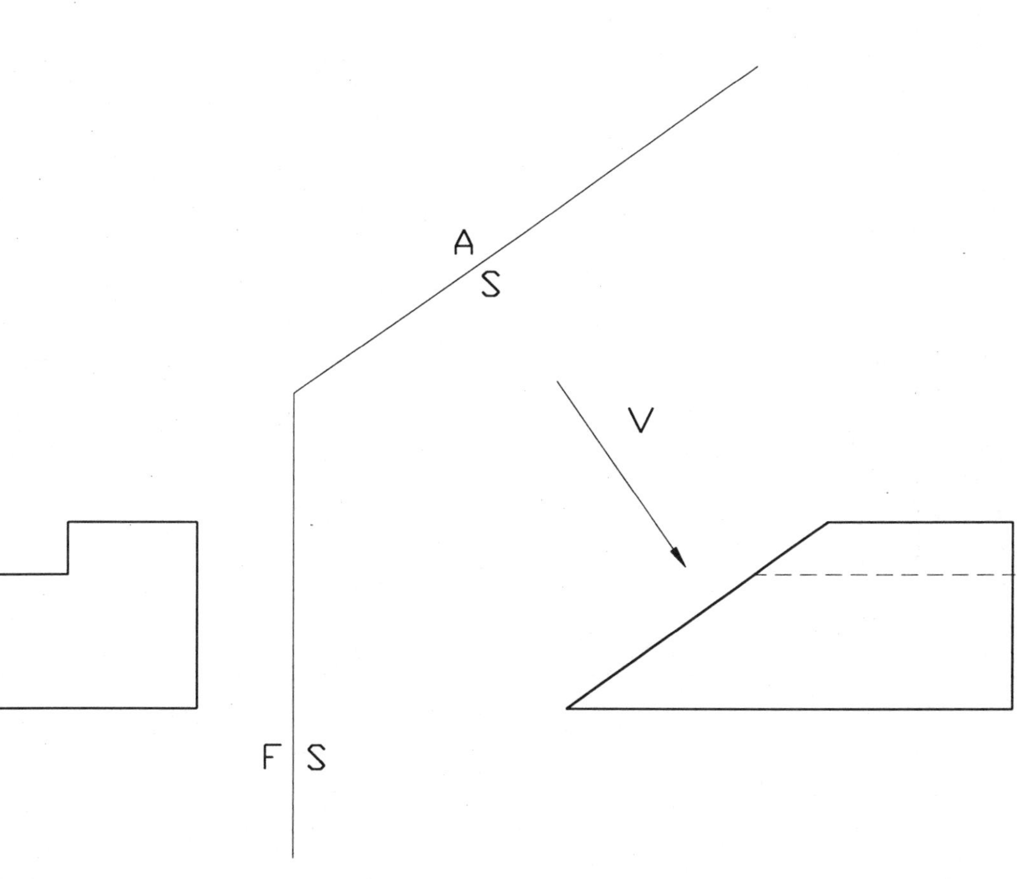

FIGURE 8–8

STEP 2 Draw construction lines perpendicularly to the sloping surface, at each end and at any other features that are to be projected into the auxiliary view. See Fig. 8–9.

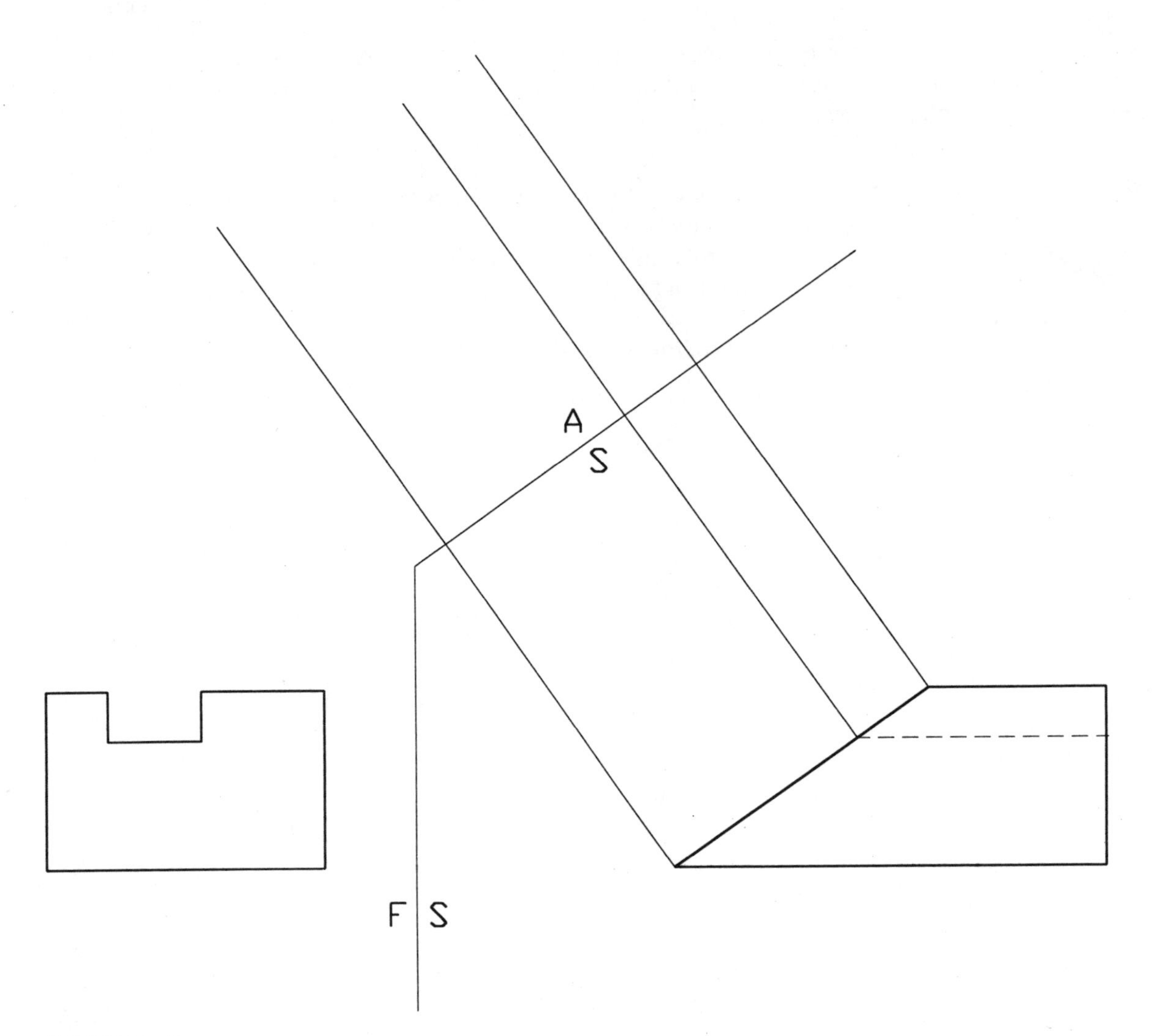

FIGURE 8–9

STEP 3 Measure horizontal distances W and X from the edges of the front view to the folding line F-S. Mark off these distances in the auxiliary view by measuring from the folding line A-S. See Fig. 8–10.

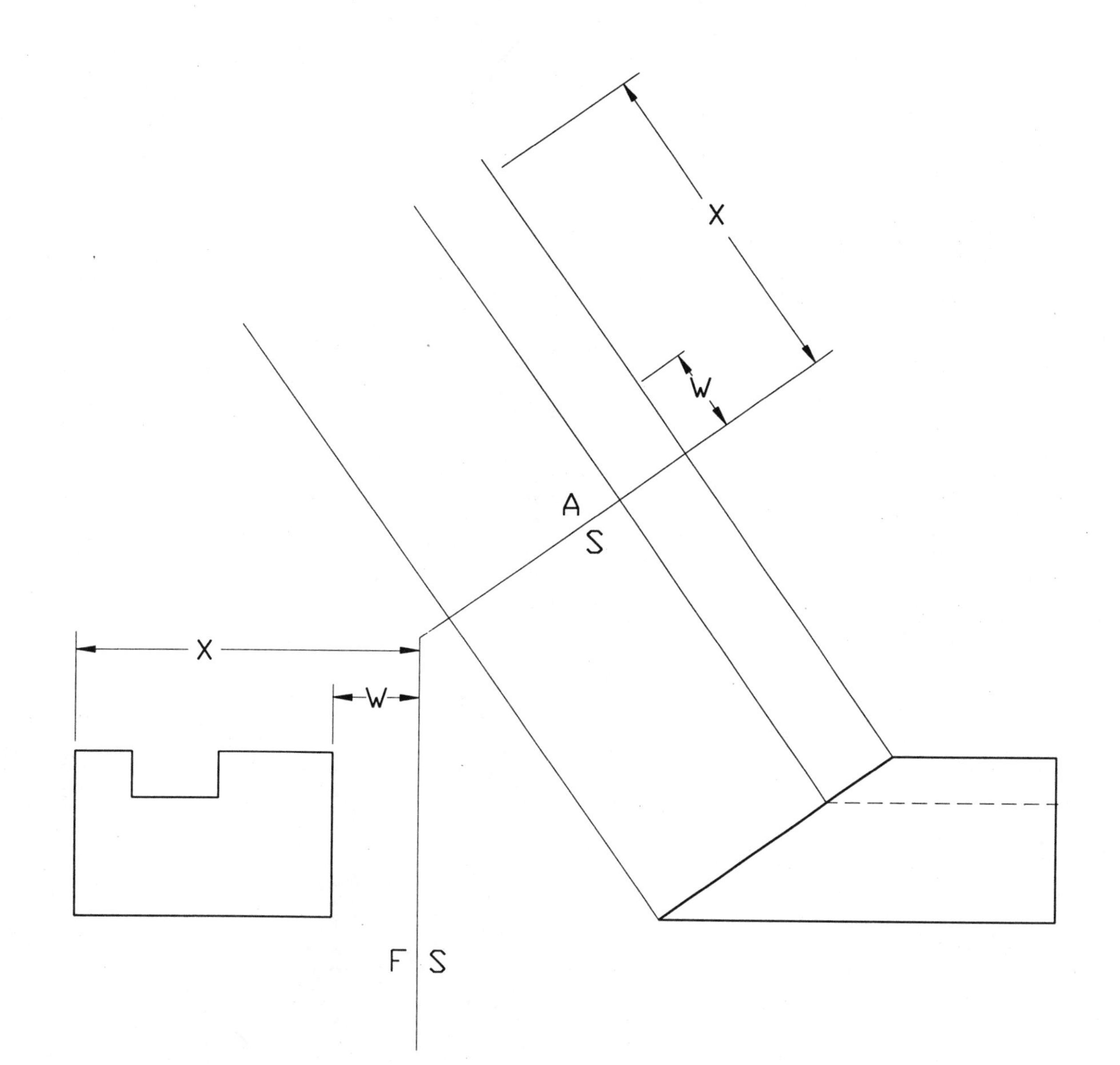

FIGURE 8–10

STEP 4 Using the distances marked off in step 3, locate some corners and lines in the auxiliary view. See Fig. 8–11.

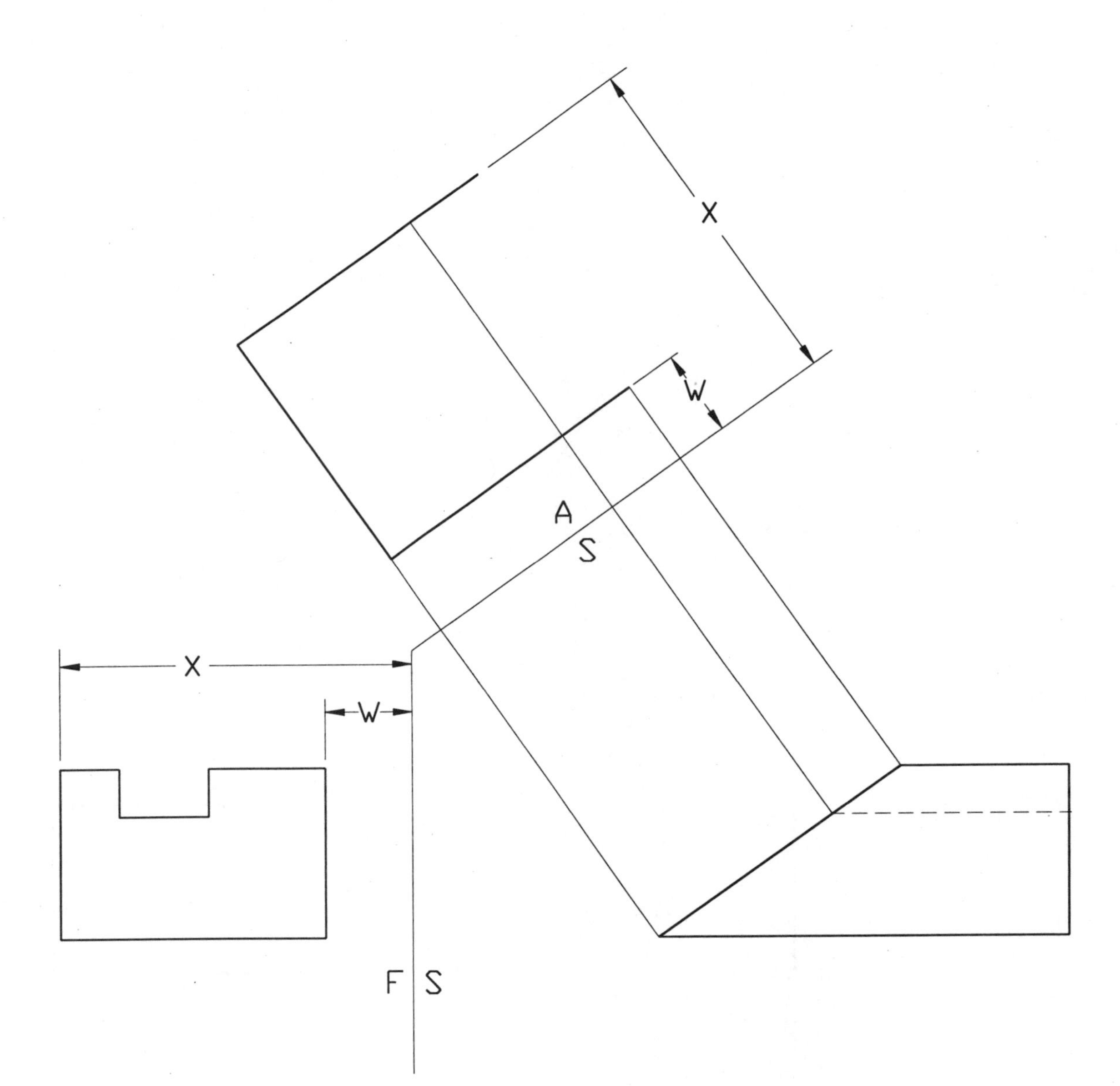

FIGURE 8–11

STEP 5 In a similar fashion, by measuring and marking off other distances from the front view, locate the remaining corners and lines. See Fig. 8–12, where distances Y and Z have been marked. Darken the required lines. If the folding lines and construction lines are too dark, erase them. With practice, you will be able to draw your construction lines lightly so that they do not have to be erased.

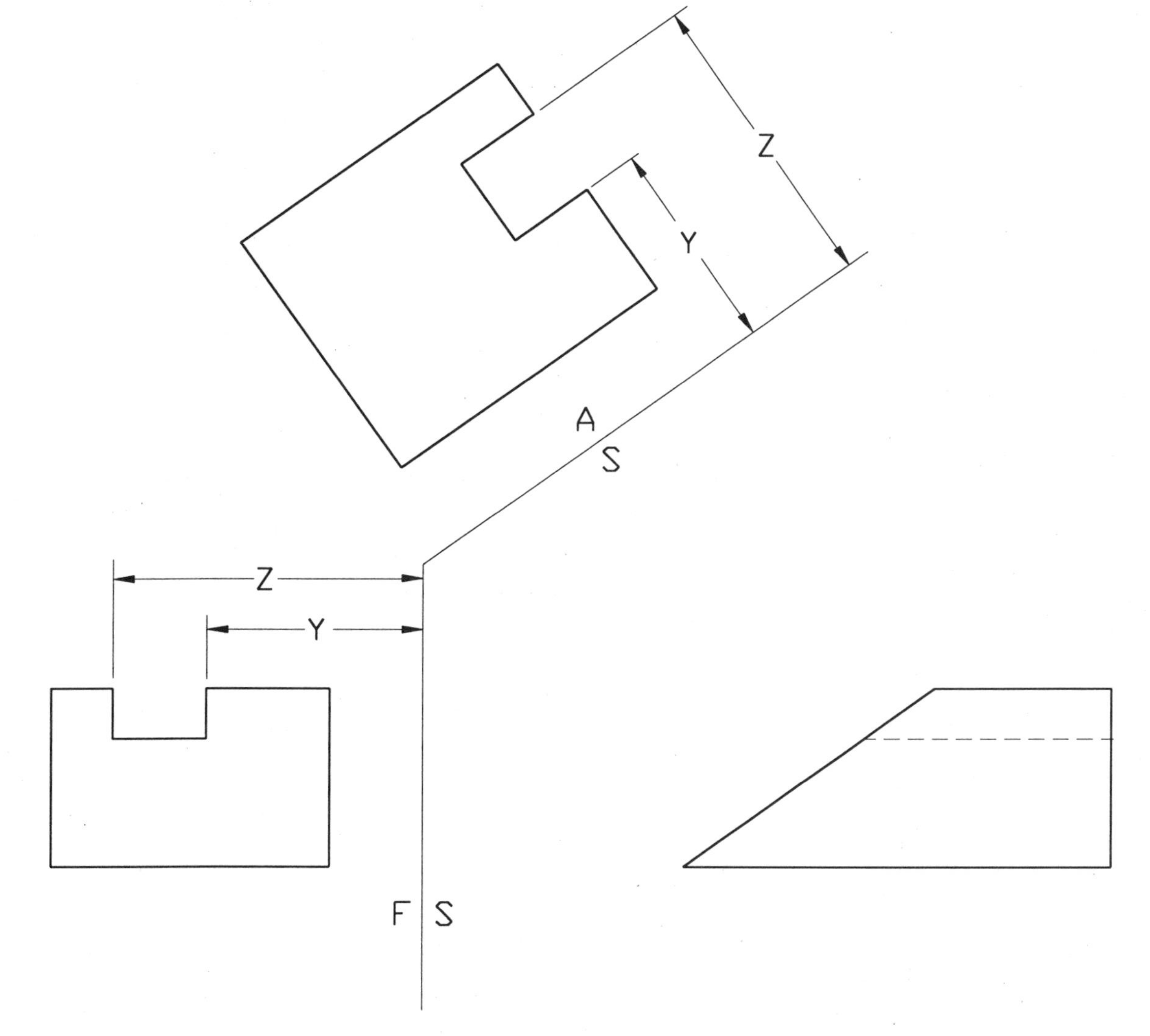

FIGURE 8–12

ACTIVITY: Remove this page from your workbook and cut out the shape in Fig. 8–13. Fold the shape so that it forms the object shown in Fig. 8–7. Now you may compare the views in Fig. 8–7 with the different surfaces on the three-dimensional object you have just made. Notice especially the sloping surface of your object that exactly matches the shape of the sloping surface of Fig. 8–7.

The drawing shown on this page is called a *development.* It is used in sheet metal applications and is studied in advanced courses.

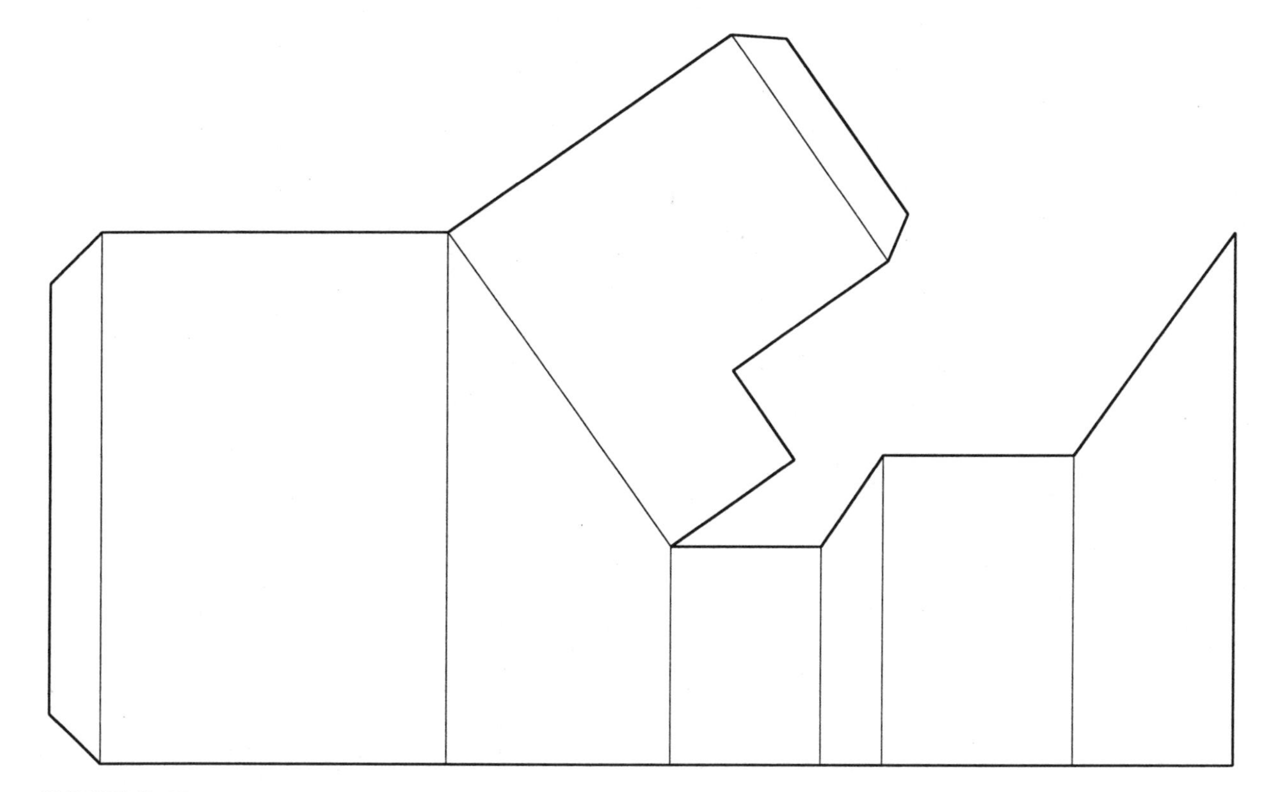

FIGURE 8–13

EXERCISES 8–1 AND 8–2

Draw an auxiliary view from the *front* view of each of the drawings shown below. You will be drawing what are called *front-adjacent* auxiliary views, since the auxiliary view is drawn perpendicularly to the front view. The numbers are provided only to assist beginning drafters. They are never shown on an industrial drawing.

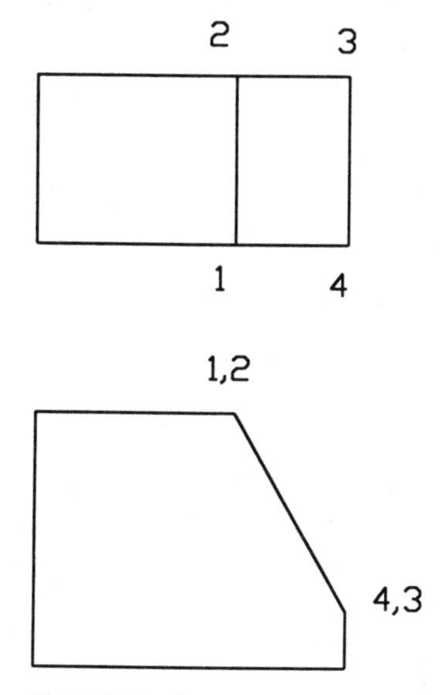

EXERCISE 8–1

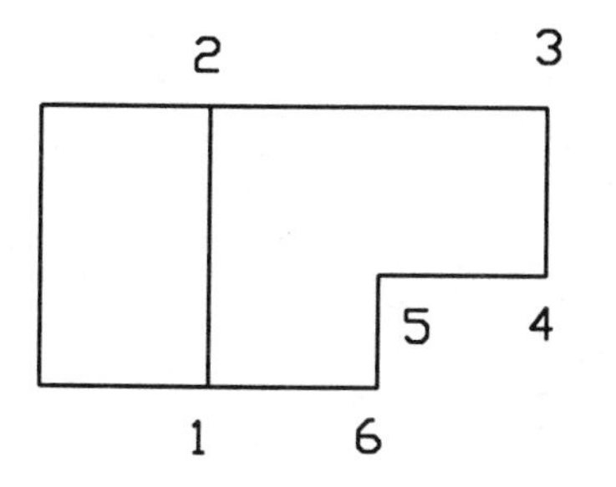

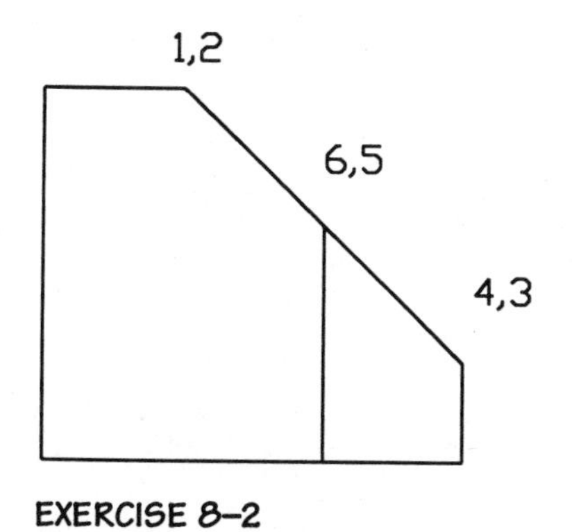

EXERCISE 8–2

DRAWN BY DATE

EXERCISES 8–3 AND 8–4

Draw a front-adjacent auxiliary view of each drawing shown below.

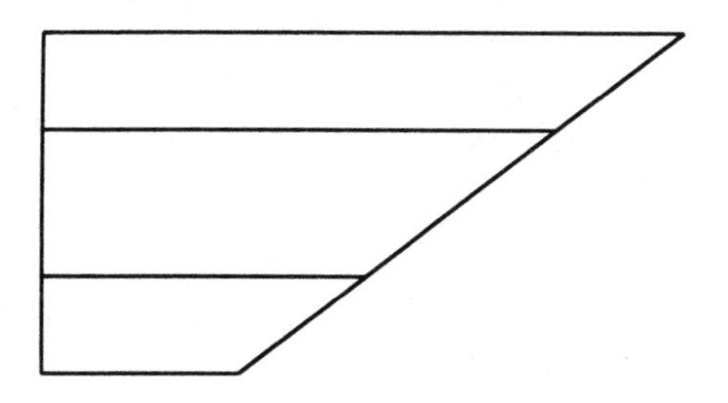

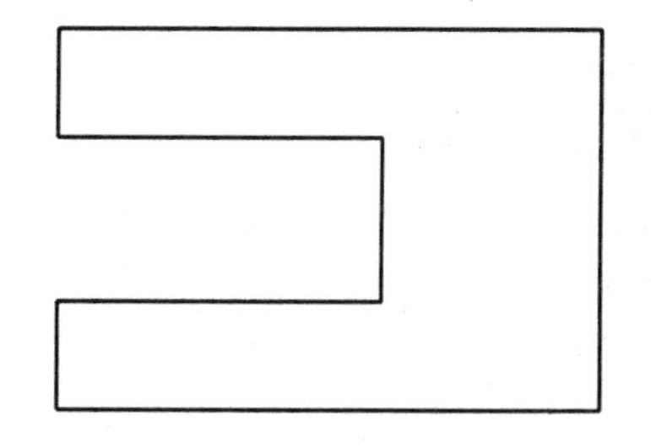

EXERCISE 8–3

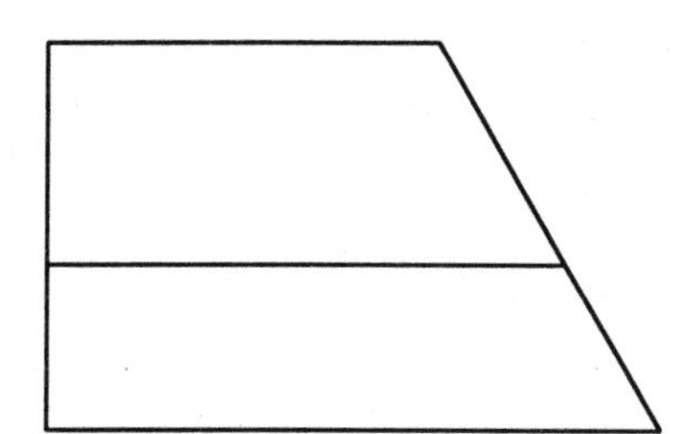

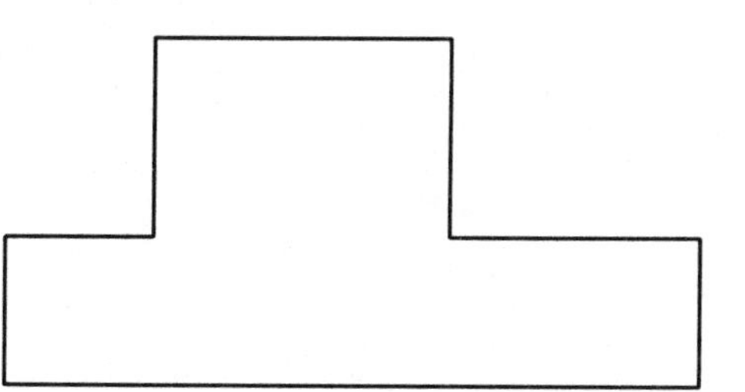

EXERCISE 8–4

DRAWN BY DATE

EXERCISES 8–5 AND 8–6

Draw an *auxiliary view* from the *side* view of each drawing shown below. Specifically, you will be drawing *side-adjacent* auxiliary views.

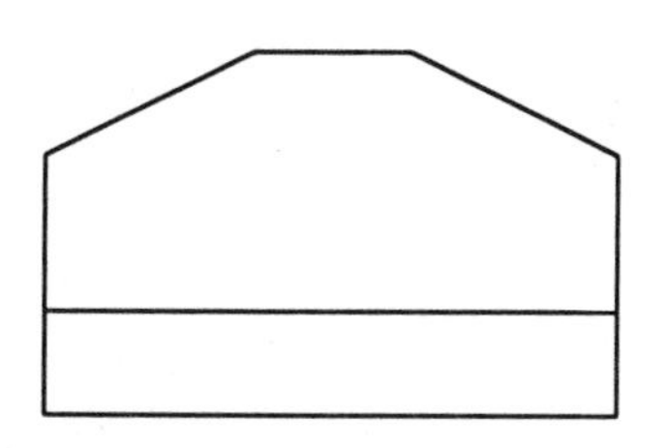

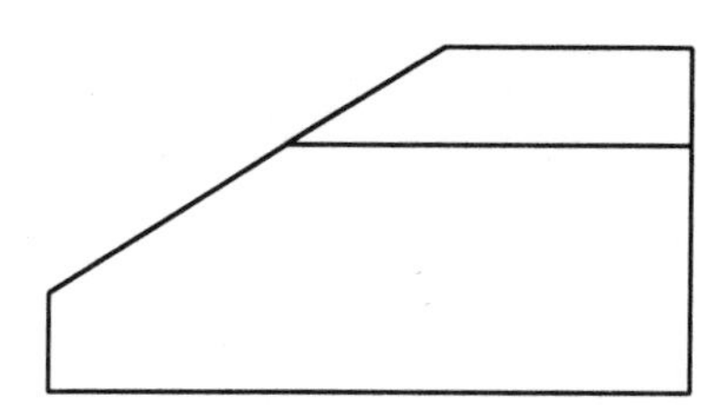

EXERCISE 8–5

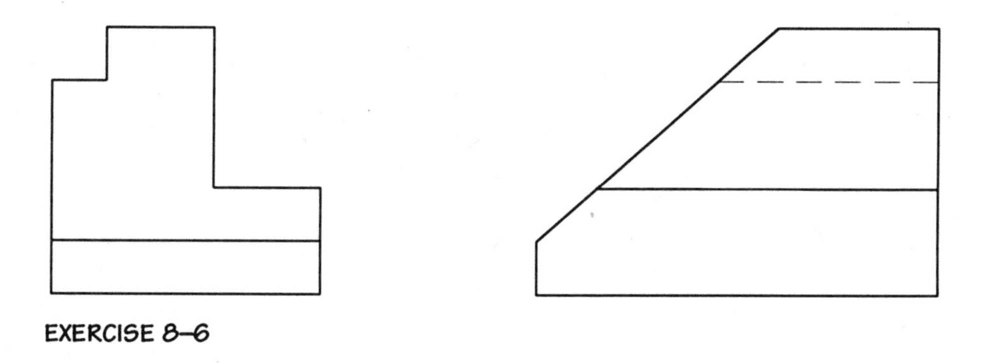

EXERCISE 8–6

DRAWN BY DATE

EXERCISES 8–7 AND 8–8

Draw an auxiliary view from the *top* view of each drawing shown below. Specifically, you will be drawing *top-adjacent* auxiliary views.

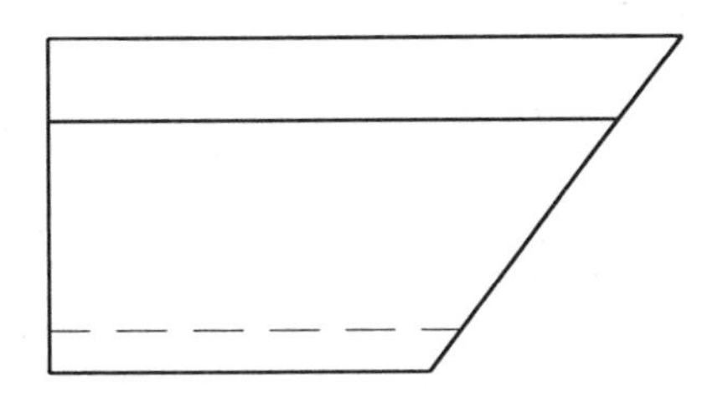

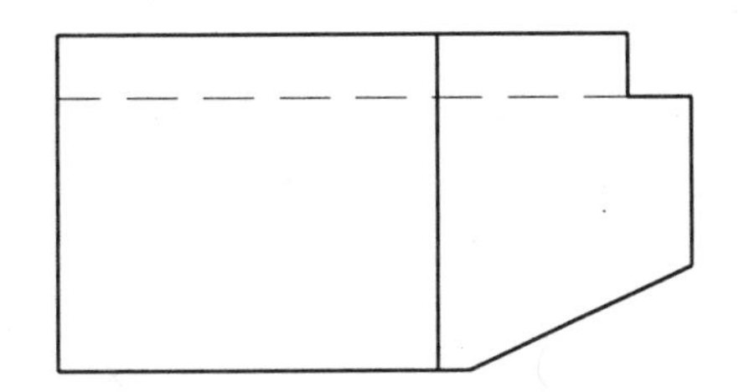

EXERCISE 8–7

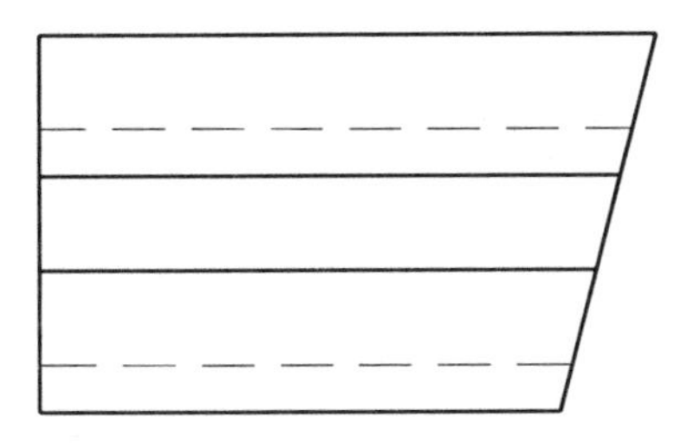

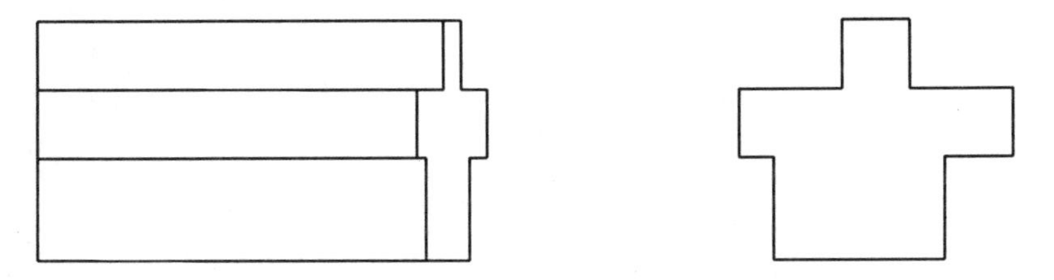

EXERCISE 8–8

DRAWN BY DATE

EXERCISE 8–9

Draw a top-adjacent auxiliary view of the sloping surface for the drawing shown below. A center line has been drawn in the front and right side views to show that the object is symmetrical. After you draw the two projection lines perpendicular to the sloping surface of the top view, draw a center line perpendicular to the projection lines as a reference. Then transfer distances 2 to 1 from the front or side view to the auxiliary view. Do likewise with distances 2 to 3, 5 to 6, and 5 to 4.

REMEMBER: The numbers are provided only to assist beginning drafters. They are never shown on an industrial drawing.

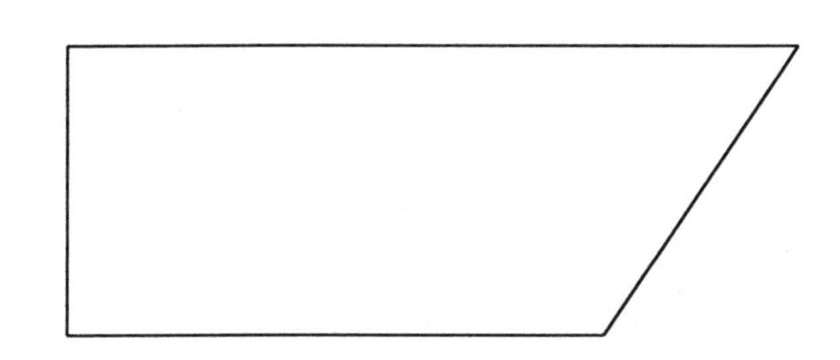

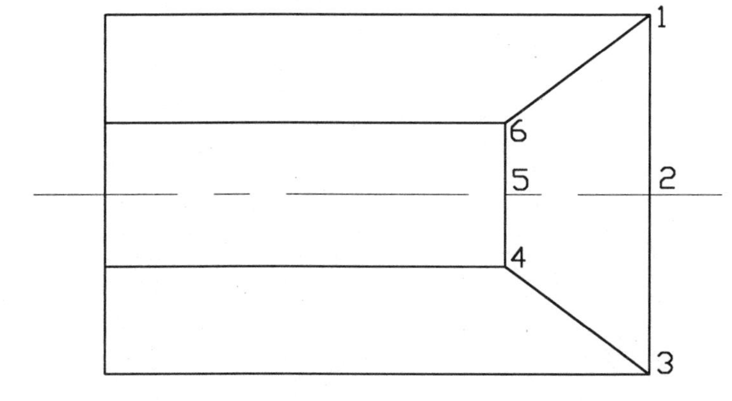

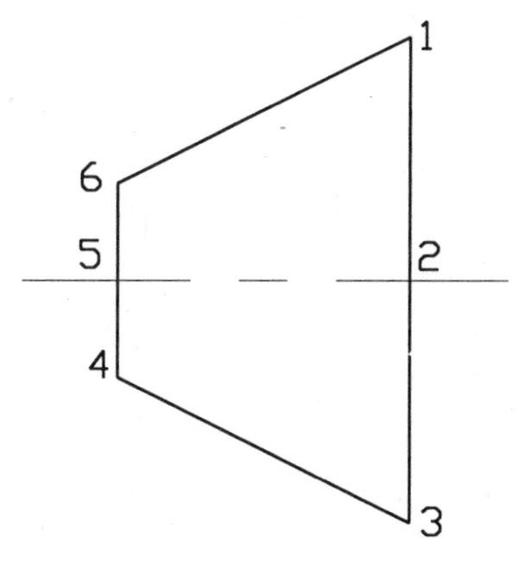

DRAWN BY DATE

EXERCISE 8–10

Draw a top-adjacent auxiliary view of the sloping surface for the drawing shown below.

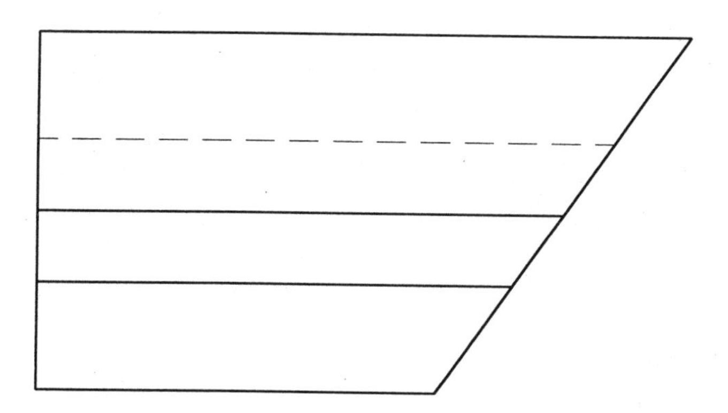

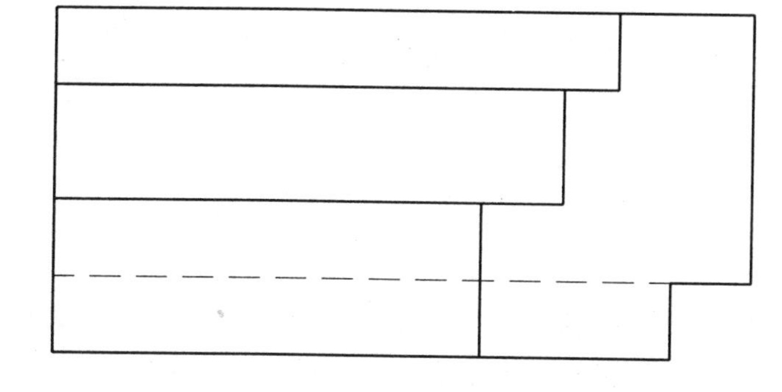

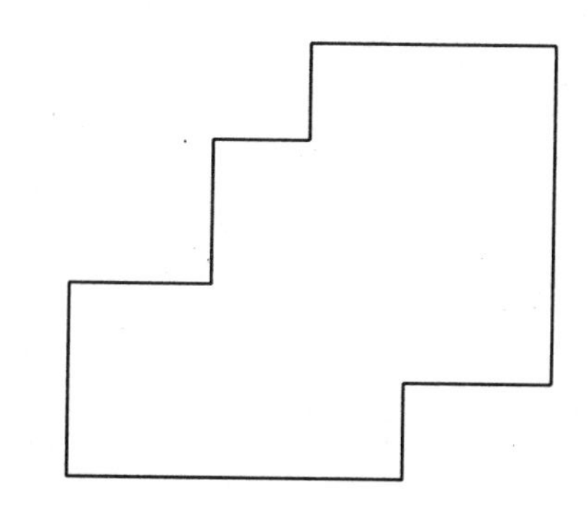

DRAWN BY ______ DATE ______

AUXILIARY VIEWS: PART 2 (CIRCULAR FEATURES)

9

OBJECTIVE

This chapter is an extension of Chapter 8. When you have finished this chapter, you should be able to draw auxiliary views of sloping surfaces that have circular features.

EQUIPMENT

You are required to have all your drafting instruments for this chapter.

AUXILIARY VIEWS OF CIRCULAR FEATURES

The following steps illustrate the procedures to draw an auxiliary view of a surface that has a circular feature. This circular feature may be a hole or a solid round sloping surface.

NOTE: If you would like to use a physical example of the drawing on this page, get a solid cylindrical object, such as a sausage, and cut one end at an angle.

STEP 1 *Identify the problem.* In Fig. 9–1, you can see the front and top views of the object. Given a sloping surface with a circular feature, the problem is to find the true shape of the circular feature (the whole sloping surface). You may think of this view as a piece of paper cut to fit onto the sloping surface. As with any auxiliary view problem, you must look in a direction perpendicular to the sloping surface and draw what you see.

STEP 2 Divide the view where you can see the curve of the circular feature (in this case, the top view) into a certain number of parts, say 12. Using your 30°–60° triangle (set square) held firmly against the parallel straightedge, draw *light* lines intersecting the circumference as if they went through the center point. See Fig. 9–2.

STEP 3 Using your triangle and parallel straightedge, project these points of intersection to the view where the circular feature is seen as a sloping edge view (in this case, the front view). See Fig. 9–3.

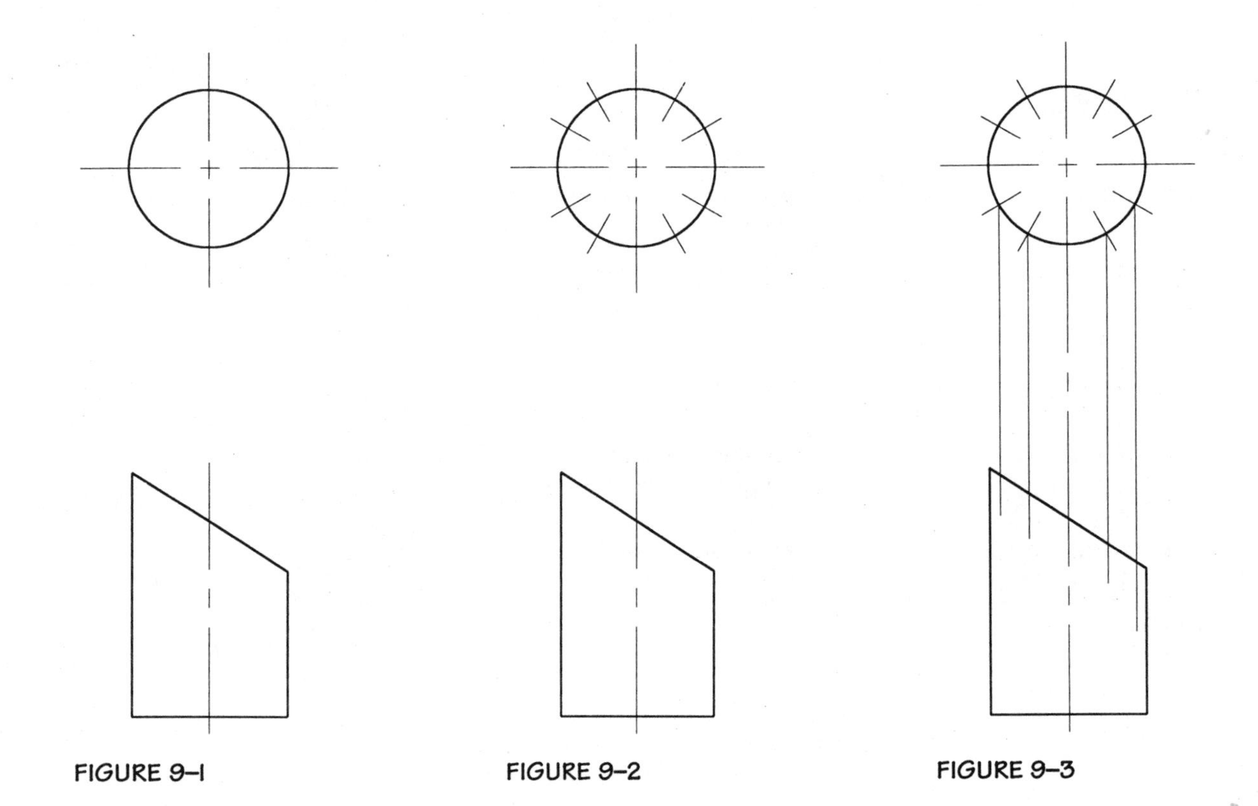

FIGURE 9–1 FIGURE 9–2 FIGURE 9–3

STEP 4 Draw *lines of sight*, or projection lines, *perpendicular* to the sloping edge view at the intersections made in step 3 and at each end of the sloping surface. See Fig. 9–4.

STEP 5 Draw a center line for the auxiliary view perpendicular to the lines of sight, at any convenient distance from the sloping surface. You should plan your work so that the auxiliary view does not touch the sloping surface or any other view. See Fig. 9–5.

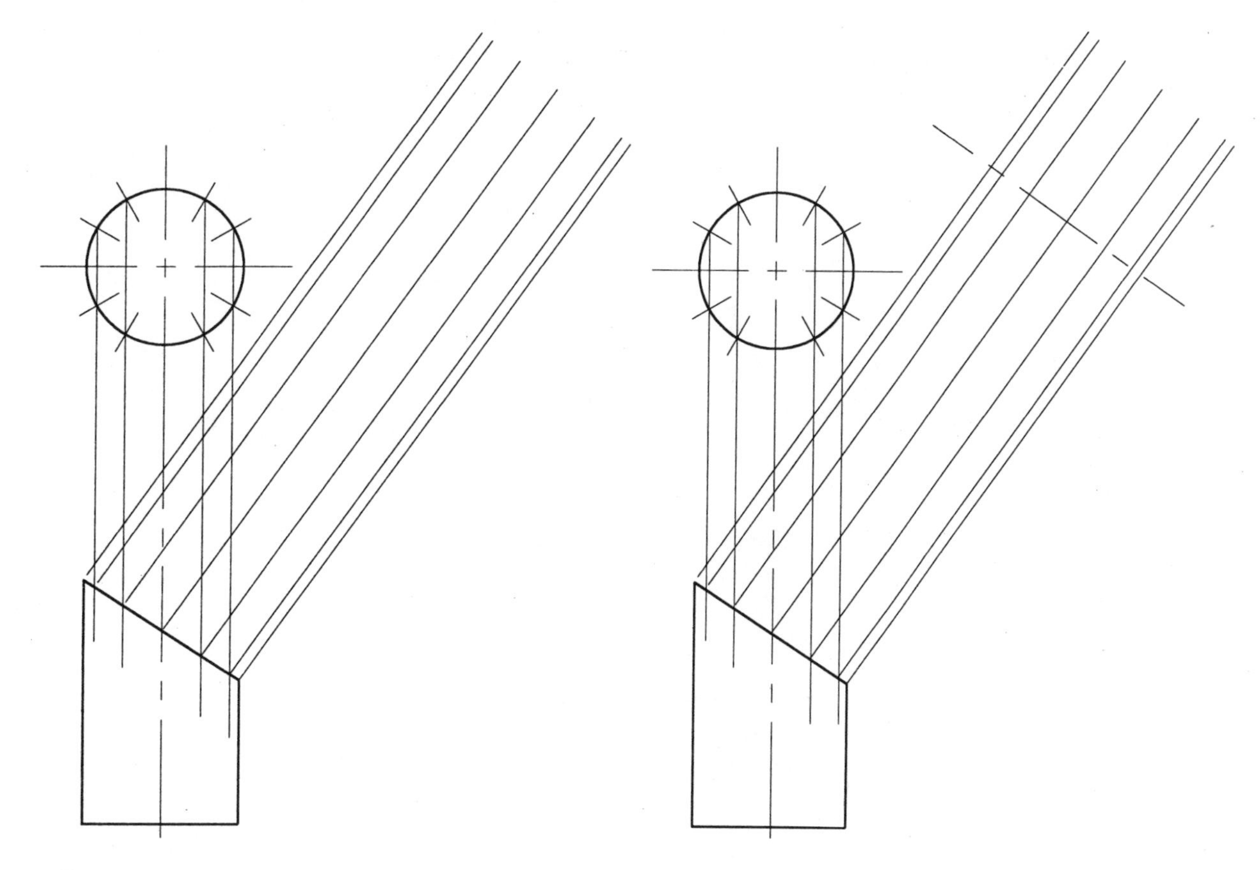

FIGURE 9–4

FIGURE 9–5

STEP 6 Number the points of intersection in the view where we see the curve of the circular feature, in this case, the top view. Use the same numbers to label the corresponding points in the view where we see the curve as a sloping edge view.

Find points 1 and 7 in the auxiliary view by measuring distances M and N in the top view of Fig. 9–6 and transferring these measurements to the auxiliary view. (In our example, distances M and N are equal.) Dividers are useful here to transfer the distances, or you may use a scale or even a compass. Normally you draw light lines to locate these points, as illustrated in the figure. Locate and label points 4 and 10 in the auxiliary view where the center line intersects the projection lines.

NOTE: To make the diagram easier to read, the vertical projection lines have been erased in Fig. 9–6.

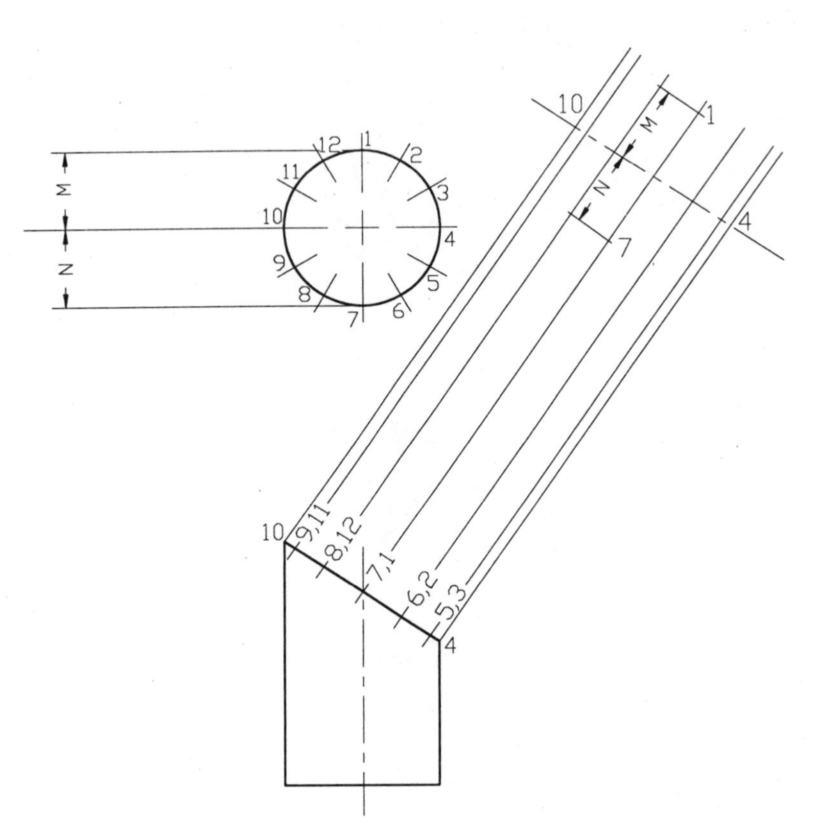

FIGURE 9–6

STEP 7 Locate the remaining points on the auxiliary view by transferring the appropriate distances from the top view. Distances P, Q, R, and S, as illustrated in Fig. 9–7, will assist you in finding points 12, 8, 3, and 5, respectively. As previously mentioned, dividers are helpful here to transfer the distances. (You may use a compass if you do not have dividers.) You may also use a scale.

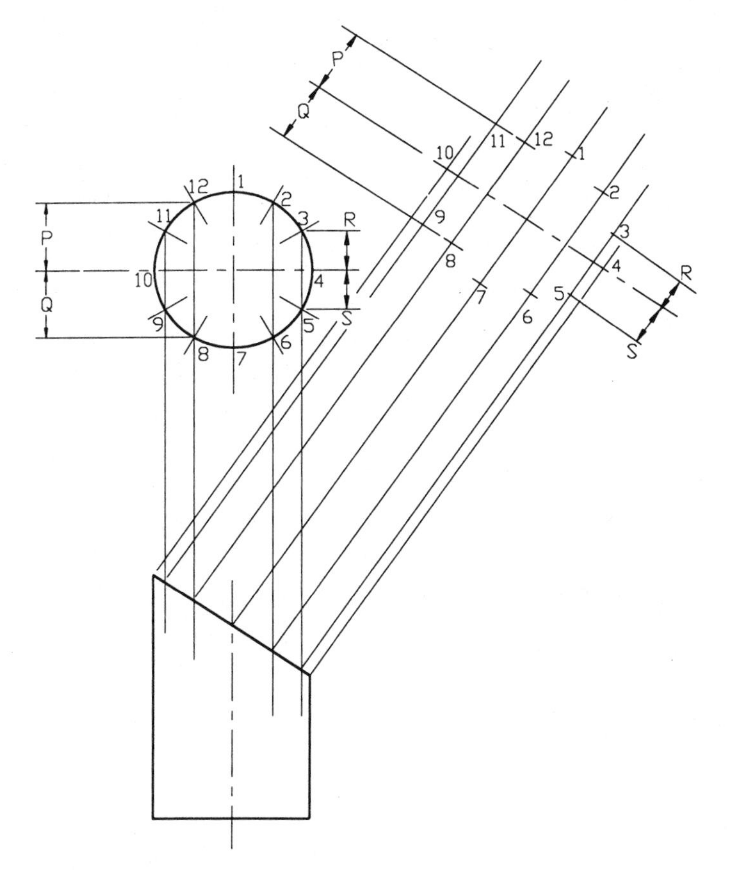

FIGURE 9–7

STEP 8 Join the points with a French curve. Drawing with a French curve is more of an art than a science. Using any portion of the inside or outside edge of the French curve, move the curve so as to join as many points as possible. A minimum of three points should be joined at one time. Work lightly at first, and darken the lines when you are satisfied with the results. Erase all unneeded lines. Do not draw the curve freehand!

As mentioned in Chapter 8, you may wish to show center lines or projection lines connecting the auxiliary view with its corresponding principal view. Figure 9–8 shows the center line between the front view and the auxiliary view to help you interpret the drawing and line up the views.

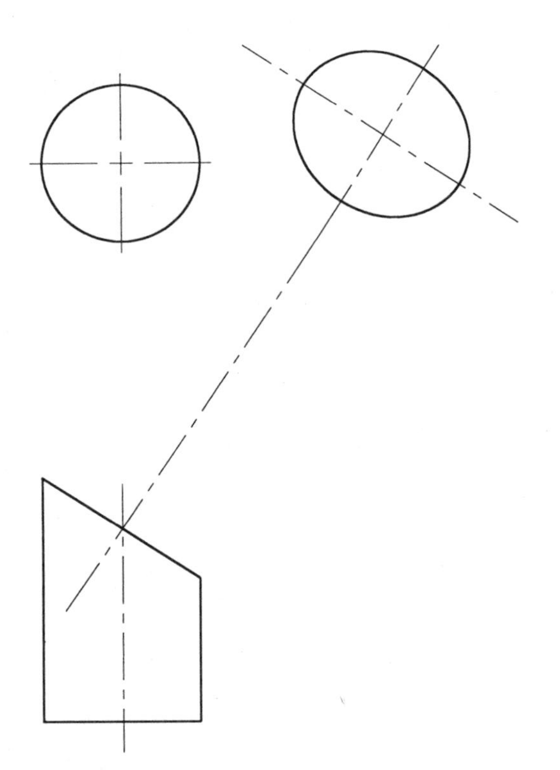

FIGURE 9–8

NOTE: Exercises 9–4 through 9–7 in this chapter do not contain any circular features. However, they do provide a review of the concepts covered in Chapter 8.

EXERCISE 9–1

The drawing below shows the front and right side views of an object containing a circular feature on a sloping surface. In the space provided, draw the front-adjacent auxiliary view (the auxiliary view drawn from the front view).

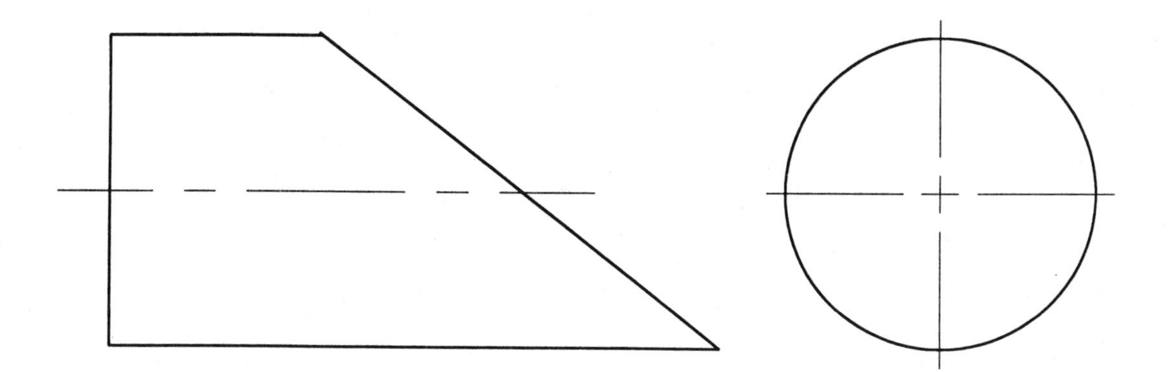

DRAWN BY DATE

EXERCISE 9–2

In the space provided, draw the necessary auxiliary view of the object shown below.

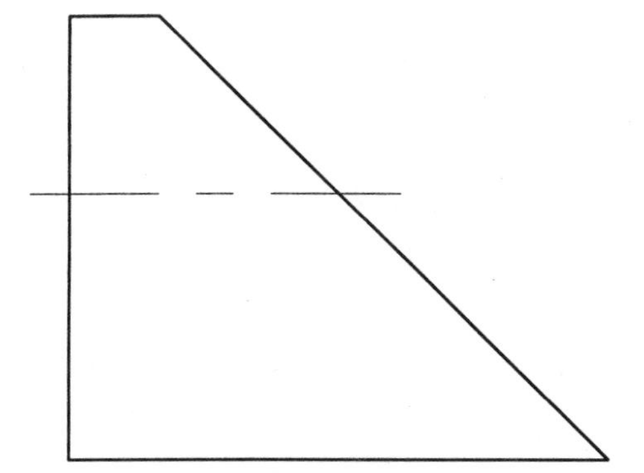

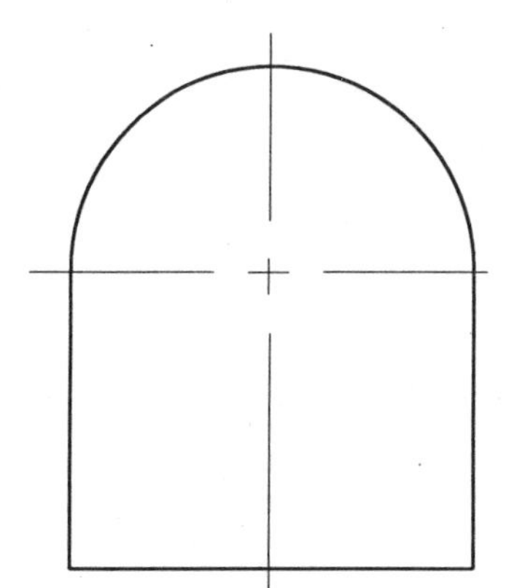

DRAWN BY DATE

EXERCISE 9–3

In the space provided, draw the front-adjacent auxiliary view of the object shown below.

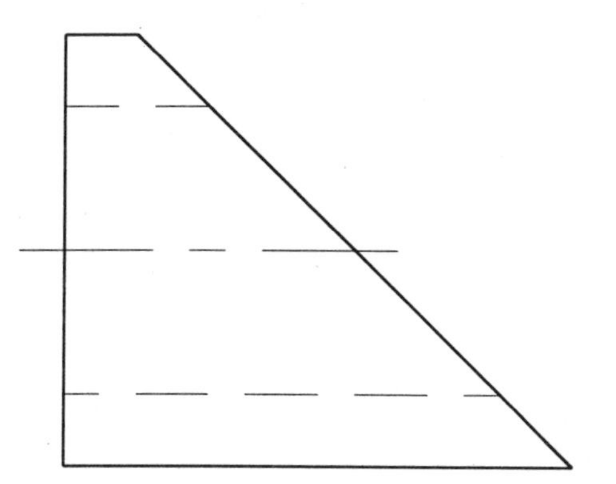

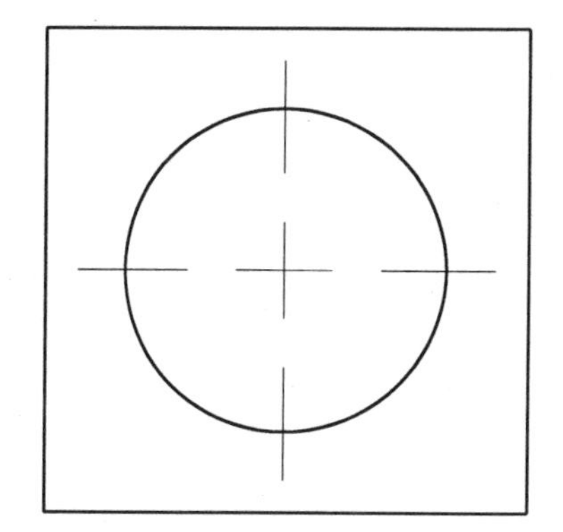

DRAWN BY ______ DATE ______

EXERCISE 9–4

In the space provided, draw the top-adjacent auxiliary view of the object shown below.

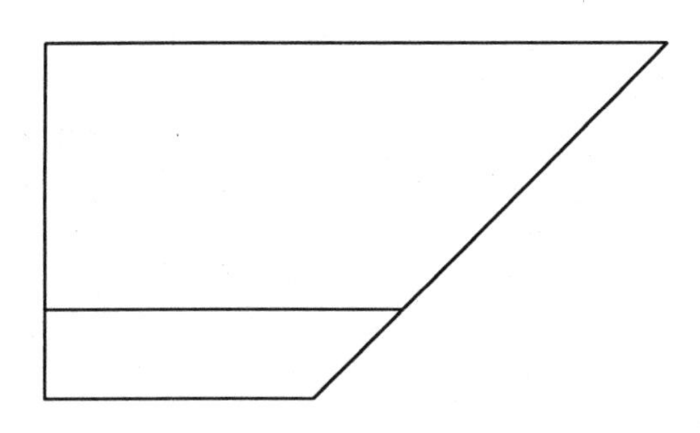

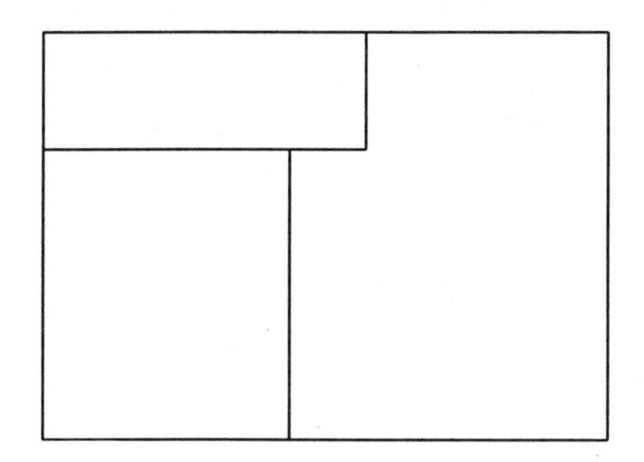

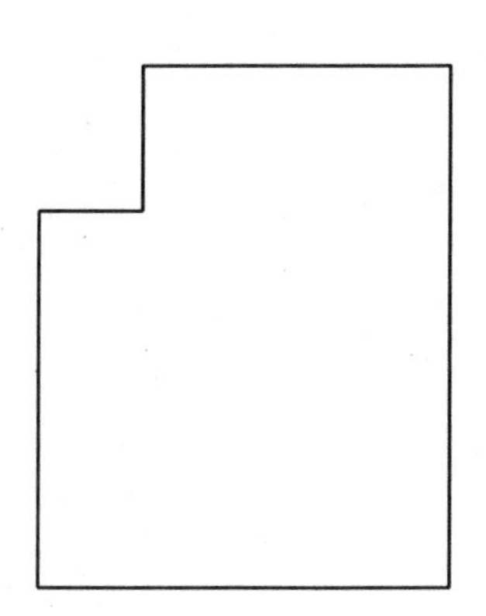

DRAWN BY

DATE

EXERCISE 9–5

In the space provided, draw the front-adjacent auxiliary view of the object shown below.

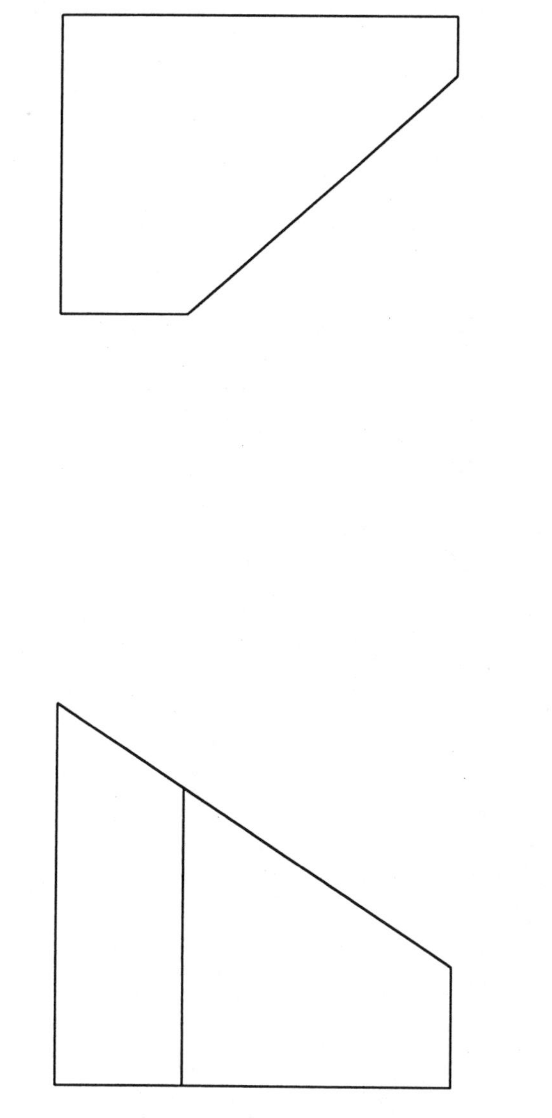

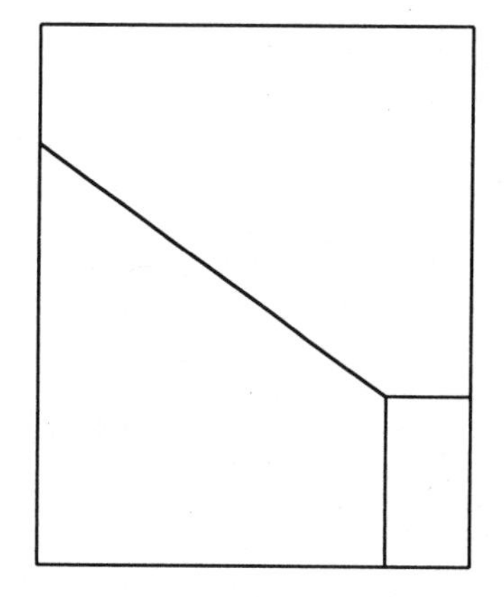

DRAWN BY DATE

EXERCISE 9–6

In the space provided, draw the side-adjacent auxiliary view of the object shown below.

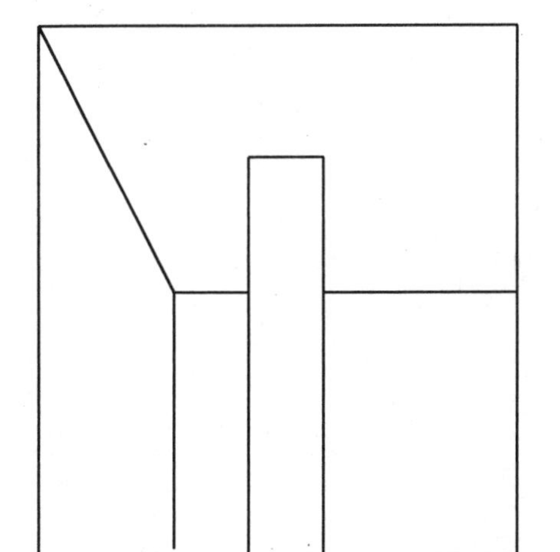

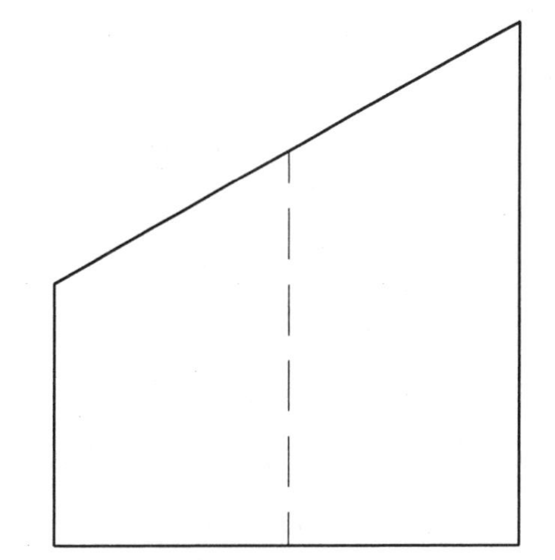

DRAWN BY _______________ DATE _______________

EXERCISE 9–7

In the space provided, draw the side-adjacent auxiliary view of the object shown below.

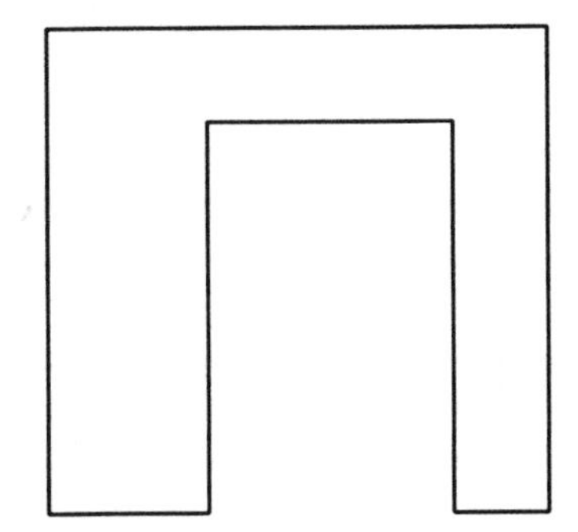

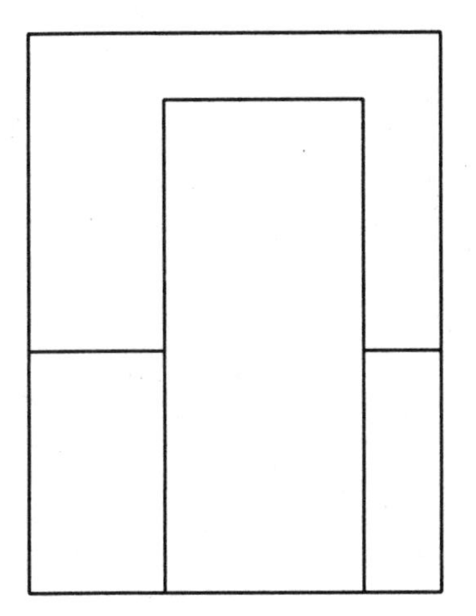

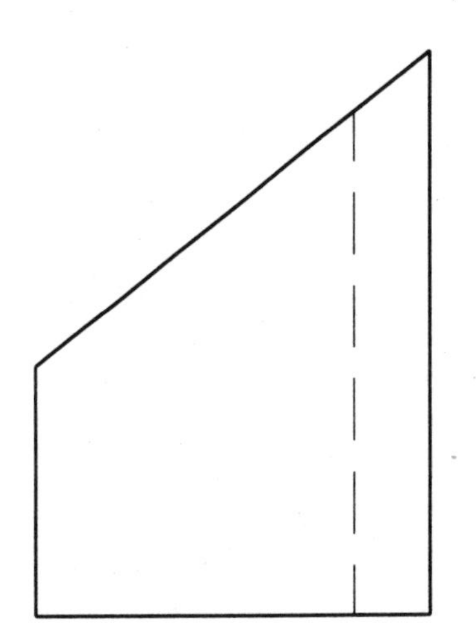

DRAWN BY DATE

SECTIONS: PART I

10

OBJECTIVE

When you have finished this chapter, you should be able to draw full-section and half-section views.

EQUIPMENT

You are required to have all your drafting instruments for this chapter.

SECTION VIEWS

A section view describes the interior of a complicated part where many hidden lines would make the drawing confusing to read. Using a section view helps to eliminate hidden lines.

The object shown in Fig. 10–1 is drawn in isometric as well as with normal orthographic views. You will observe that the right side view is not necessary to describe the part—in practice, we need only the top and front views. The side view is included for your reference. Notice that there are hidden lines in the front and side views. Study the drawing so that you fully understand what the piece looks like. This drawing does *not* have a section view. Figure 10–2 shows the same object drawn with a section view.

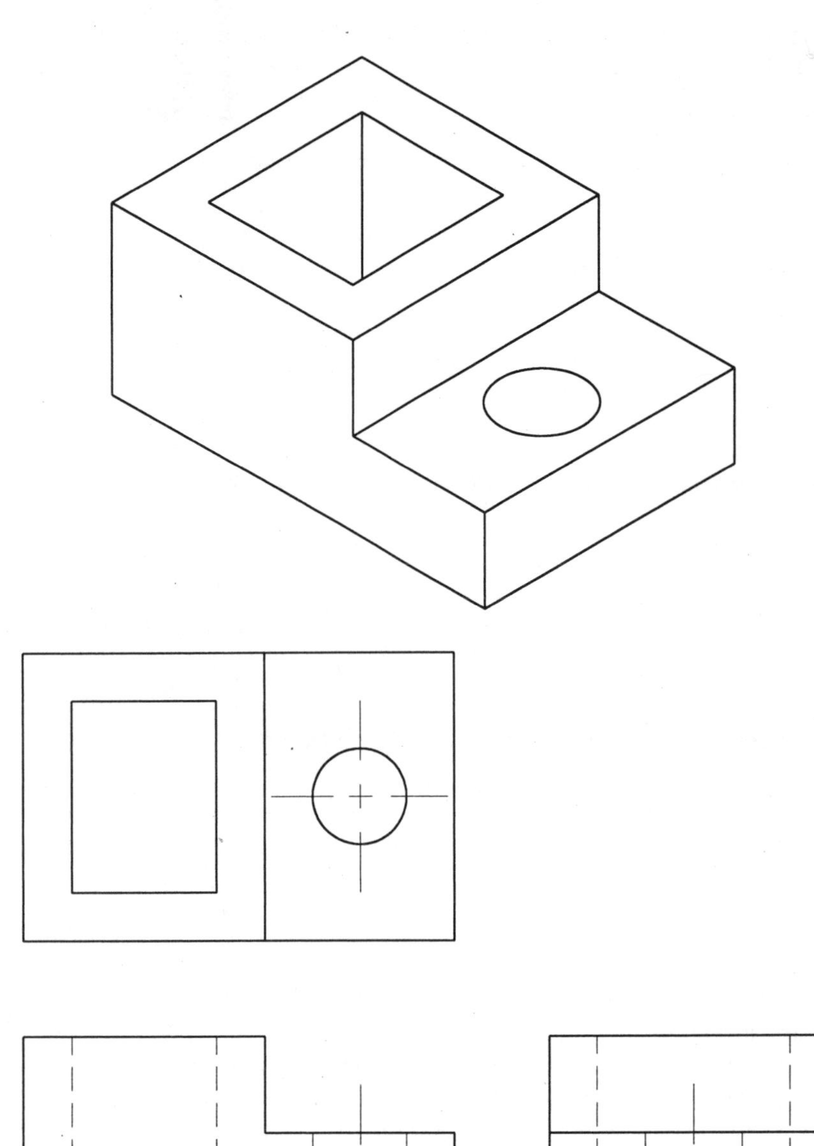

FIGURE 10–1

FULL SECTION

Figure 10–2 shows the same piece as in Fig. 10–1. This orthographic drawing, however, has the front view drawn as a *full section.* To obtain this section view, imagine that we cut the top view in half with a saw, along the thick *cutting plane line.* The purpose of the cutting plane line is to show where the cut was made. The arrowheads at the end of the line show the direction in which we are looking after making the cut. We now remove the front half of what was cut and look at what is left. We are now looking into the object. If we draw what we see, we will have a *full section.* (The cutting plane line goes fully through the object.) To tell anyone reading the drawing that this cut was made only in our imagination, we put *section lines* (also called *cross-hatching*) on the surfaces touched by the saw.

To visualize where the section lines should be placed, imagine that the part is cut on the cutting plane line. Hold the half that has the cutting plane line arrows pointing toward it and press it onto an ink stamp pad. The surfaces that pick up ink should be section-lined.

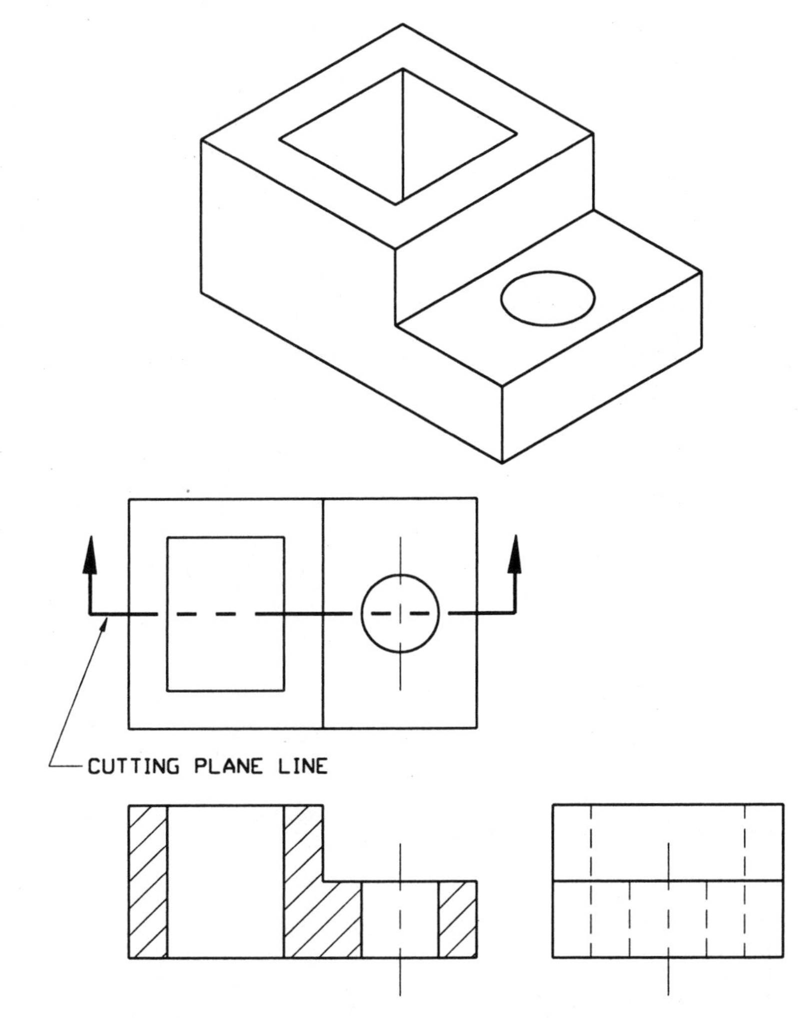

FIGURE 10–2

EXAMPLES OF FULL SECTIONS

Figures 10–3 and 10–4 show examples of full sections. The section lines are spaced about 1/4" apart. For large areas, they may be spaced farther apart, and for small areas, closer together. (Very thin surfaces may be shaded in, as in the case of a gasket.) Usually, hidden lines are omitted from a section view, unless they are needed for dimensioning or clarity. The section view may replace a top view, front view, side view, or any other orthographic view.

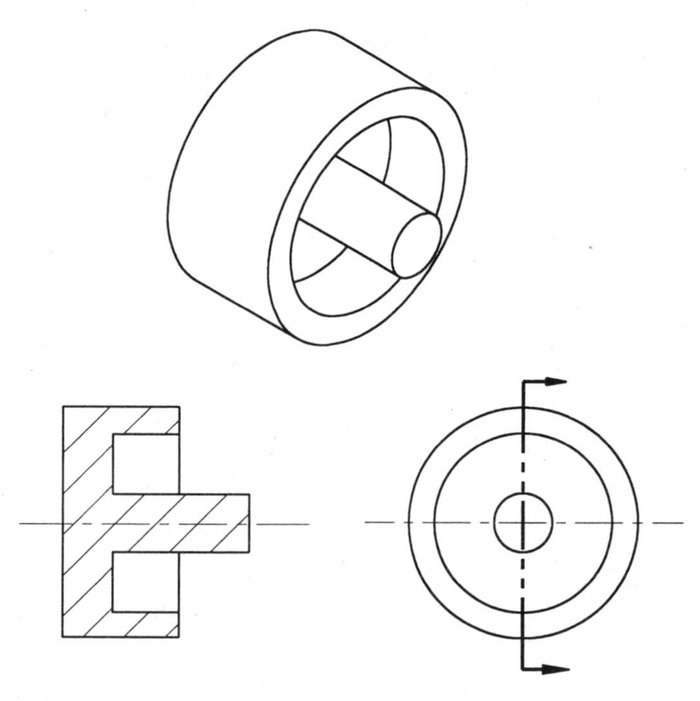

FIGURE 10–3

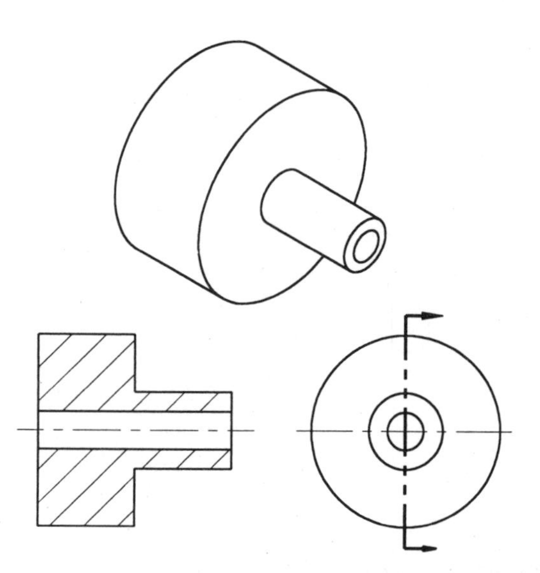

FIGURE 10–4

CUTTING PLANE LINE

There are several ways to draw a *cutting plane line.* American standards differ from Canadian and International standards. Figure 10–5 demonstrates various features of ANSI standards (see Fig. 10–6 for CSA and ISO standards). Note the following about the ANSI cutting plane line: It is a thick line, and its arrowheads are attached to the ends of the line, indicating the direction in which you are looking after the cut has been made. Two equally acceptable styles of cutting plane line are shown in parts (a) and (b). In parts (c) and (d), the line has been terminated because the drawing is complicated and would appear cluttered if the cutting plane line were to run through it. Whereas (a), (b), and (c) have the section view in the front view, (d) shows a different orientation in which the section view is in the top view.

The lengths of the long and short lines of the cutting plane line may be of any convenient length. The following guideline is suggested: Draw the long lines between 3/8" (10 mm) and 1 1/2" (38 mm) long, depending on the size of the area to be sectioned. Draw the short lines about 1/8" (3 mm) long. Space the lines about 1/16" (2 mm) apart.

The cutting plane line may be omitted altogether when a line of symmetry, or center line, coincides with the cutting plane line, or when the location of the cutting plane line is obvious. In Fig. 10–5, the cutting plane lines could have been omitted.

This workbook uses the ANSI standard cutting plane line.

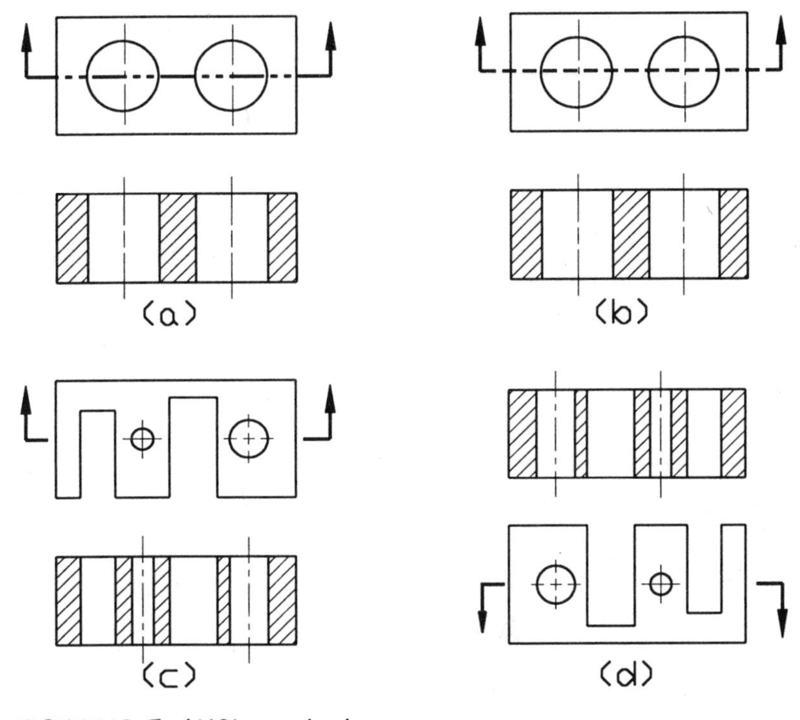

FIGURE 10–5 ANSI standards

CSA (Canadian) and ISO (International) standards specify a different kind of cutting plane line in which the ends of the line and any corners are thick and the rest of the line thin. Notice that the thin part of the line looks the same as a center line. Also note how the arrows are placed. Although arrow placement differs from that shown in Fig. 10–5, it follows the same principle of indicating the direction in which you are looking after making the cut. Figure 10–6 illustrates several features of the cutting plane line according to these standards.

Study the illustrations to familiarize yourself with the differences, and follow the standards that conform to your institution.

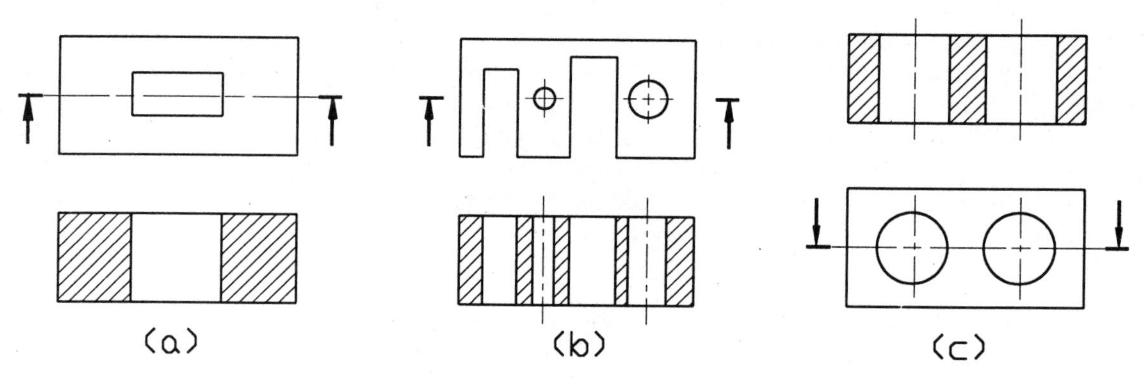

FIGURE 10–6 CSA and ISO standards.

SECTION LINES: METHODS

Section lines, or cross-hatching, should be evenly spaced and thin. If possible, draw lines at a 45° angle with your triangle. If necessary, the lines may be drawn at a 30° or 60° angle (also using your triangle), sloping to the left or the right. Figure 10–7 illustrates some common errors in drawing section lines.

CORRECT

INCORRECT
LINES MUST TOUCH BORDER

INCORRECT
LINES ARE TOO CLOSE

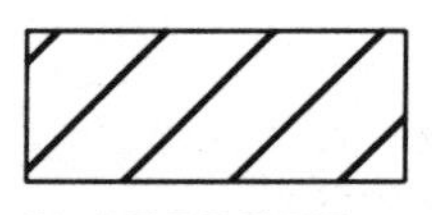

INCORRECT
LINES ARE TOO THICK

INCORRECT
LINE SPACING IS NOT EQUAL

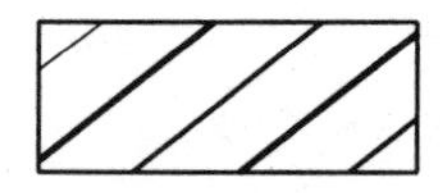

INCORRECT
LINE THICKNESS VARIES

INCORRECT
LINES ARE TOO FAR APART

FIGURE 10–7

In certain cases, angles other than 30°, 45°, or 60° may be used. Figure 10–8 illustrates incorrect and correct methods of cross-hatching. The 45° section lines in Figs. 10–8(a) and (c) do not stand out because they are parallel to a sloping edge. In parts (b) and (d), the angle of the section lines has been changed so that the section lines are no longer parallel to the sloping object line. They therefore stand out better.

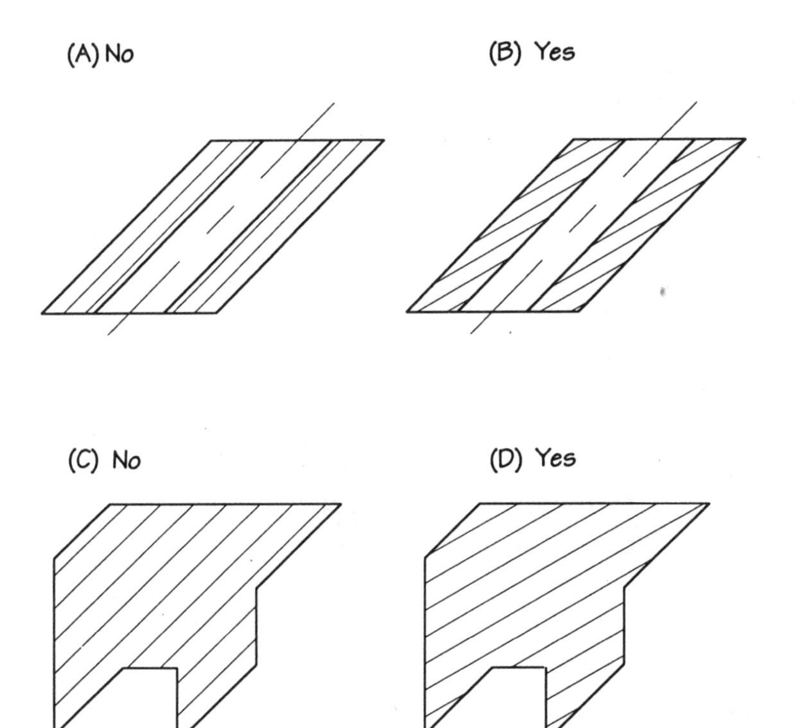

FIGURE 10–8

SECTION LINES: STYLES

Different styles of section lines are used to represent different materials. Some examples are given in Fig. 10–9.

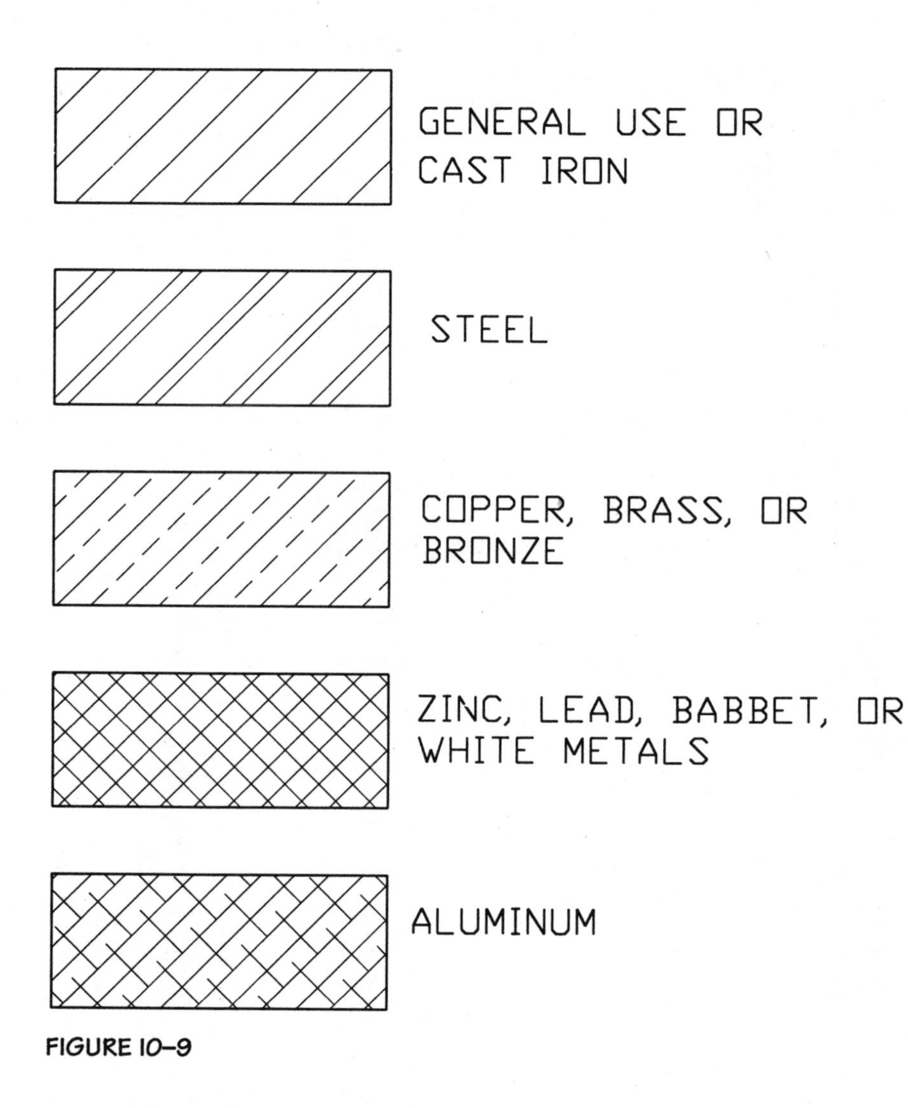

FIGURE 10–9

When you have more than one section view on a drawing, it is a good idea to identify each with letters, such as A-A, B-B, and so on, about 1/4" (6 mm) high. In Fig. 10–10, it is not really necessary to use letters, since we can identify the views through orthographic principles, but as drawings get more complicated, you will see the need to identify sections.

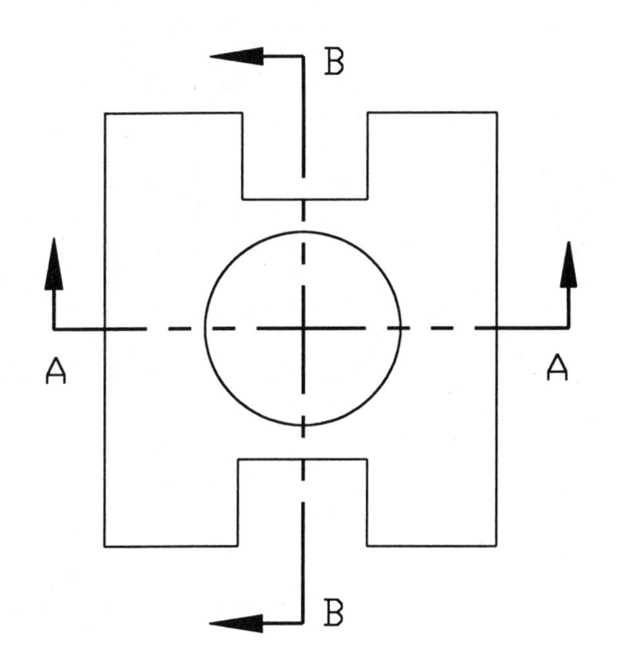

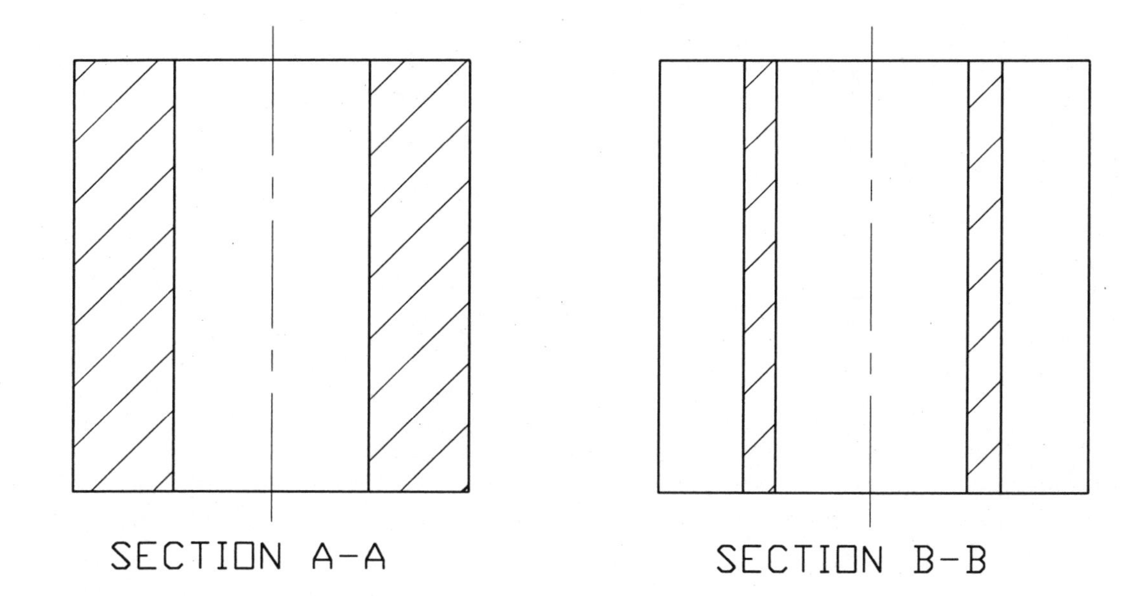

FIGURE 10-10

HALF-SECTION

A *half-section* is used when you want to show both the inside and the outside of an object in the same view. It is usually used with objects that are symmetrical (have a line of symmetry), such as round objects. To obtain this type of view, imagine that the object is cut *halfway* through by a cutting plane line, which then exits the object at 90°. Remove the part cut out (a quarter of the object) and, in the direction of the arrow, look at what is left. This will be a half-section.

Figure 10–11 shows an example of a half-section. Notice that there is no solid line in the front view dividing the sectioned half from the unsectioned half, but a center line instead. The imaginary knife would cut through the part along the cutting plane line. Notice that only one arrow is used in half-sections.

As in full-section drawings, the cutting plane line may be omitted if the object is symmetrical and the location of the line is obvious.

Hidden lines are omitted in the nonsectioned half unless they are required for clarity or for ease in dimensioning.

Note that you need the front view to visualize the part in Fig. 10–11. You cannot know the shape if given only the top view.

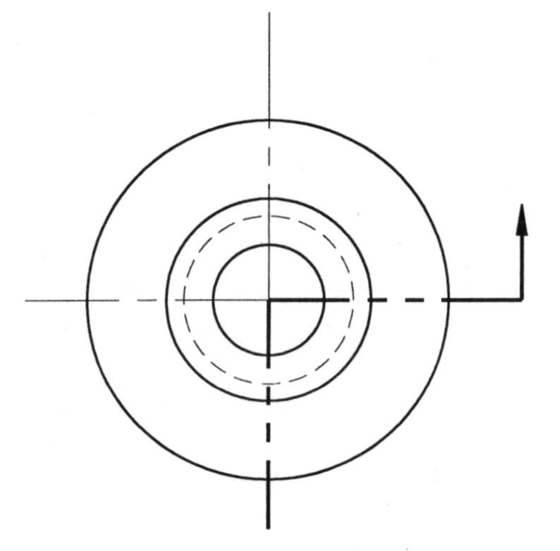

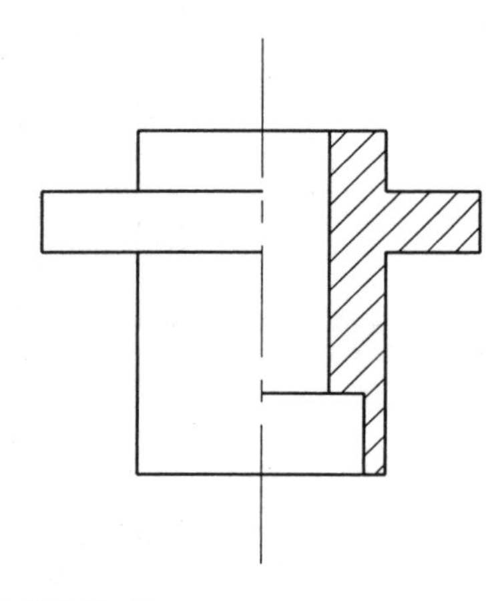

FIGURE 10–11

Figure 10–12 shows another example of a half-section. You will see once again that in the front view, hidden lines were not drawn on the left portion. Check the standards of your company or instructor to decide when to add hidden lines to a half-section. (Remember, they are usually added only for clarity or to aid in dimensioning.)

By studying the top and front views in this figure, can you imagine what the piece looks like?

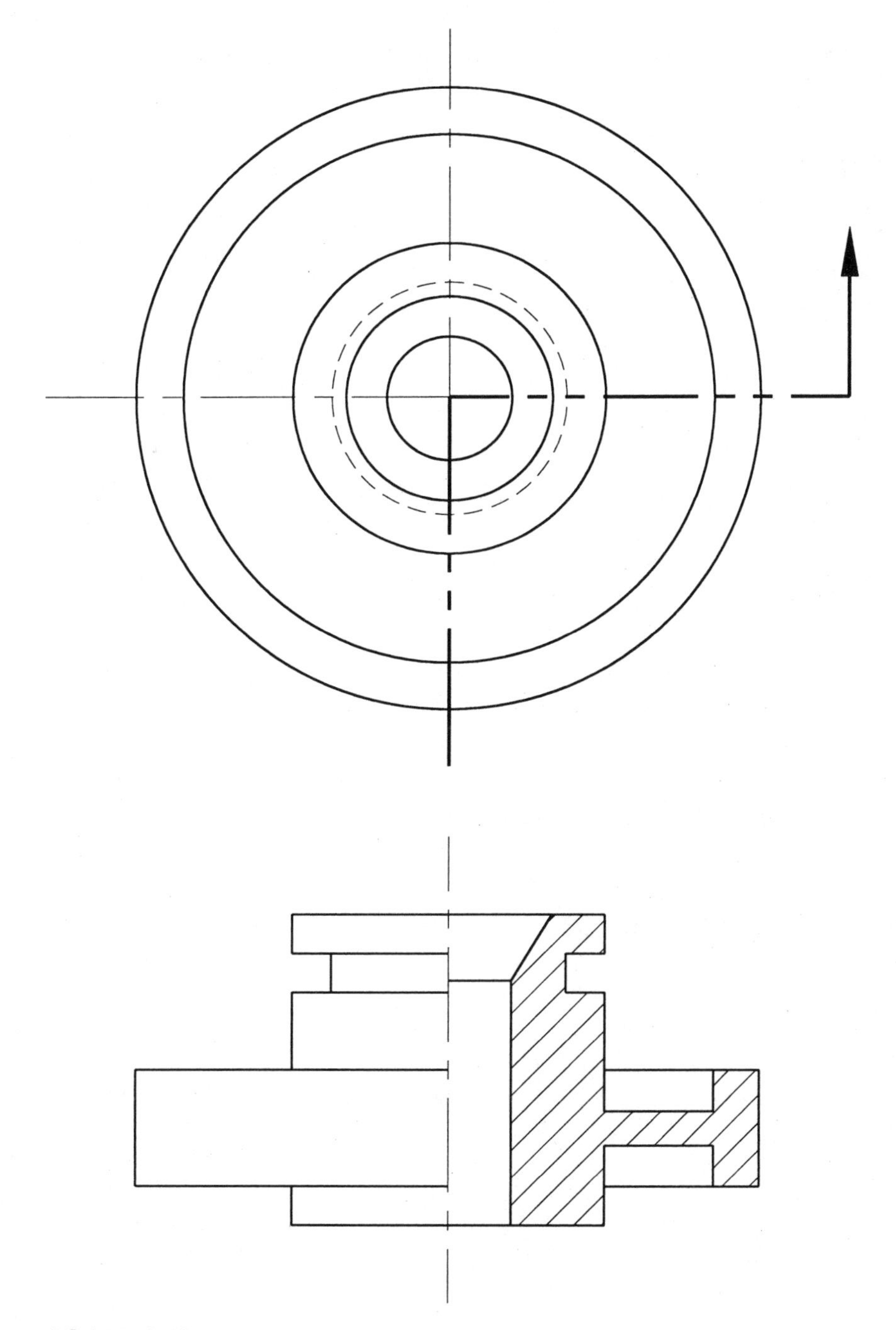

FIGURE 10–12

Remove the following exercise pages from your workbook so that you can do the exercises directly on the page.

EXERCISE 10–1

Modify the front view below to make it a full section.

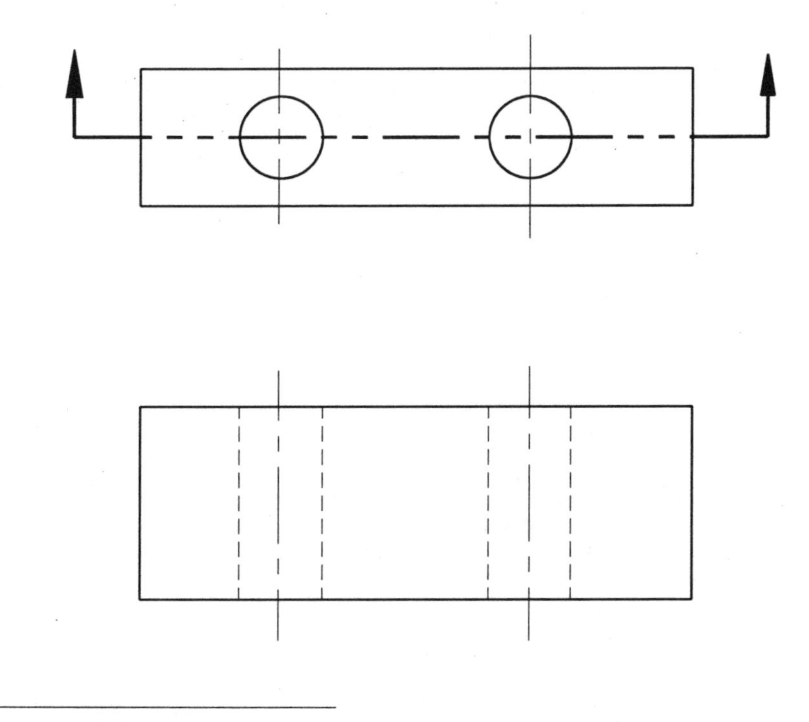

EXERCISE 10–2

In the figure below, modify the right side view to make it a full section. Add the cutting plane line to the front view.

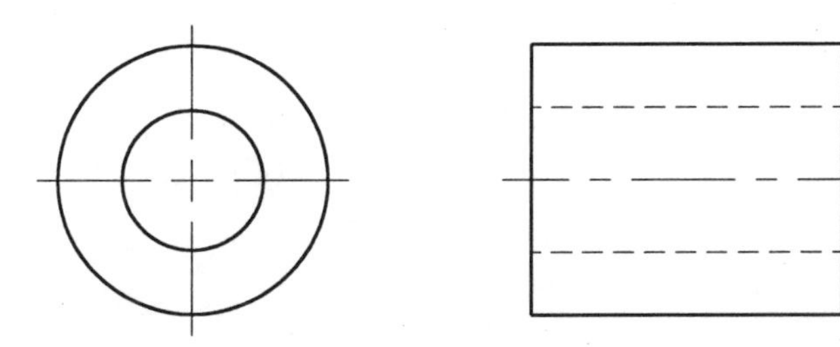

DRAWN BY DATE

EXERCISE 10–3

In the figure below, a front view and right side view are given. Draw a left side view as a half-section.

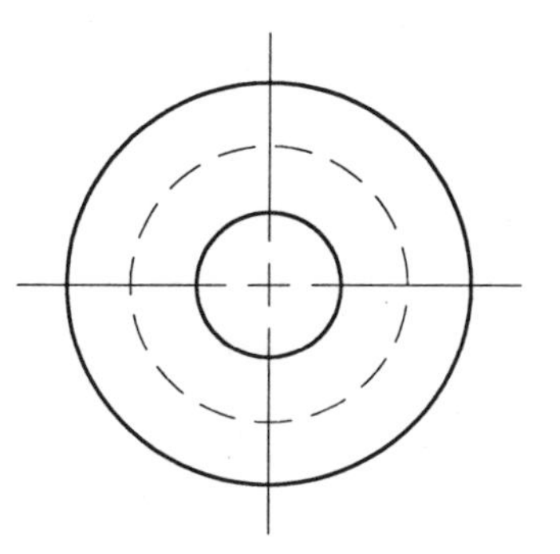
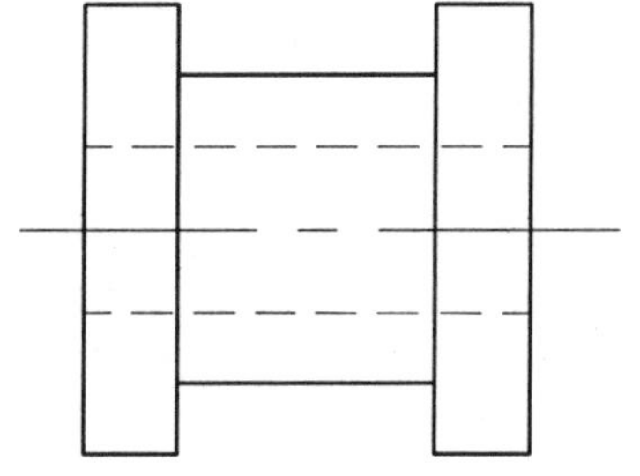

EXERCISE 10–4

In the figure below, the front and left side views are given. Draw the right side view as a full section.

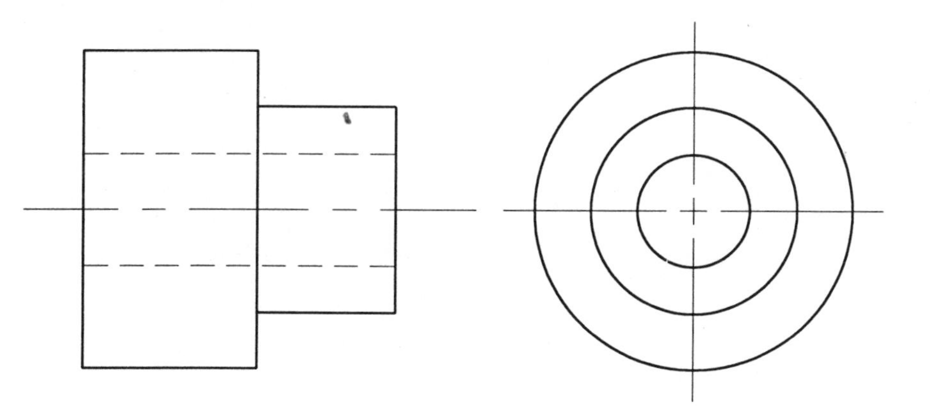

EXERCISE 10–5

The figure below shows the front and right side views of an object. Draw sections A-A and B-B on this sheet.

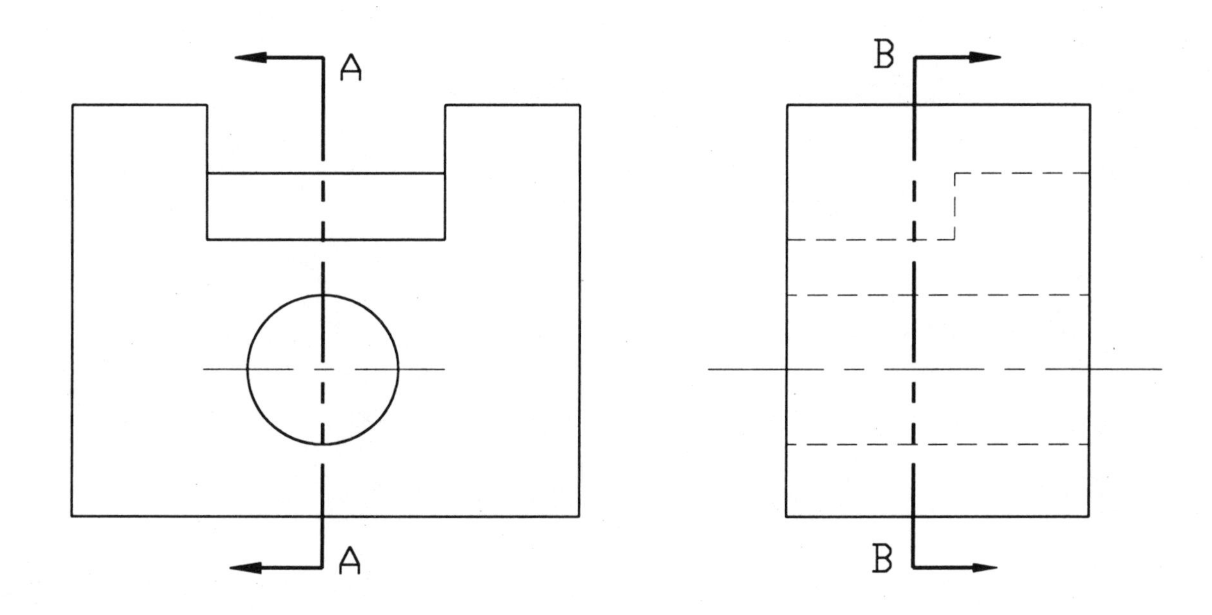

DRAWN BY DATE

EXERCISE 10–6

Complete the missing half-section view in the figure below.

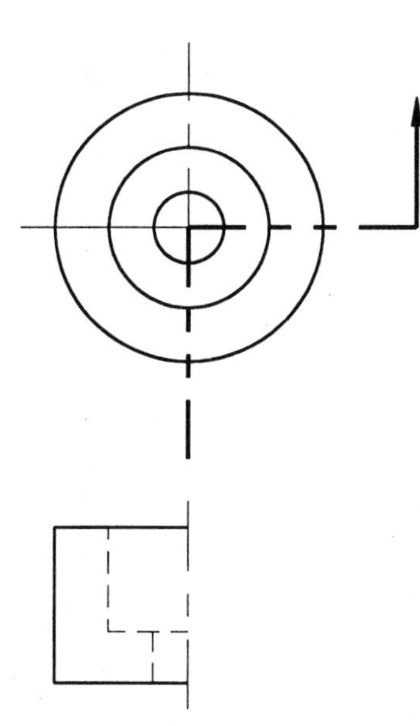

EXERCISE 10–7

Complete the missing half-section view in the figure below.

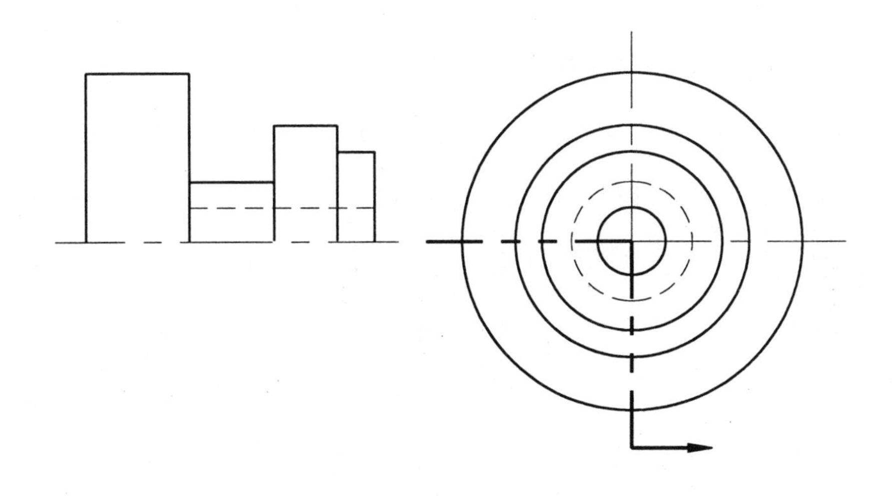

DRAWN BY DATE

EXERCISE 10–8

In the figure below, complete the front view as a half-section.

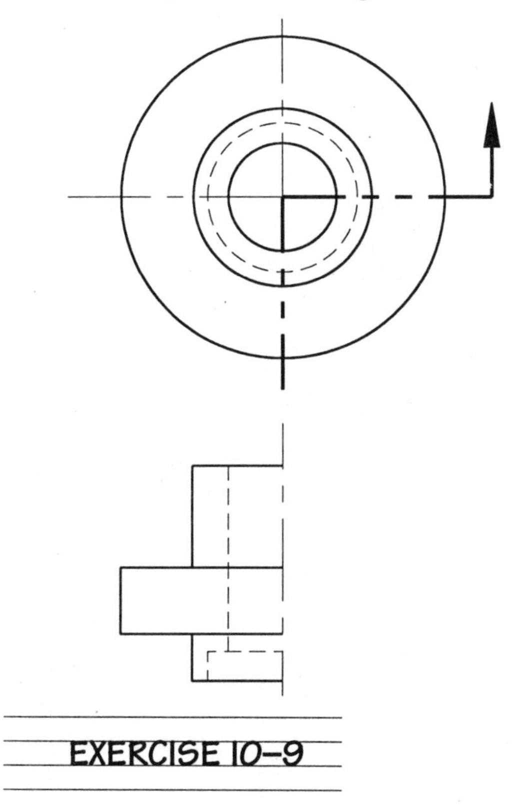

EXERCISE 10–9

In the figure below, complete the front view as a half-section.

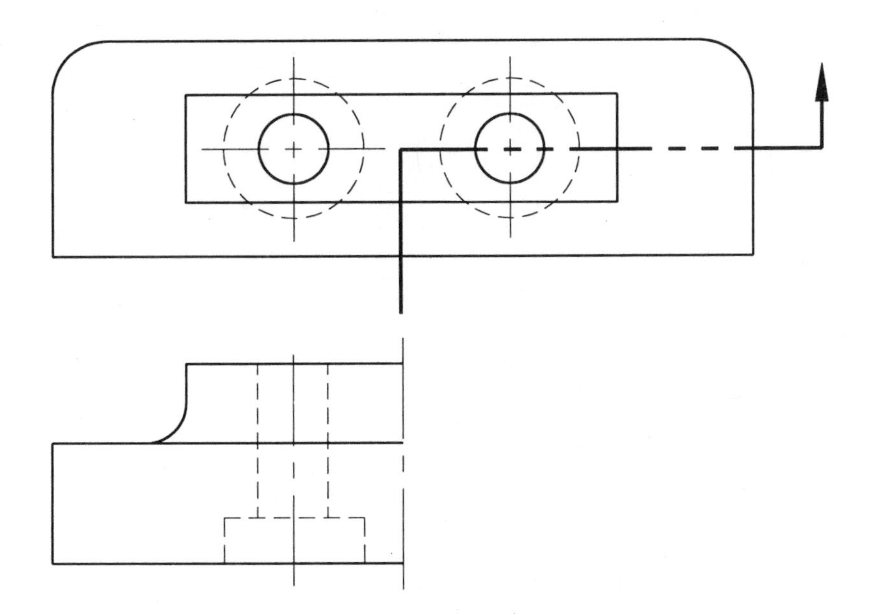

DRAWN BY ____________________ DATE ____________

EXERCISE 10–10

Draw the missing front view of the object illustrated below. This view will contain the half-section.

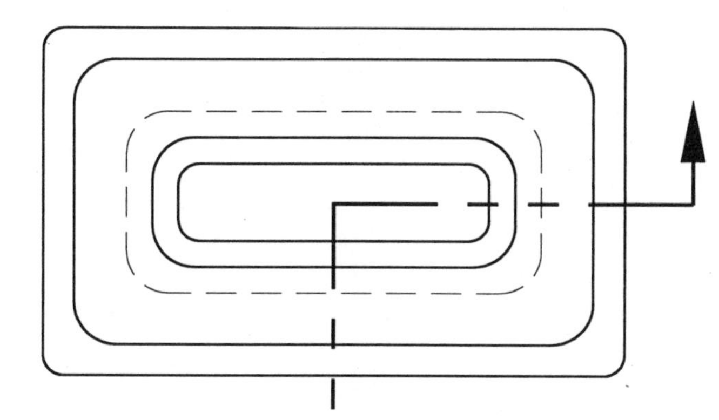

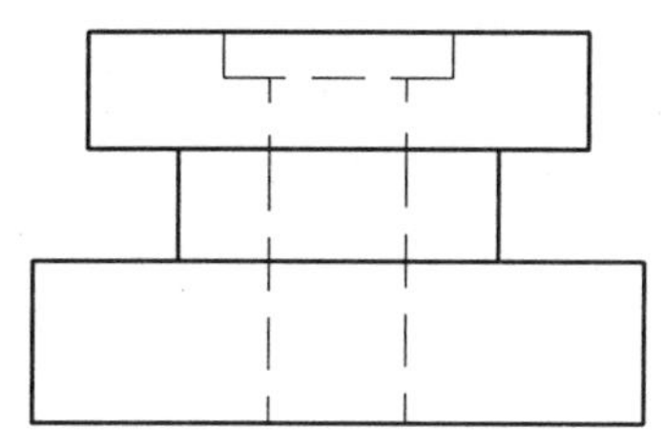

DRAWN BY DATE

EXERCISE 10–11

Draw the missing front view in the figure below. It will be a full section view.

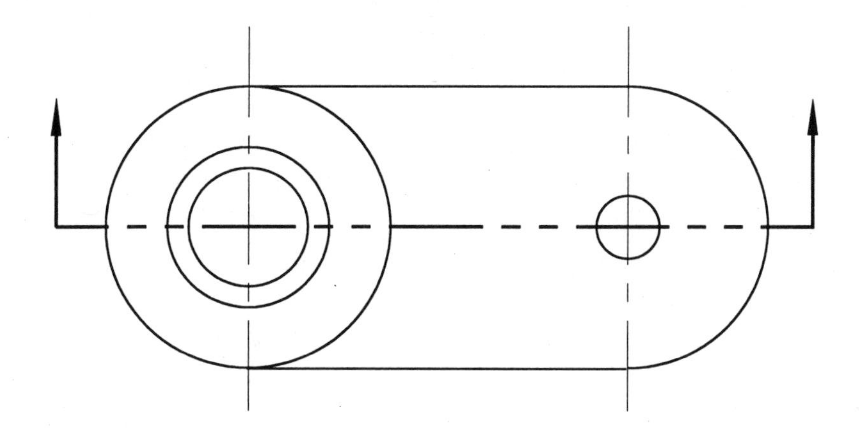

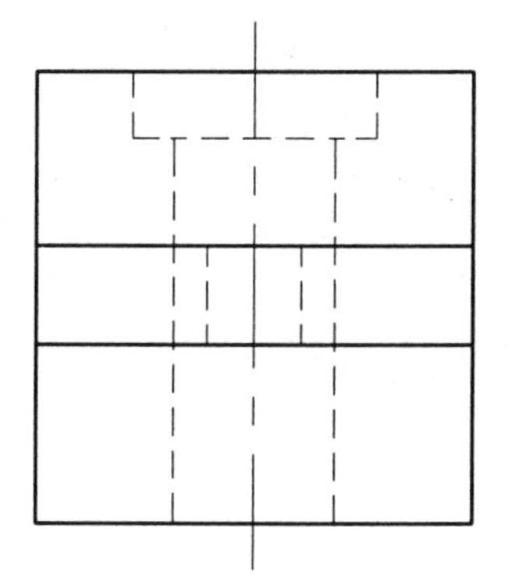

DRAWN BY DATE

SECTIONS: PART 2

11

OBJECTIVE

When you have finished this chapter, you should understand the purpose of, and be able to draw, more advanced types of section views.

EQUIPMENT

You are required to have all your drafting instruments for this chapter.

OFFSET SECTIONS

An *offset section* shows internal features that are not in a straight line. The cutting plane line goes through the center of each internal feature that you want to show. Figure 11–1(a) shows an ANSI standards cutting plane line. Notice that the arrows indicate the direction in which you are looking. In the front view, there is no solid line showing the corners of the offset cut, because we only *imagine* the cut. Also notice that the center lines and section lines are thin and that the object lines and cutting plane line are thick. (The cutting plane line is sometimes drawn thicker than the object line.)

Figure 11–1(b) shows how CSA and ISO standards specify the offset section cutting plane line. Notice that the line is thick at each corner of the cut and that the thick lines are joined by thin lines. The thin lines may be omitted for clarity.

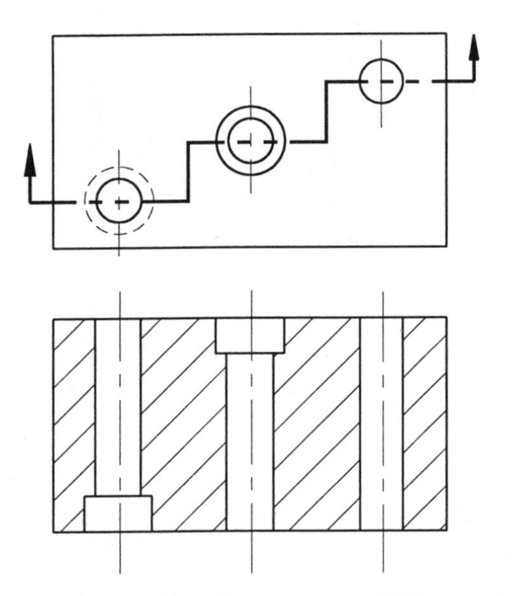

FIGURE 11–1(A) **Offset section: ANSI standard**

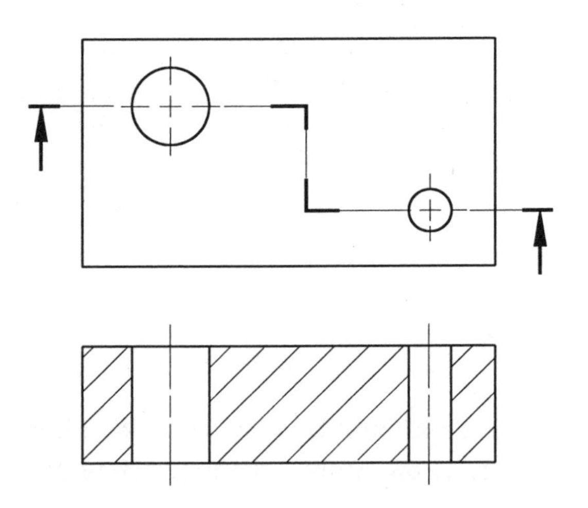

FIGURE 11–1(B) **Offset section: CSA and ISO standard**

REMOVED SECTIONS

A *removed section* is similar to a full section, except that it can be placed anywhere on the drawing. (One exception in its placement is that a section view should never be shown in first-angle projected position in a third-angle projection drawing. This could lead to misinterpretation.) The removed section is identified by a label, such as A-A or B-B. Many removed sections can be placed on the same drawing. We show only what the knife touches, not what is behind the plane of the cut. Figure 11–2 shows the front and right side views of an object, together with a removed section view.

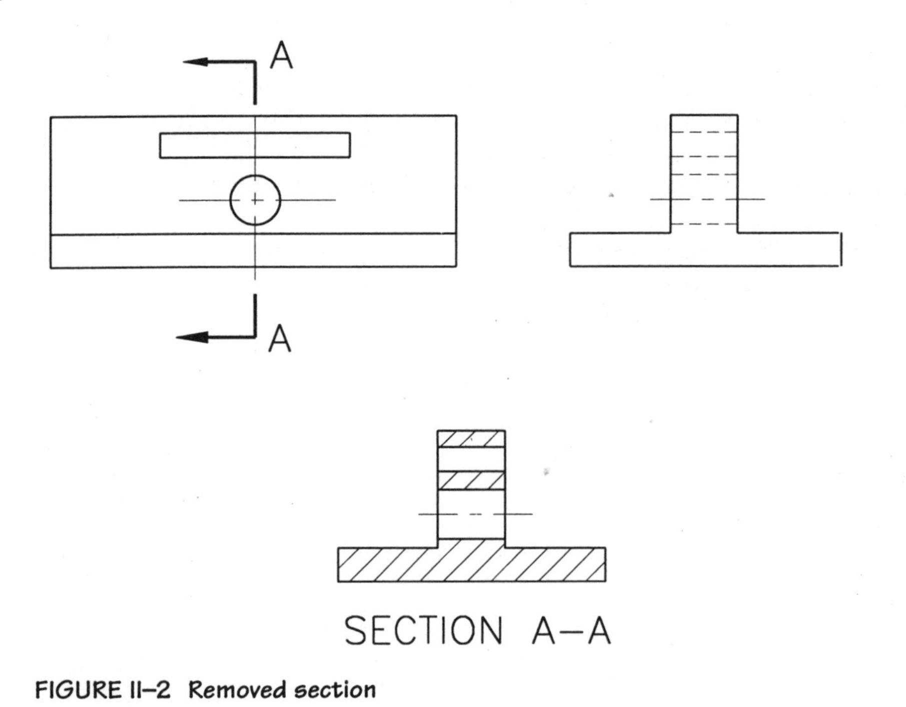

FIGURE 11–2 **Removed section**

BROKEN-OUT SECTIONS

A *broken-out section* is used to show certain internal features in a small area of a part where it would be a waste of time to draw a full section. It may be used when the part is not symmetrical and a half-section would therefore be inappropriate. There is no cutting plane line. A *break line* separates the broken area from the nonbroken area. Long break lines are thin, straight lines with notches, as shown in Fig. 11–3(a). Short break lines are thick, freehand, and irregular, as in Fig. 11–3(b).

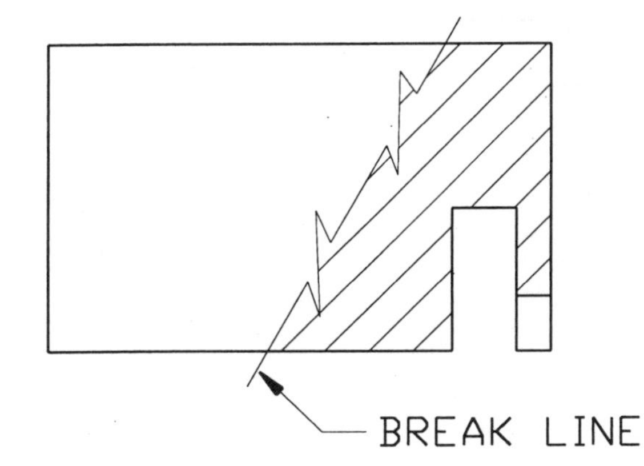

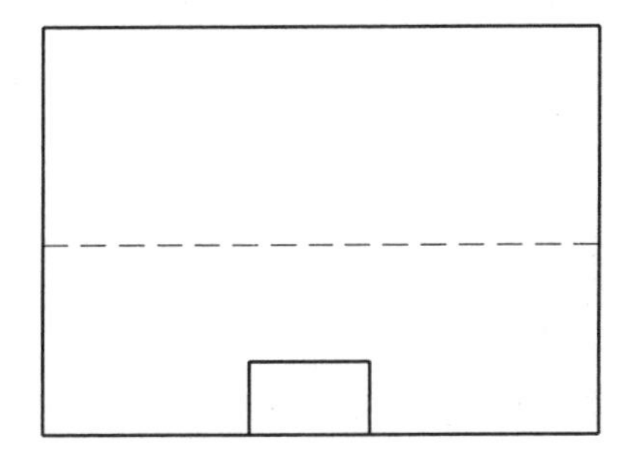

FIGURE 11–3(A)

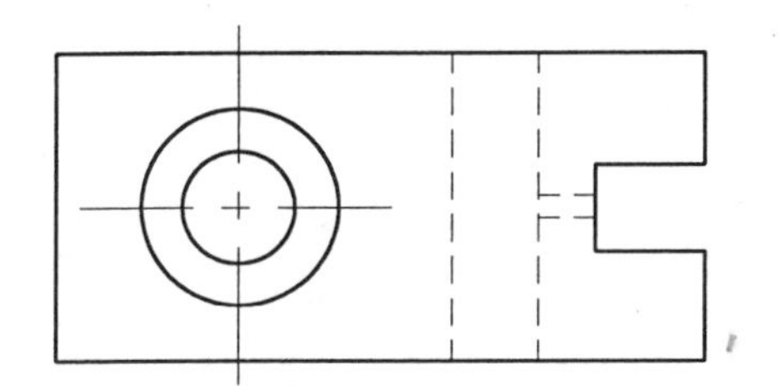

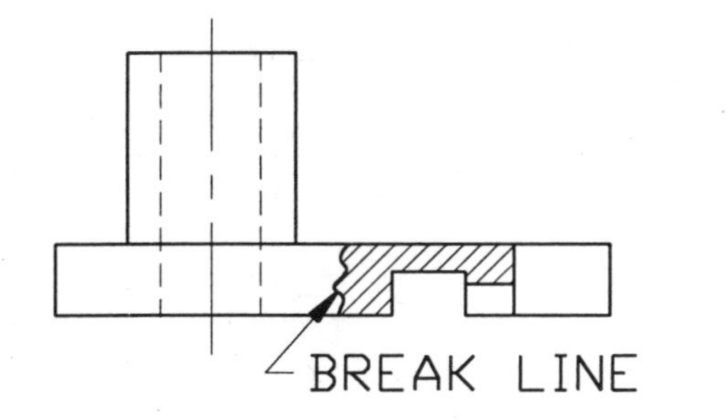

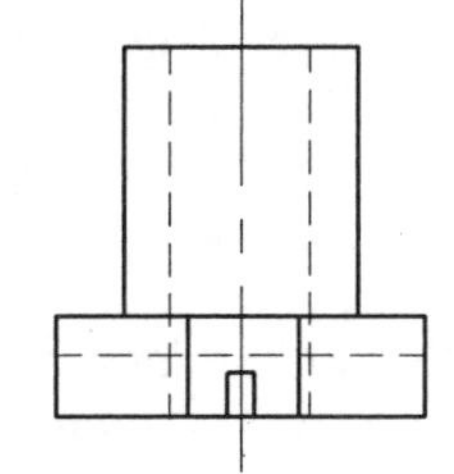

FIGURE 11–3(B)

REVOLVED SECTIONS

Figure 11–4(a) uses two views to describe the object. Figure 11–4(b) shows how we can save space by using only one view in which the plane of the section view has been revolved 90° and superimposed on the front view. This illustrates a method of drawing a revolved section using break lines. The object line is not touching the revolved section. This method is acceptable by all three standards (ANSI, CSA, and ISO). (Figures 11–5b and c on the next page illustrate two other ways to draw a revolved section.)

You do not have to label your views as revolved views.

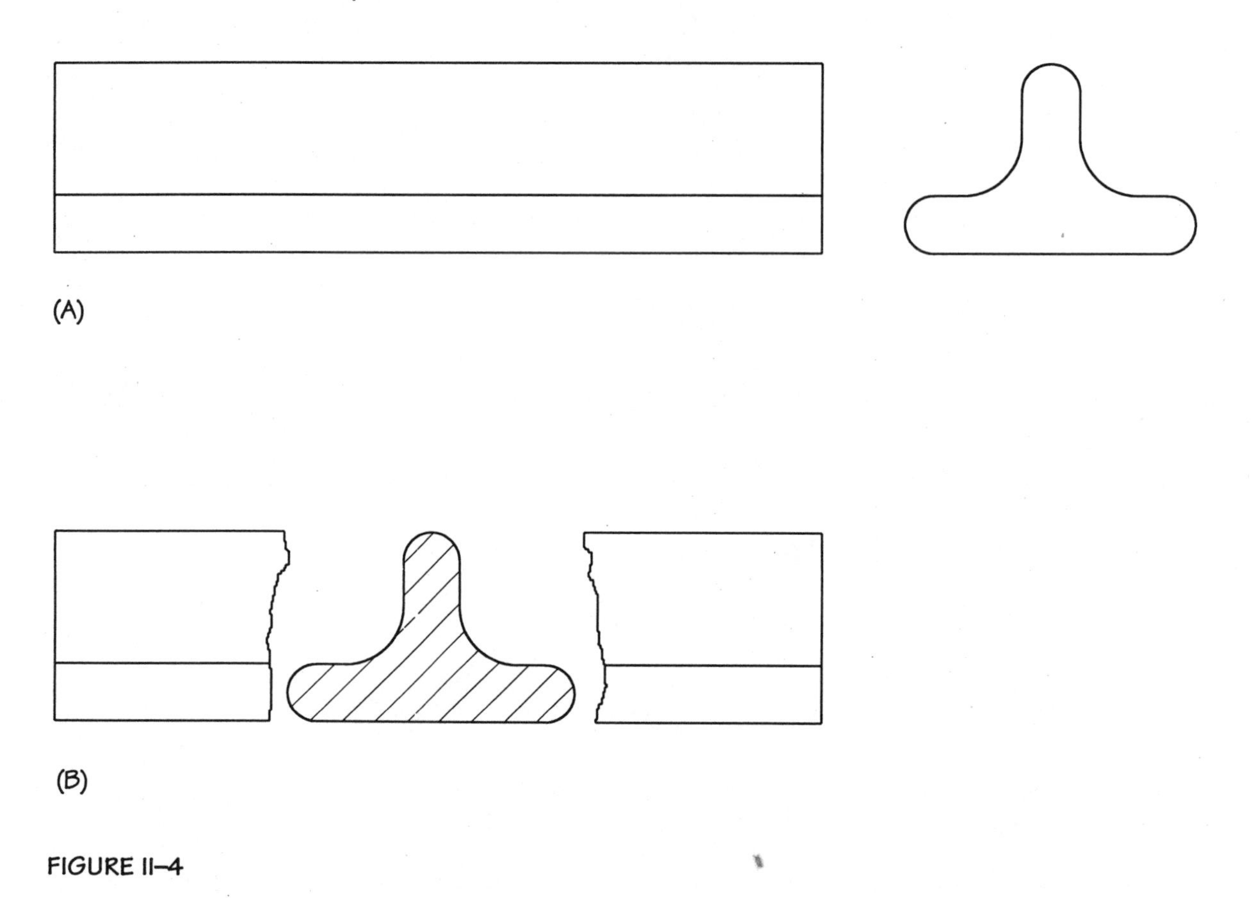

FIGURE 11–4

In Fig. 11–5(a), once again two views are used to describe an object. Figures 11–5(b) through (e) show another method of drawing revolved sections in which the object line is touching the revolved section. ANSI standards prescribe that the outline of the section view be thick, as shown in parts (b), (d), and (e). CSA and ISO specify that when the object line touches the revolved section, the outline of the revolved section is to be thin, as shown in part (c).

Figures 11–5(d) and (e) illustrate other important aspects of revolved sections. In part (d) we see that two different cross sections can be shown on the same view by inserting two revolved sections. Notice that the smaller part of the object is round in its cross section, whereas the larger piece is oval or elliptical. In part (e), as in the other examples, the revolved section is *always* tangent to the object at the point where it is rotated. This is an important rule to remember, especially when the object tapers, as in part (e).

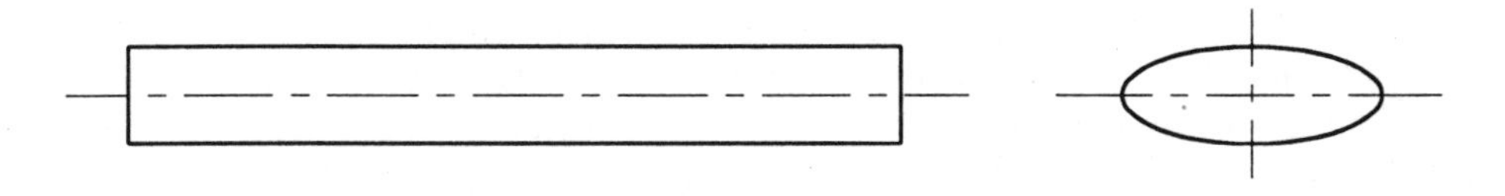

(A)

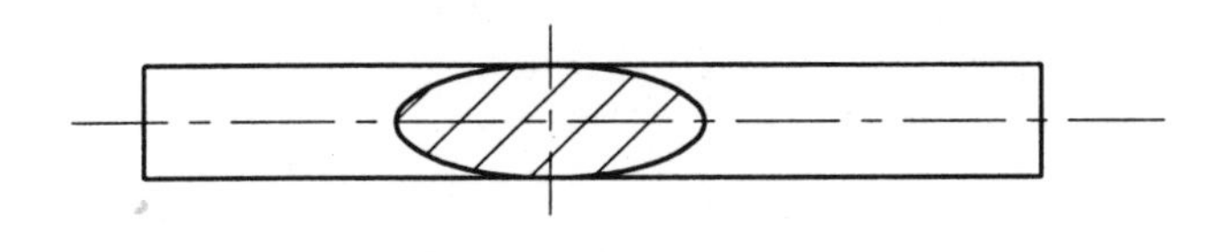

(B)

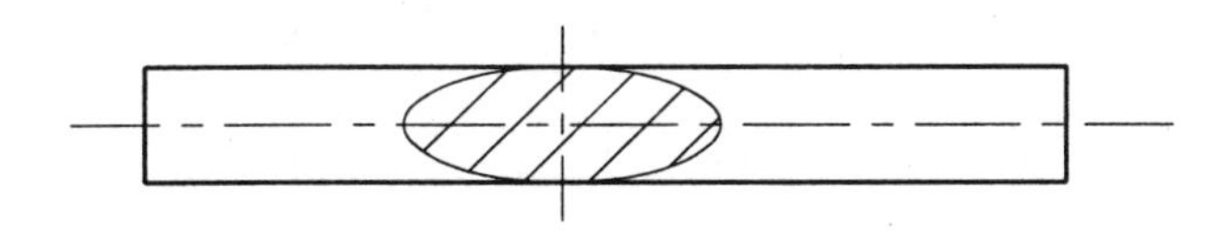

(C)

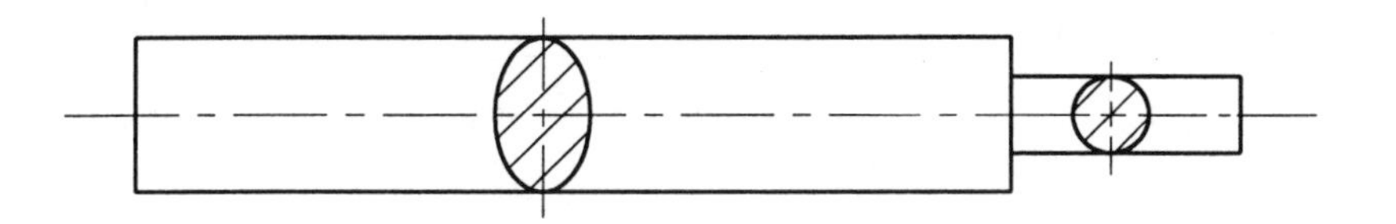

(D)

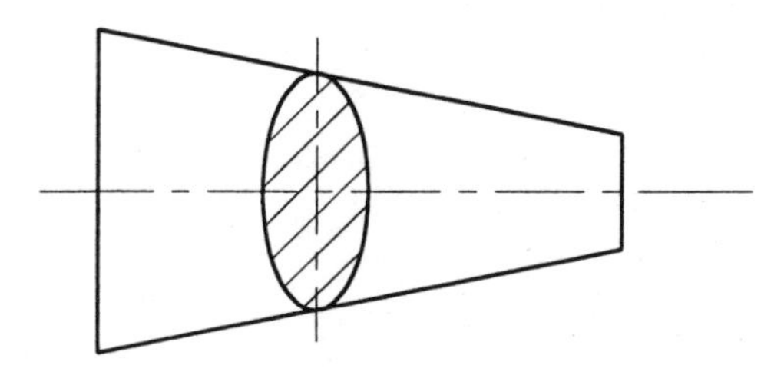

(E)

FIGURE 11–5 **Revolved sections**

WEBS IN SECTION

Figure 11–6 shows a part similar in shape to a train wheel. In this simplified drawing, the wheel has an outer rim, which would be in contact with the track, and an inner hub with a hole in it. The hub is connected to the rim by a web. The web is section-lined because it is solid all around the wheel. This drawing is an example of a full section. It is presented here to provide a contrast with the drawings in Fig. 11–7.

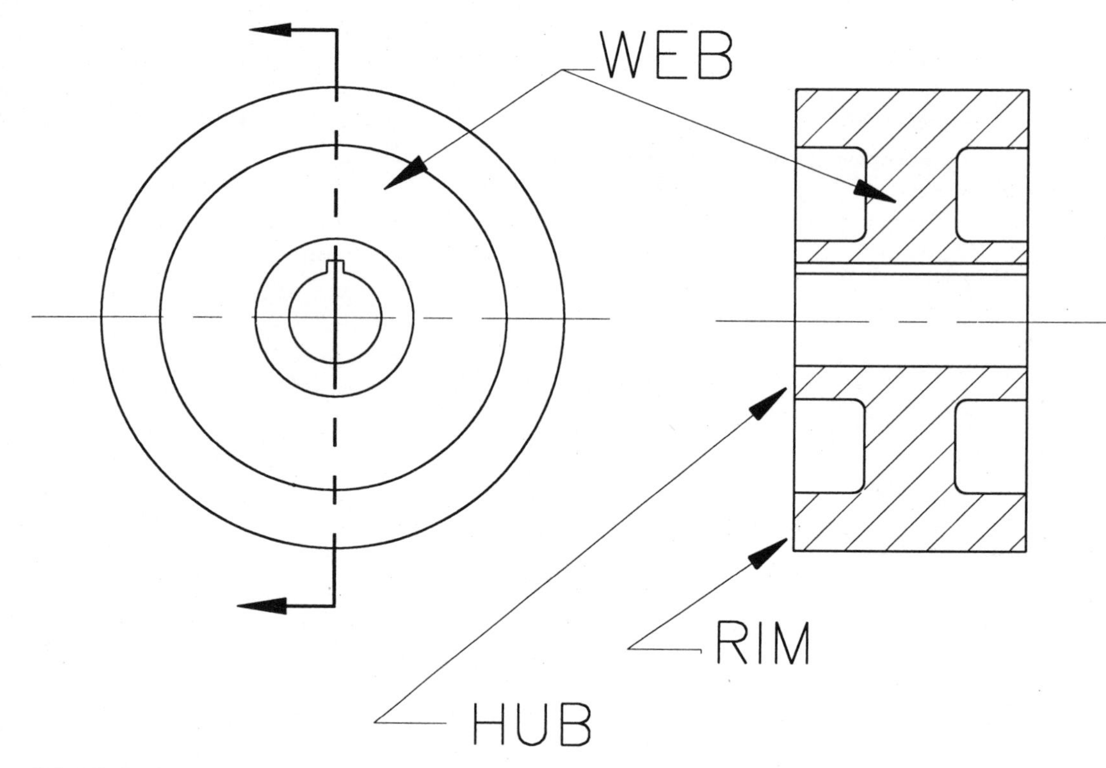

FIGURE 11–6

RIBS AND SPOKES IN SECTION

(CONVENTION 1)

Drafting conventions, or conventional practices, are methods that have been generally agreed upon for drawing certain features because they save time and improve clarity. These conventions do not follow the rules of orthographic projection that you have learned so far. One such convention is how ribs and spokes are sectioned.

Figure 11–7(a) shows a top view and a full-section front view as we might think it should be drawn. Because the cutting plane line passes through the ribs, they have been section-lined. Two problems arise from this. First, no distinction has been made between the relatively thin ribs and the cylindrical shape around the center. Second, anyone reading the drawing is given the mistaken impression that the ribs are solid all around the object (as the web was in Fig. 11–6). Figure 11–7(b) shows the preferred way of drawing ribs in section. Because the ribs are relatively thin and not continuous all around the object, they are not section-lined, even though the cutting plane line passes through them. CSA standards provide an alternative: The ribbed section may have double-spaced section lines, as shown in Fig. 11–7(c).

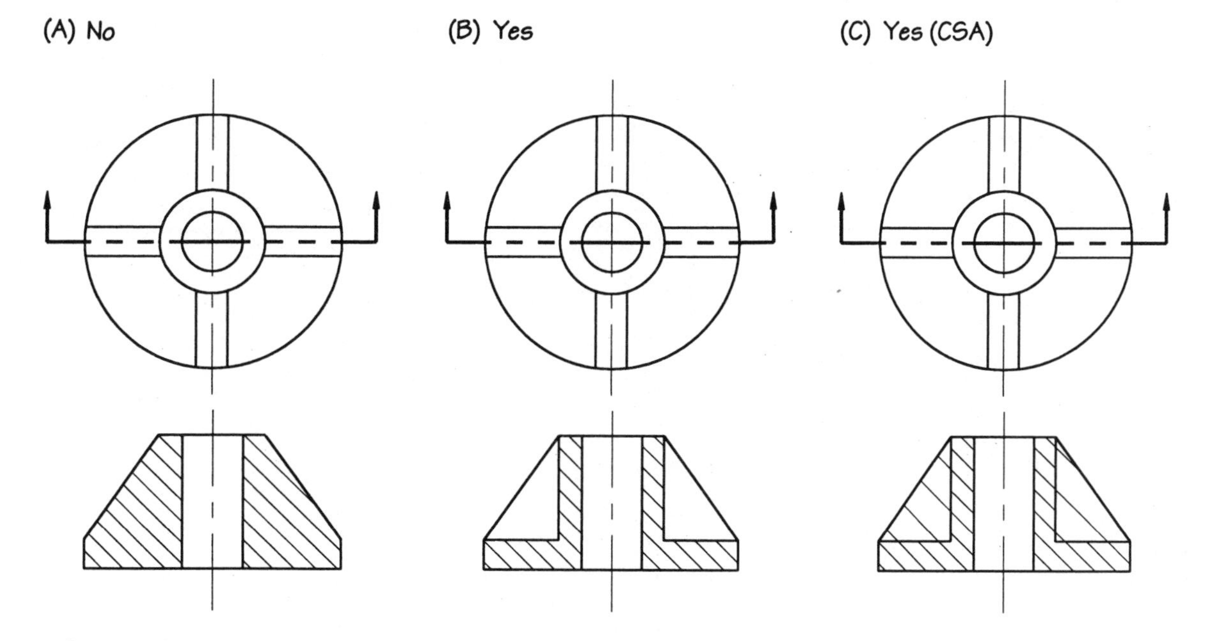

FIGURE 11–7 Ribs In Section

In Fig. 11–8, we have spokes joining the inner hub and outer rim. Because the spokes are not continuous, they are drawn using the same convention that we used for ribs; we do not section-line them. Section-lining spokes, as was the case with ribs, would give a false impression of solidity.

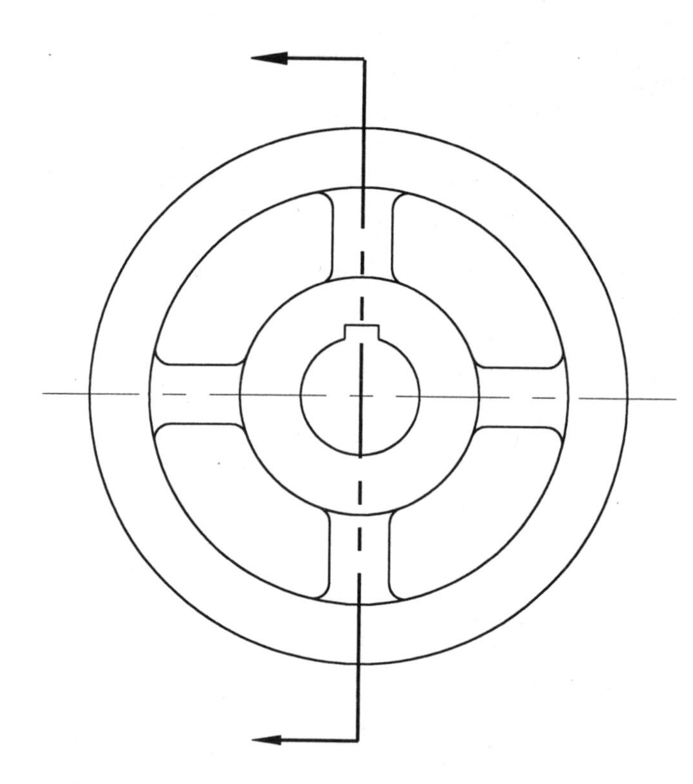

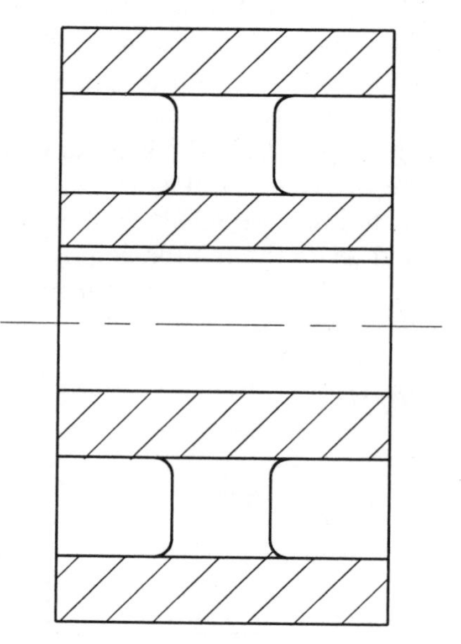

FIGURE 11–8 **Spokes In Section**

ALIGNED FEATURES CONVENTION

(CONVENTION 2)

A second drafting convention is to draw a section view as if the cutting plane and the features were rotated to a position at which they are easier to draw. The section view in Fig. 11–9(a) is orthographically correct but difficult to draw and to read. The view shown in Fig. 11–9(b) is preferred. To simplify the drawing, we rotated the cutting plane line 45° so that it passes through one of the small holes. With the new *aligned* view, it looks as if the hole is perpendicular to our line of sight when we look at the object from the front. We rely on the reader to look at *all* views and understand common drafting conventions. If the plane had not been rotated, the small holes would have appeared to be closer to the center hole than they actually are. By aligning the features, we are able to see the true distance between the hole and the edge as well as the distance that separates the small hole from the center hole. The drawing shown in Fig. 11–9(b) is an example of a half-section aligned view.

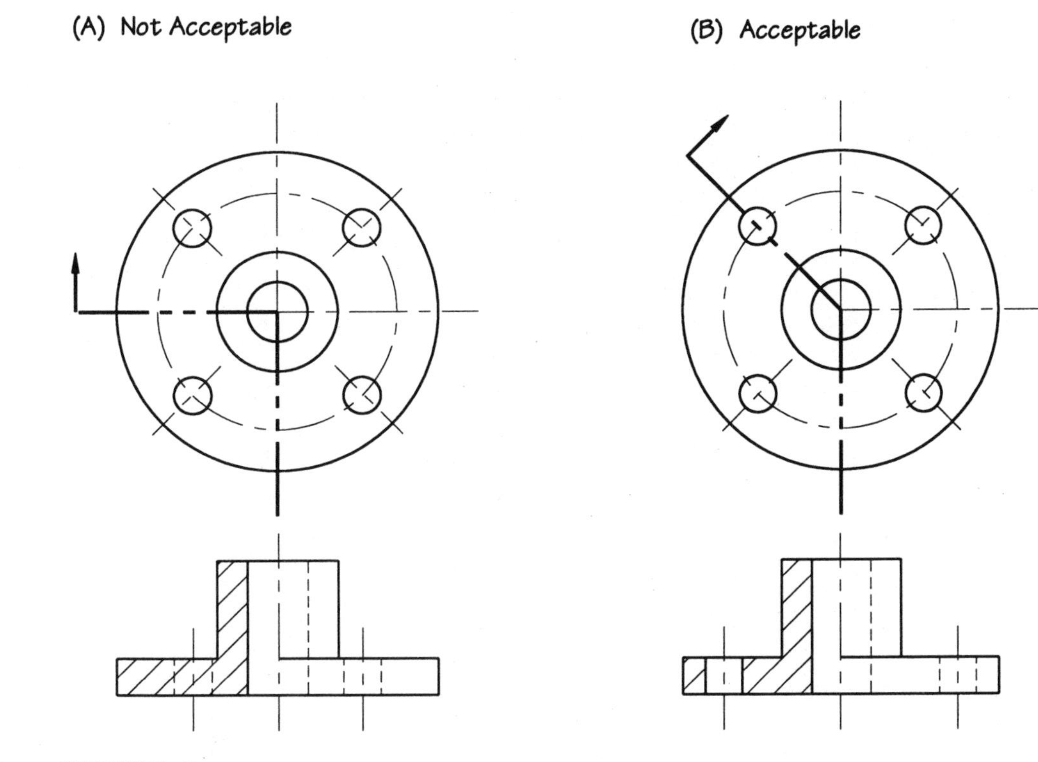

FIGURE 11–9

Figure 11–10 shows spokes in section. (It looks similar to Fig. 11–8.) Once again, we align the cutting plane line with one of the spokes for clarity of presentation in the section view.

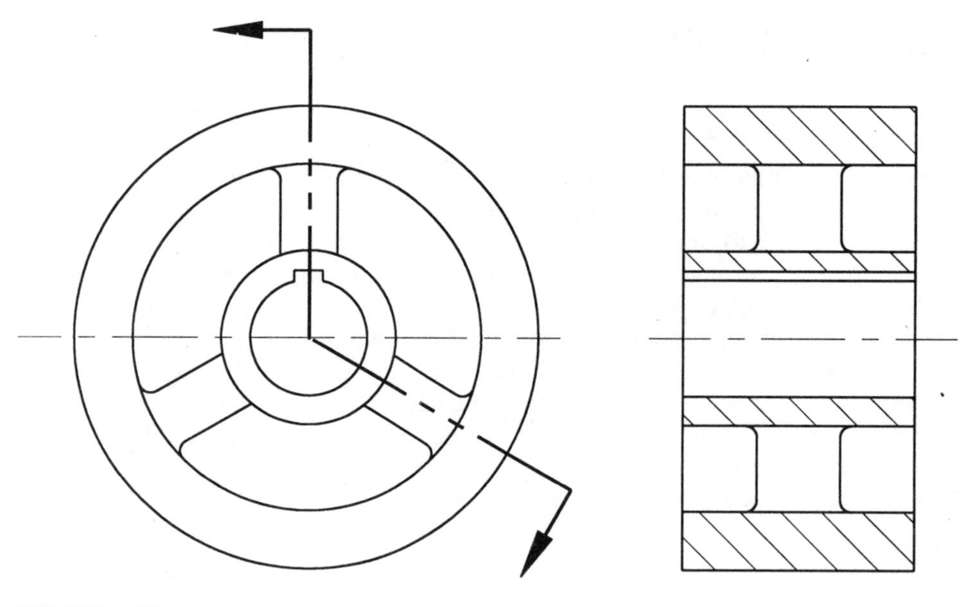

FIGURE 11–10

The object shown in Fig. 11–11 presents a new problem if we wish to draw a section view of it. It is not enough to align the cutting plane line with one particular feature, because there are two types of important features (small circles and ribs) on different planes. In this case, we rotate the features so that they are in line with the cutting plane line. In Fig. 11–11, the holes were rotated to a "natural" position to avoid distorted features. This convention is followed even though the holes and ribs now look as if they are aligned (they are not).

Note that the ribs are not section-lined, as we learned in Convention 1.

Remember to look at *all* views when studying a drawing.

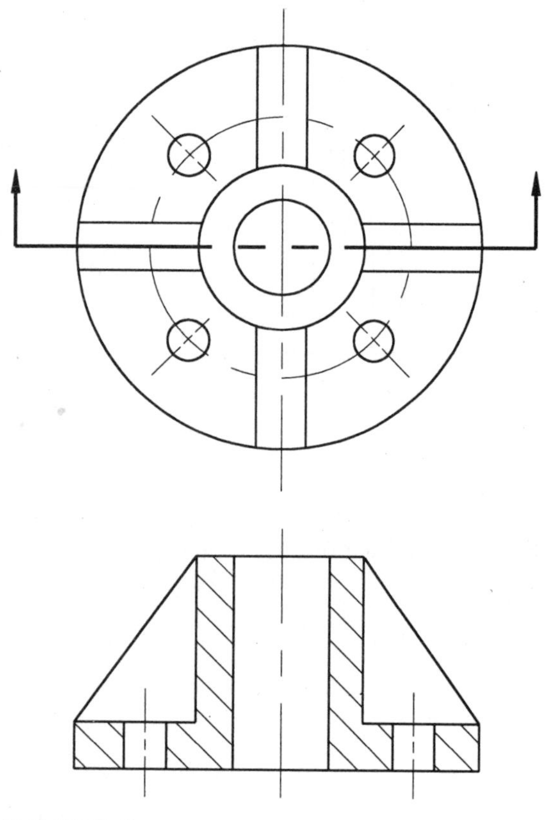

FIGURE 11–11

By studying the half-section view in Fig. 11–12, we see that the aligned features convention applies not only to section views but also to regular orthographic views and their hidden lines.

(A) Not Acceptable

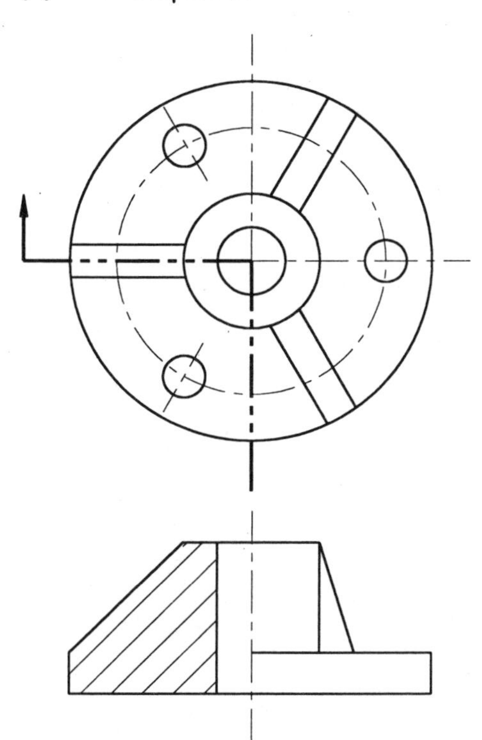

(B) Acceptable

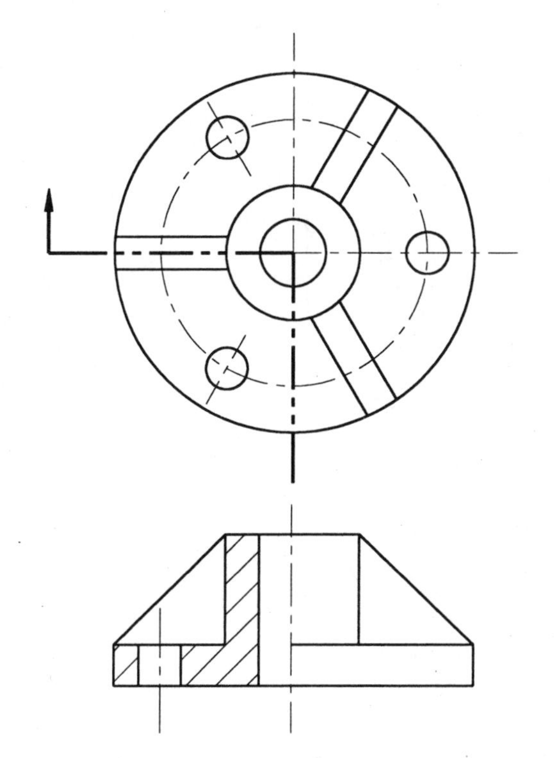

FIGURE 11–12

ALIGNED SECTION LINES

(CONVENTION 3)

Notice that in Figs. 11–13(a) and (c), if you put a straightedge parallel to the section lines, they do not line up between one part of the object and another. This practice is incorrect. In Figs. 11–13 (b) and (d), all the section lines line up with each other. This is correct. Complex parts must have section lines that are aligned even if they are not continuous. Aligned section lines simplify interpreting complex drawings.

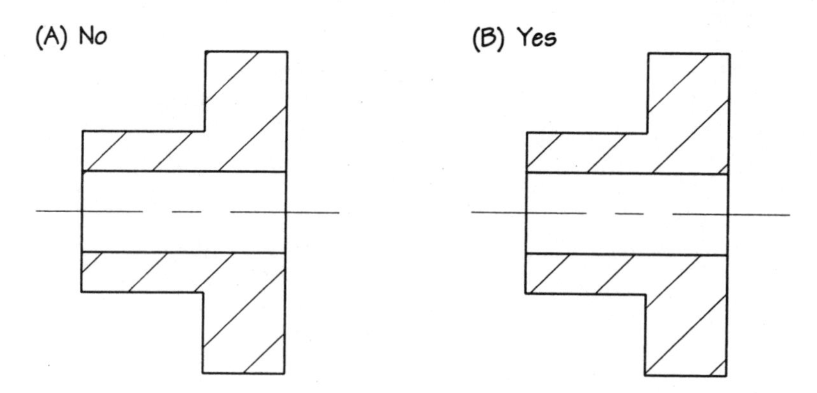

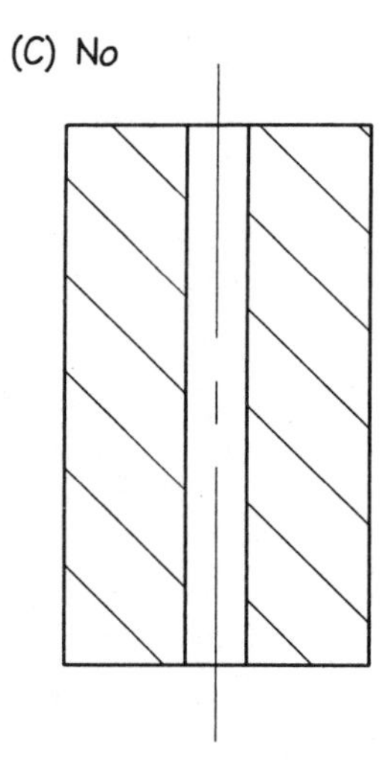

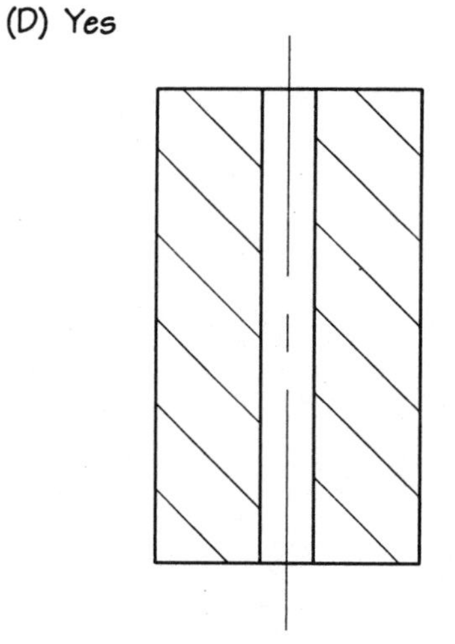

FIGURE II–13

ASSEMBLY VIEWS IN SECTION

So far in our study of drafting, we have learned to draw one piece at a time. These working drawings of one piece are called *detail drawings*. Another type of drawing is used to show how different parts are assembled. It is called an *assembly drawing*.

An assembly view draws several different parts together as they would look in their assembled condition. Because an assembly drawing usually shows different parts fitting inside each other, a section drawing is often used. Each piece is identified by a different number in a "bubble," or circle, as shown in Fig. 11–14(a) on the next page. Assembly drawings are the subject for a second-level course but are mentioned here briefly because they often use section views.

We have already looked at three drafting conventions. Following are four more drafting conventions that apply particularly to assembly drawings.

Convention 4: Do not draw section views of fasteners such as bolts, nuts, or pins.

Convention 5: Do not draw section views of round bars.

Convention 6: Use different angles or spacing to differentiate between adjacent pieces made of the same material. When two different sectioned parts are drawn beside each other, as in the orthographic assembly view in Fig. 11–14(a), we differentiate them by changing the spacing or the angle of the section-lining. Different angles of section-lining are illustrated in pieces 1 and 4 in the figure.

Convention 7: Use different styles of section-lining to differentiate between adjacent pieces made of different materials. This is illustrated with pieces 3 and 4 in Fig. 11–14(a). Examples of these different styles of section lines are illustrated in Fig. 10–9 of Chapter 10.

Figure 11–14(b) illustrates an exploded isometric drawing of the assembly drawing in part (a). Can you relate the parts numbered in the orthographic assembly to the parts in the exploded isometric drawing?

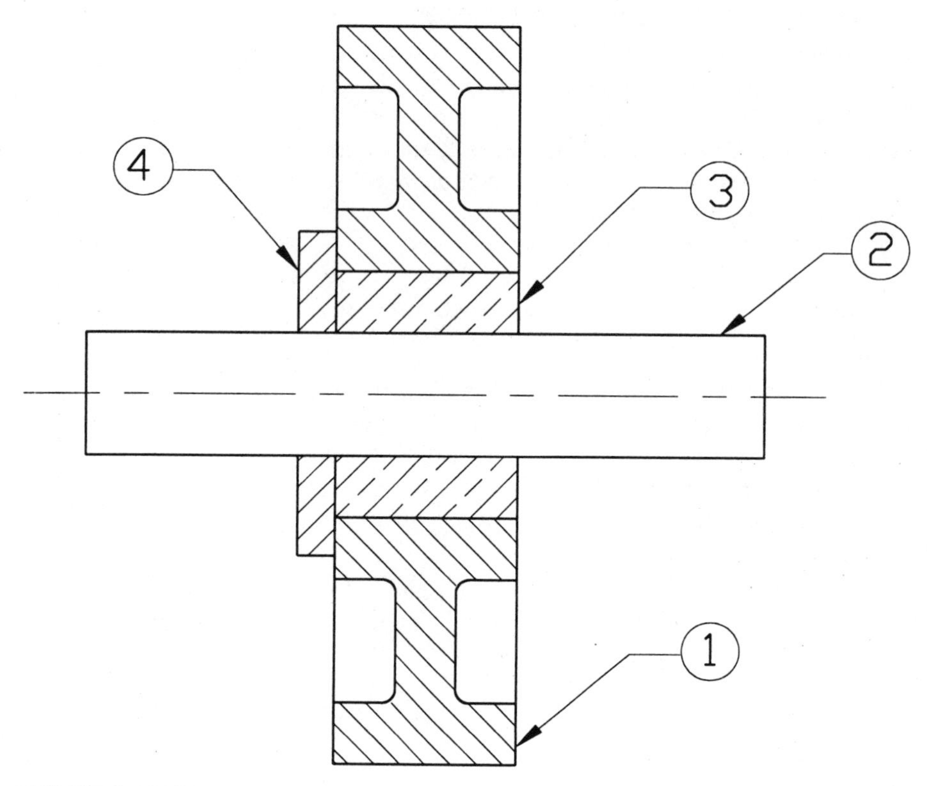

FIGURE 11–14(A)

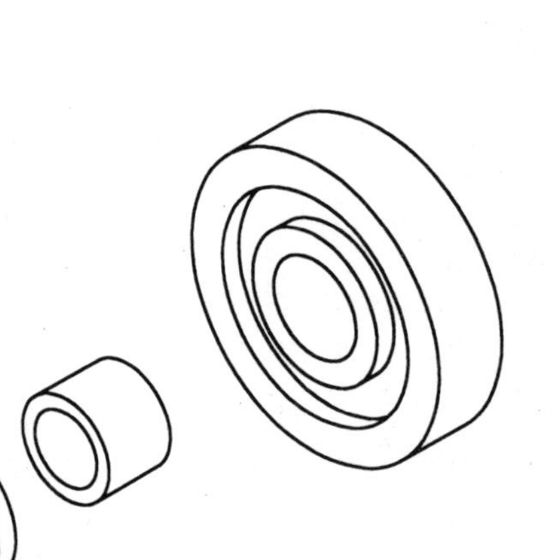

FIGURE 11–14(B)

EXERCISE 11-1

Modify the front view in the figure below to make it a full section. Do not forget to add the cutting plane line.

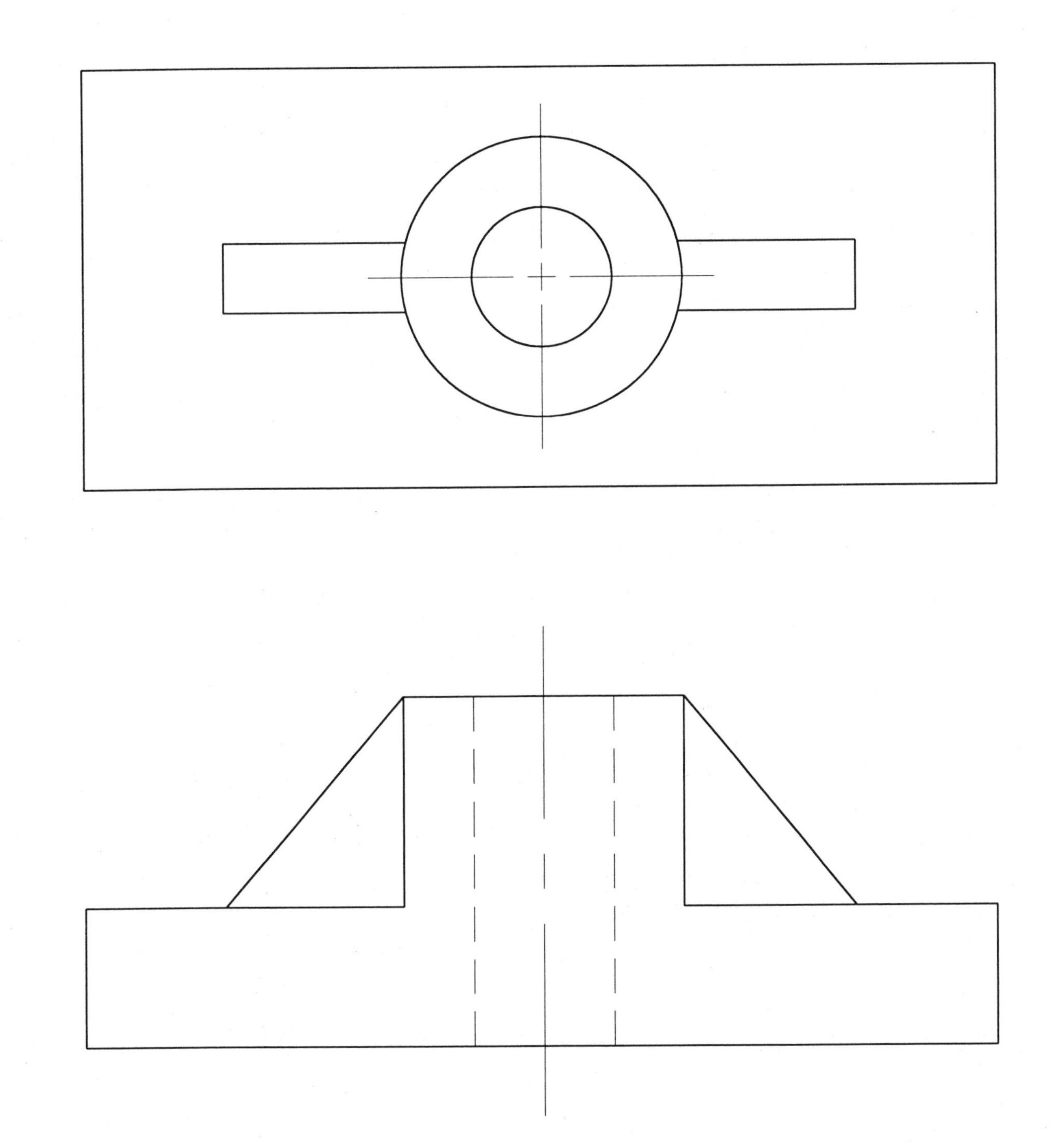

DRAWN BY ____________________ DATE ____________

EXERCISE II-2

Complete the front view of the figure below by adding section lines and any missing center lines, hidden lines, and object lines.

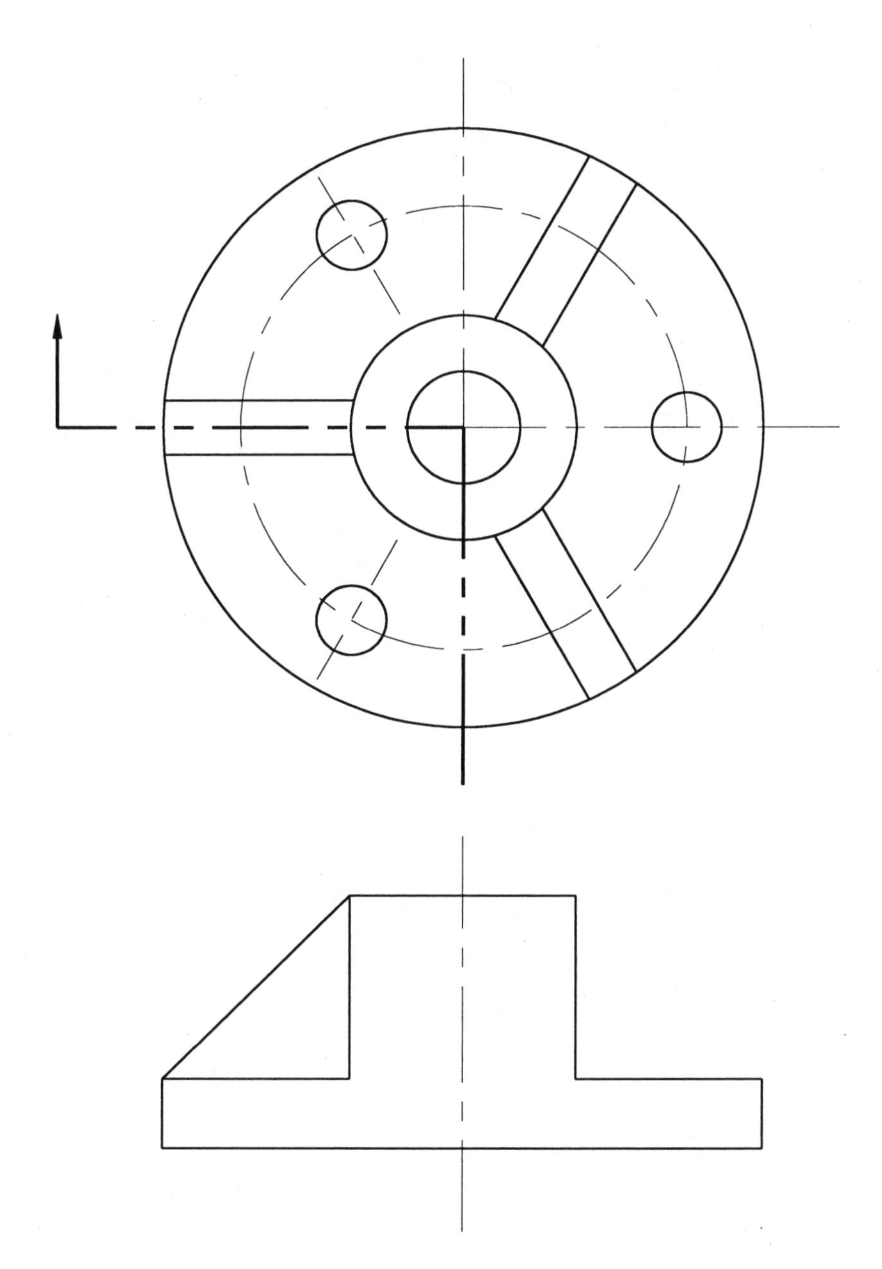

DRAWN BY ______________________________ DATE ______________

EXERCISE 11–3

Modify the front view in the figure below to make it a full section. Note that there is one rib on the left side of the object and that there are two on the right.

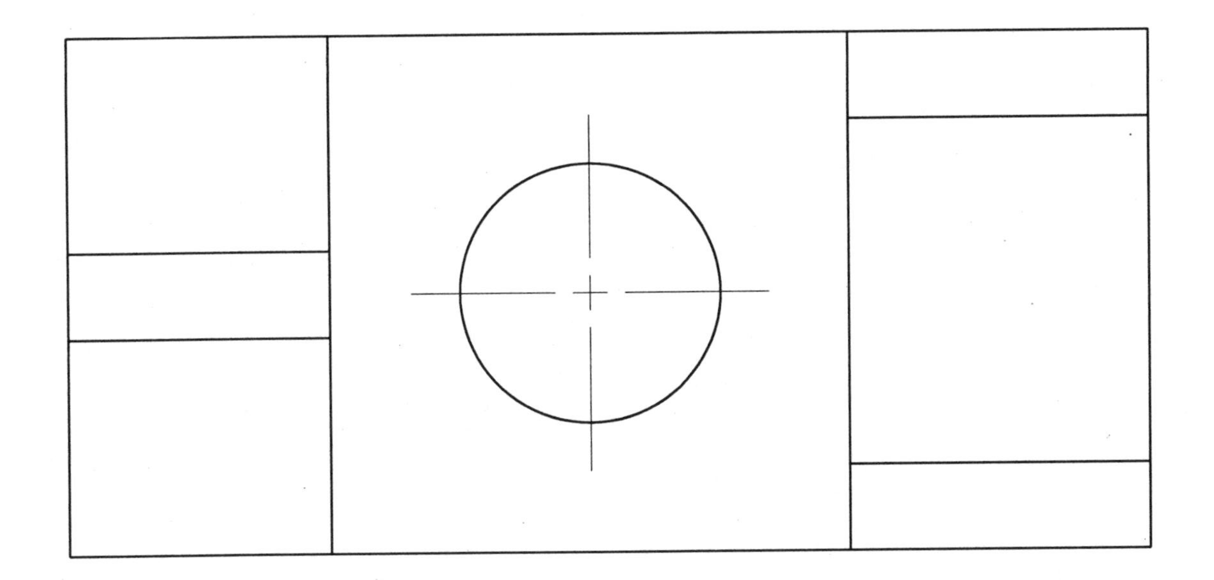

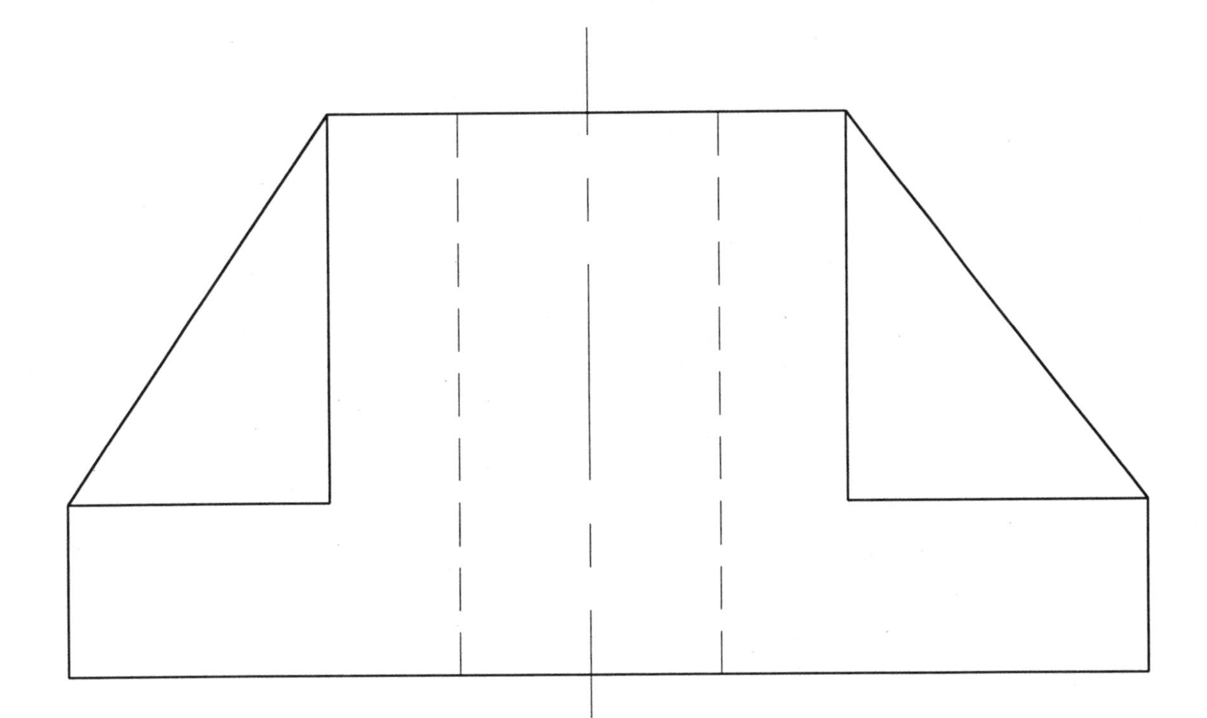

DRAWN BY ____________________ DATE ____________

EXERCISE II-4

Modify the front view in the figure below to make it an offset section. You will have to add the cutting plane line.

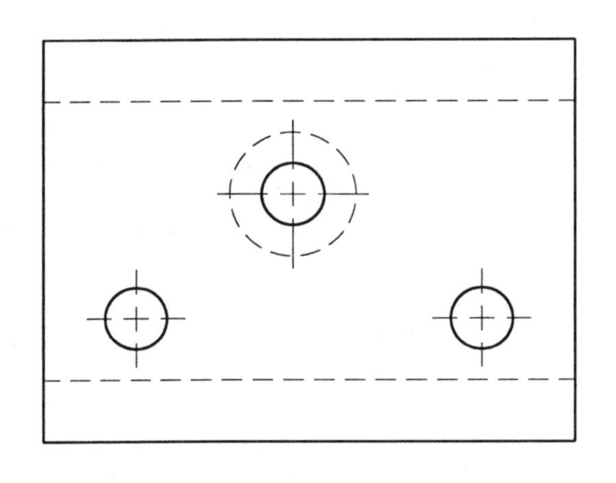

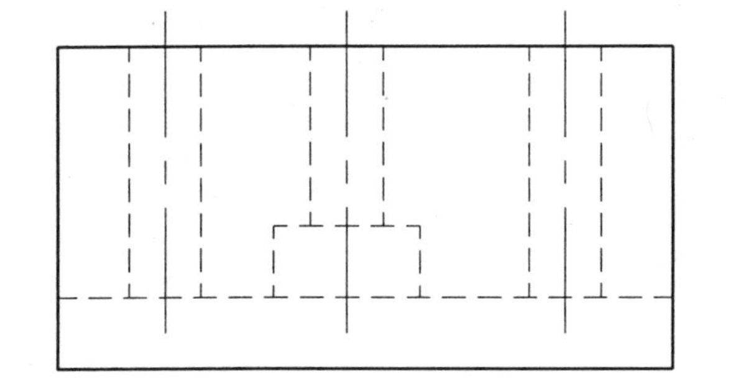

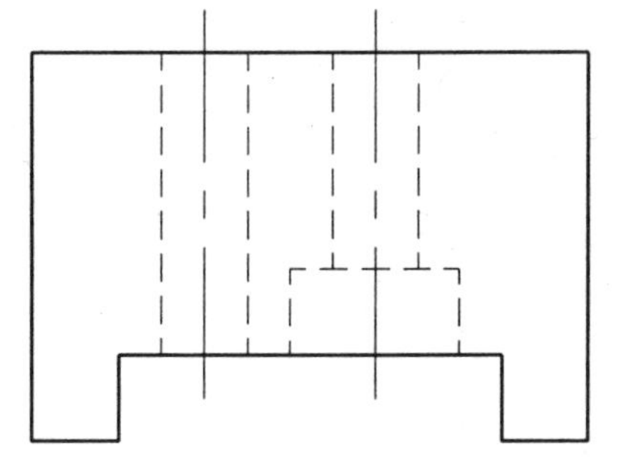

DRAWN BY DATE

EXERCISE II–5

Draw the removed section A-A in the figure below.

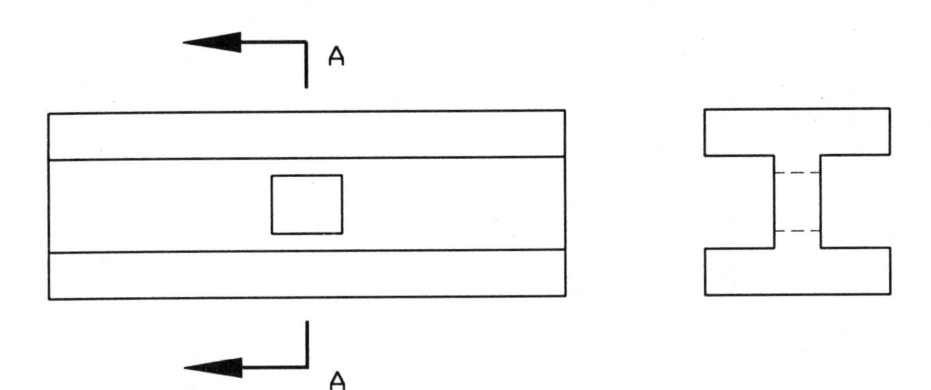

EXERCISE II–6

Redraw the figure below as a revolved section.

DRAWN BY DATE

EXERCISE 11–7

Modify the front view of the figure below to make it a full section.

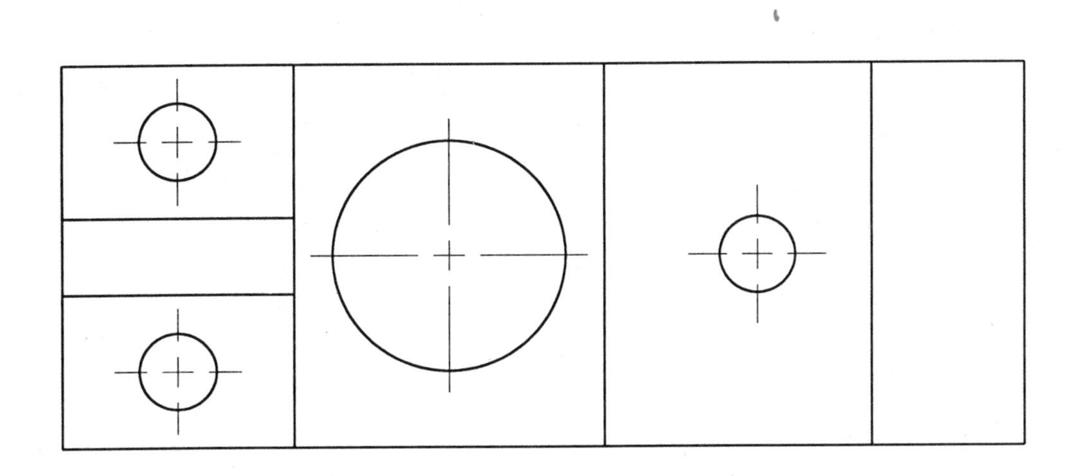

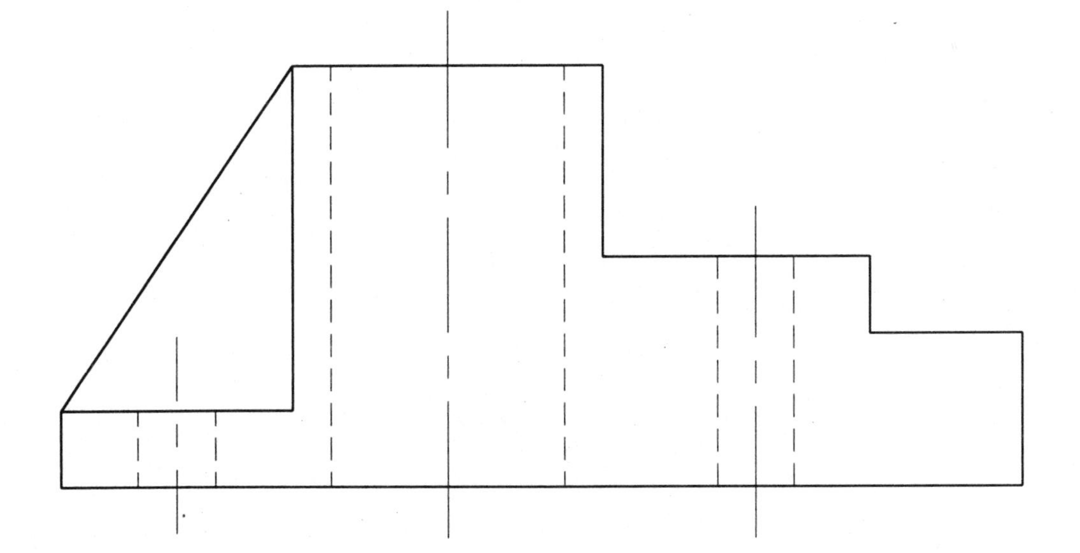

DRAWN BY DATE

EXERCISE 11-8

Draw a right side view of the figure as a full section. Add a cutting plane line.

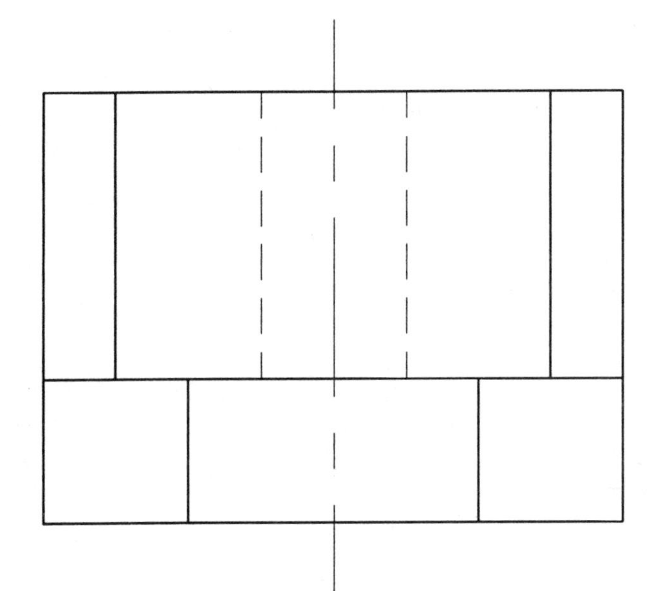

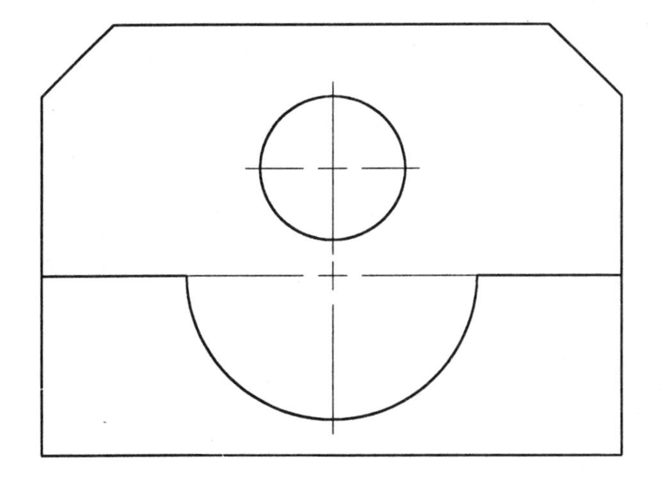

DRAWN BY DATE

DETAIL DRAWINGS

12

In this chapter, you will use all the information from the previous chapters to make drawings that could be used in the industrial workplace.

OBJECTIVE

When you have finished this chapter, you should be able to make detail drawings that include all information necessary to fabricate a part.

EQUIPMENT

You are required to have all your drafting instruments for this chapter.

NOTES

a. A *detail drawing* completely describes one part or object with orthographic views. It gives complete information (such as size and shape) necessary to manufacture one piece. You must include all dimensions, all views necessary to describe the object, and all notes pertaining to the manufacture of the part, such as surface finish and tolerances. Use only the minimum number of orthographic views. Draw only the parts to be made. Do not draw parts that are bought, such as bolts and nuts, unless special machining is required after the purchase.

(In contrast to a detail drawing, an *assembly drawing* gives all the information necessary to assemble the individual components of a mechanism. The assembly drawing includes overall dimensions, mounting dimensions, and the identification of each part. You will study assembly drawings in more advanced drafting courses.)

b. More than one detail may appear on the same sheet of paper.

c. Draw a dark line to separate one detail from the next. Arrangement of the details is subjective, but try to conserve space on the drawing paper.

d. On your detail drawing, you must verify that you have worked to your company standards and that you have included

- all dimensions, properly located,
- the scale of the drawing (and that you have made the drawing to scale),
- form and linear tolerances (you will learn about these in advanced drafting courses),
- a statement about the surface finish, if applicable (you will learn about finishes in advanced drafting courses),
- specifications about the material to be used to make the part,
- the title and date, your name, and a drawing number.

Figure 12–1, on the following page, is an example of a detail drawing from industry. It is reproduced with the permission of Dynacast Inc. Note that some drafting standards used on this drawing may vary from what you have learned in this book. The important thing to remember is to always work to your company standards.

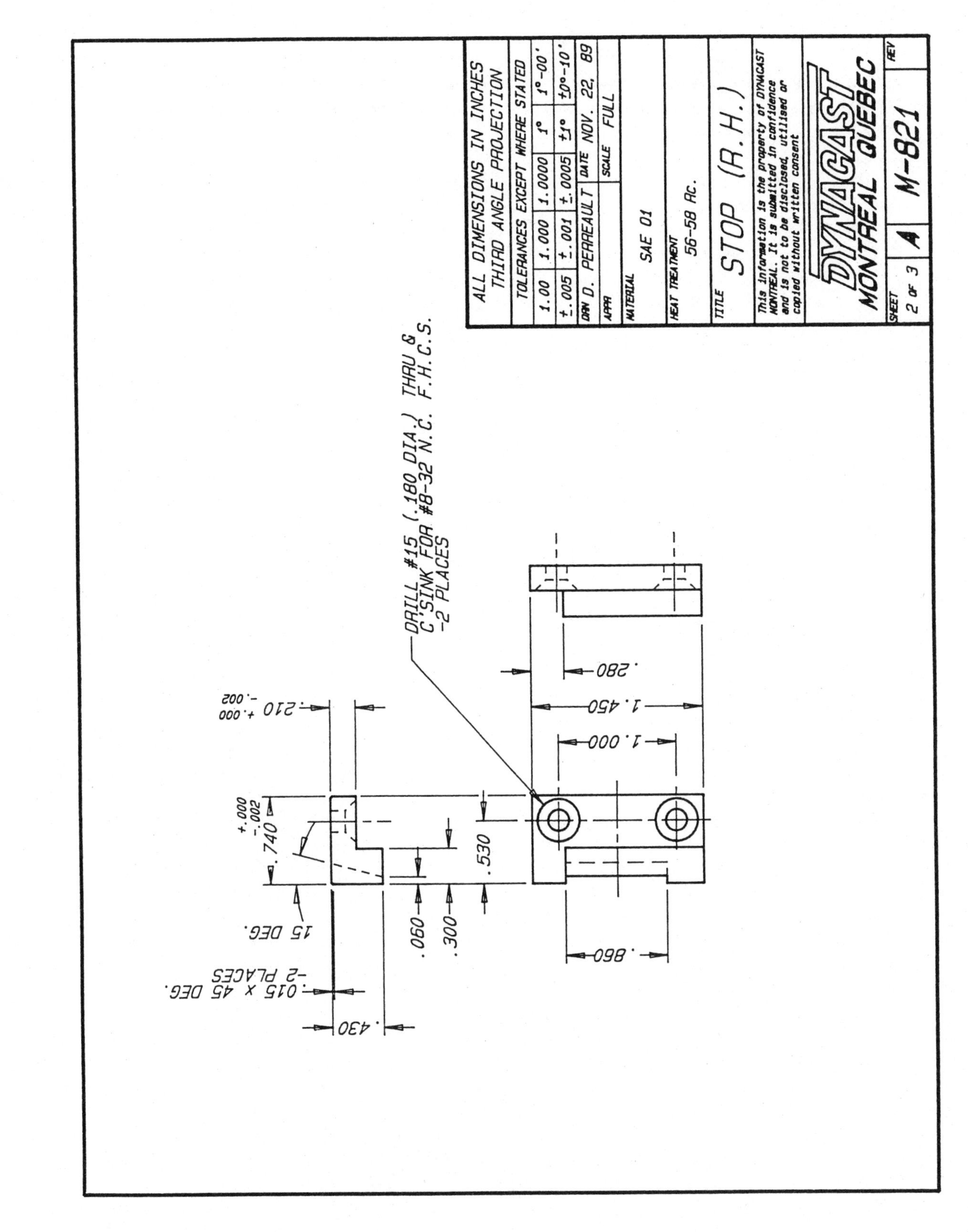

FIGURE 12–1 (Courtesy of Dynacast Inc.)

EXERCISE 12–1

Make a detail drawing of the *angle stop* shown below, including an auxiliary view of the sloping surface. Use your metric or inch scale to determine sizes. Measure to the closest 1/16" (or nearest millimeter).

The drawing on this page (and the next two pages) is an isometric pictorial and is presented here as an exercise. It does not show any dimensions. You must completely dimension your detail orthographic drawing, as discussed in Chapter 7; the tradesperson should not have to do any calculations to locate features. To measure this drawing, use your metric or inch scale according to your choice of standards. Use a full or reduced scale. Center the drawing only "by eye." Do not use detailed calculations to center it, as was done previously with simpler drawings, because you now have experience in drafting, and the sloping surfaces would make it difficult to center the drawing vertically.

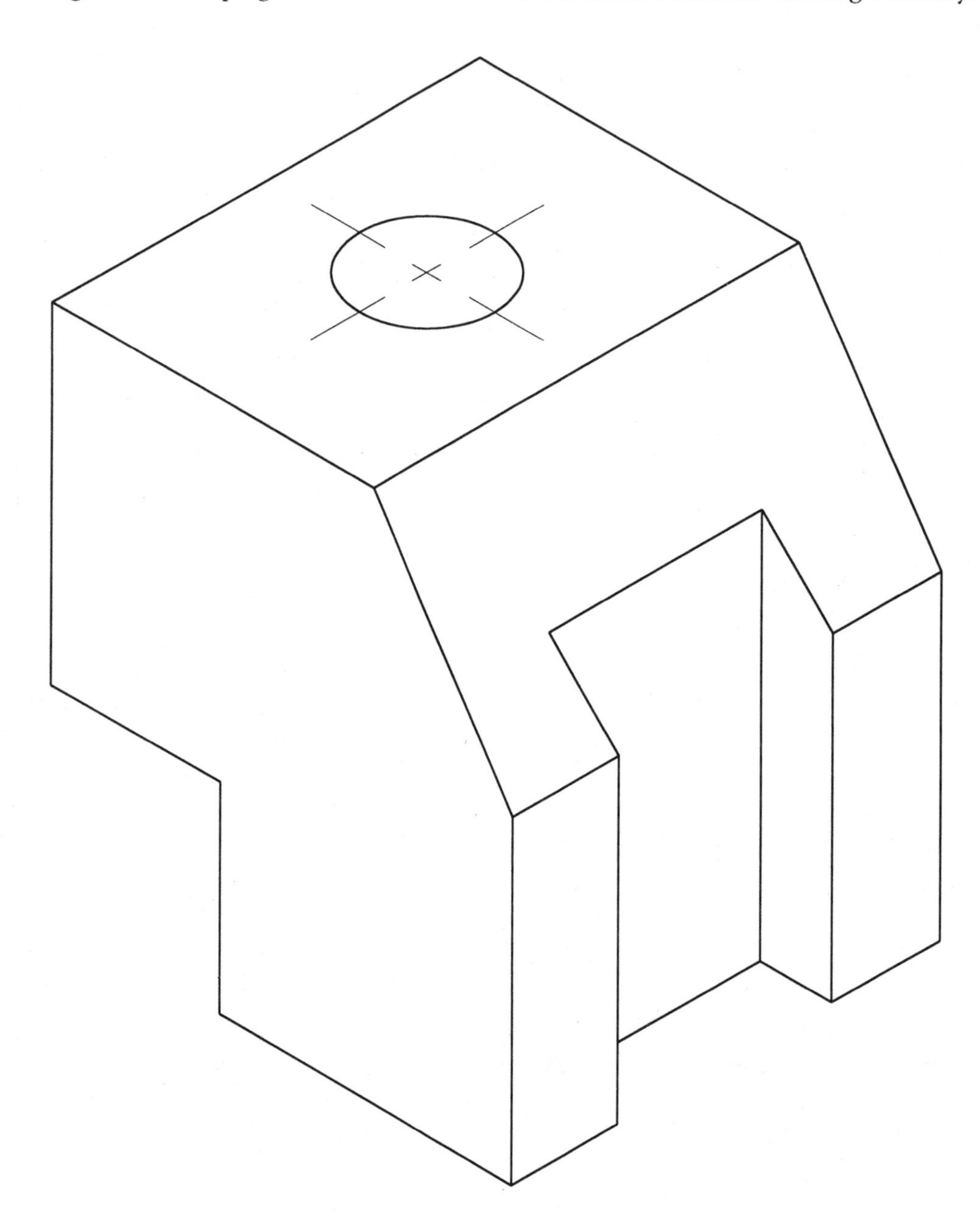

EXERCISE 12-2

On a C size sheet (metric A2), draw an orthographic detail drawing of the part shown below, remembering to include all the information listed on page 312. Note that you will need a front view, a partial top view, and two partial auxiliary views. Only a partial top view is required, since a complete top view of the sloping surface would be distorted.

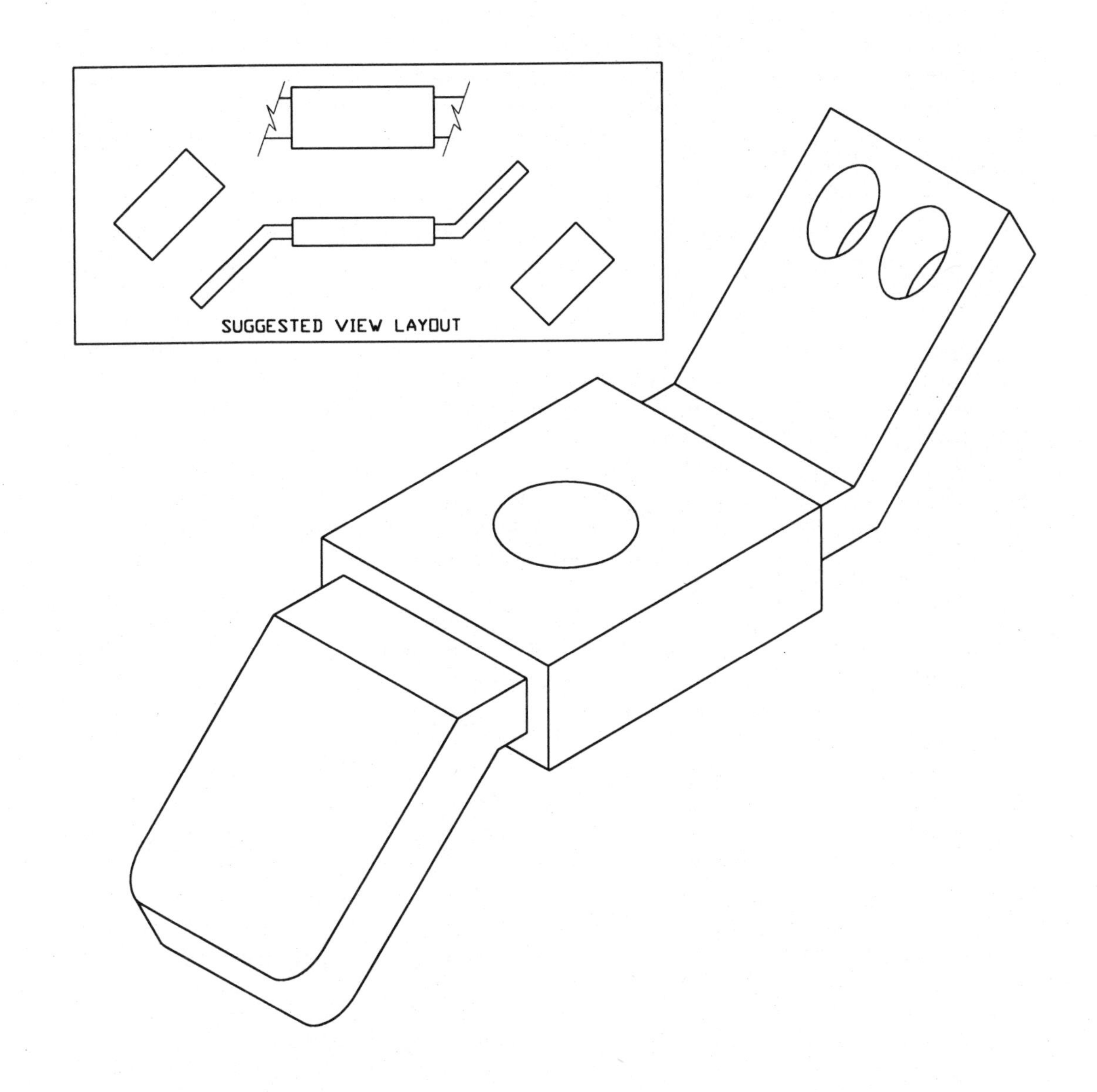

EXERCISE 12–3

On the next page are five objects in isometric views. Study these pieces so that you can visualize how they fit together to form a roller. Make detail drawings of the parts. In fact, you will need only four detail drawings (not five as you may think at first) because two pieces are identical. Scale the views on this sheet with your inch or metric scale. This drawing is half-scale. Make your drawing full scale. Use match fit—for example, 1" hole for 1" shaft—and do not worry about tolerances. They will be studied in a more advanced drafting course. Fit all the detail drawings onto one C size sheet (metric A2). Note that the sheet should not be divided into four equal parts, because some details will need more space than others. The shaft, for example, needs only one view.

OPTIONAL CHANGES:

1. Draw the wheel so that it has four spokes and a hub instead of being solid.

2. Add *spot-faces* to the two holes in the end supports. (See page 328 in the Appendix.)

3. *Chamfer* the ends of the shaft at 1/16" × 45° (1 mm × 45°). (See page 323 in the Appendix.)

4. In a drawing office, experienced drafters are promoted to *checkers*. The checker's job is to verify every detail of the drawing, including dimensions, views, and standards. The checker works with a blueprint (or copy) of the original. He or she verifies each dimension and then puts a yellow line through that dimension to indicate that it has been checked and is correct. If the dimension is wrong, the checker puts a red line through the dimension and writes the correct one. This process of yellow for correct features and red for incorrect ones is used for all drawing details. As an optional step in this exercise, you could exchange your drawing with a fellow student's and check each other's work.

NOTE: Each student will have slightly different sizes when measuring the following objects. There is no one correct size in this case.

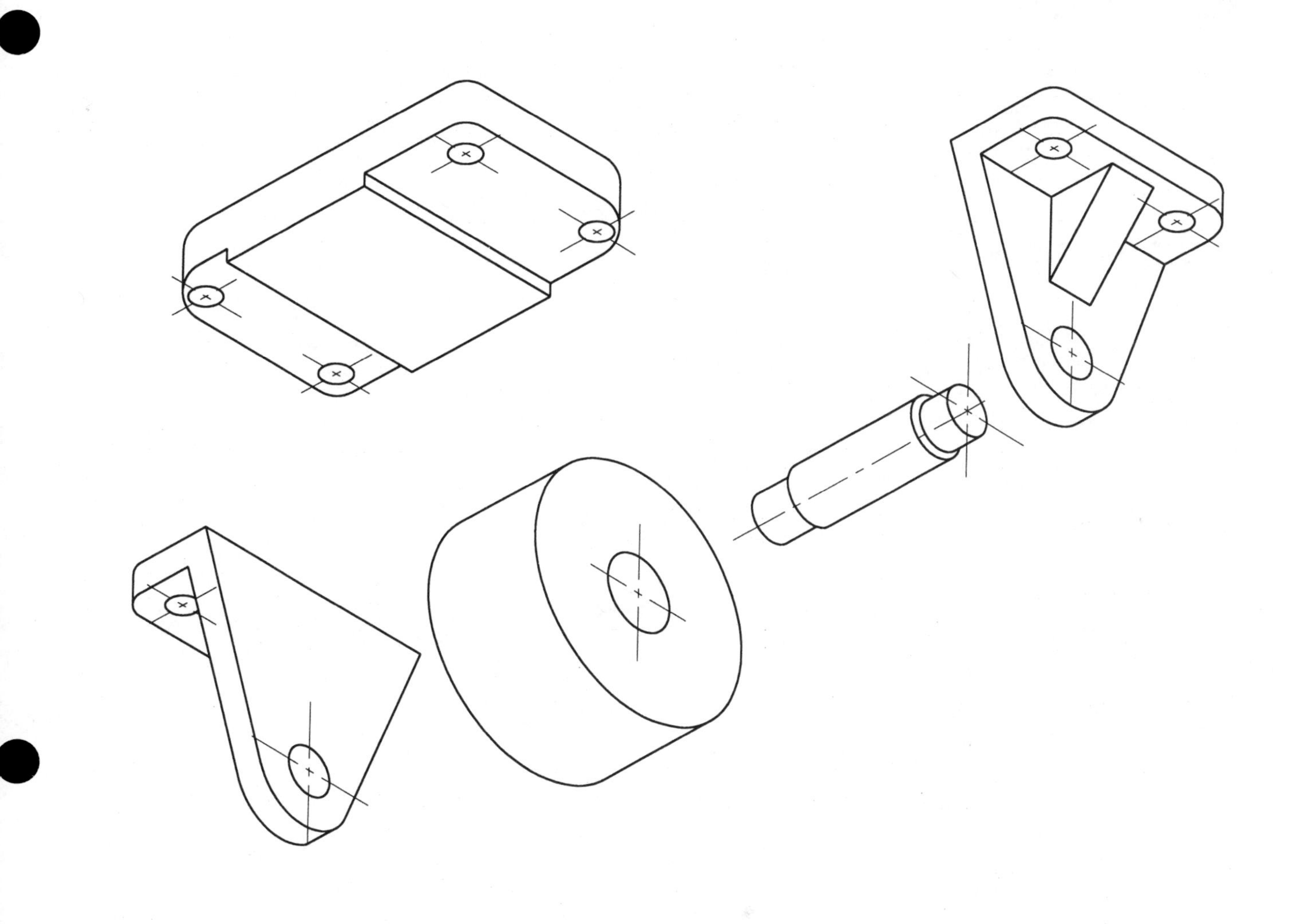

DESIGN EXERCISES

Up until now in our study of drafting, we have learned to make detail drawings of parts that have already been designed. A more advanced job in drafting is to design the shape of the part as well as to make the detail drawing. This page contains some exercises that will challenge your design skills.

EXERCISE 12–4

Design and make detail drawings of a pencil holder. It should be capable of holding at least 10 regular wooden pencils. In its simplest form, it could take the form of a cube with 10 drilled holes in which the pencils will stand vertically. Try to be creative and produce an interesting design.

EXERCISE 12–5

Design and make a detail drawing of a paper towel holder. Your design could be a freestanding device to hold the paper towels in a vertical position or a design whereby the towels would be horizontal.

EXERCISE 12–6

Design and make detail drawings of a candle holder.

EXERCISE 12–7

Design and make detail drawings of a bread cutting board.

BLUEPRINTS

Blueprints are copies of a drawing made with white lines on a dark blue background. This process is no longer used. Copies of a drawing are now made with blue lines on a light or white background and, technically speaking, are whiteprints. However, they are still popularly known as blueprints, and for this reason we will use the term *blueprint.*

A blueprint is a copy of the original drawing. The original drawing represents many hours of labor and would therefore cost additional money should it become lost and have to be redrawn. Therefore it is usually kept securely in a drawing office, and only the copy or blueprint is circulated out of that office. It would be a good idea to make blueprints of your drawing, since they will show you whether your line quality is dark enough. Light construction lines will not reproduce on a blueprint.

Many different processes are used in making blueprints. One commonly used method, the ammonia process, is described here.

The ammonia process uses paper coated with a light-sensitive material that is often yellow in color. Light will quickly destroy this yellow coating. For this reason, keep the blueprint paper in a dark location.

Turn the blueprint machine on and set the speed according to instructions. You may have to experiment with the speed adjustments to find the proper setting.

Place a piece of blueprint paper of the appropriate size *behind* the drawing so that the yellow side is touching the back of the drawing. This means that you will see the drawing properly, and you will see the yellow through the drawing.

Put the drawing and blueprint paper together into the bottom roller, which will expose the sheets to a strong light.

When the two papers exit the machine, separate them and feed the blueprint (now looking like a white piece of paper with yellow lines) into the top roller. This exposes the blueprint to ammonia gas, which changes the color of the lines from yellow to blue.

If you want a darker copy, set the machine to a faster speed. If you want a lighter copy, set the machine to a slower speed.

To conserve paper, do *not* make multiple copies if one is "a little too light" or "a little too dark" or "the paper was not straight."

Before you leave the blueprint room, turn off the machine and make sure that the supply of blueprint paper is located where no light or ammonia fumes will destroy the coating.

APPENDIX

USING YOUR DRAFTING EQUIPMENT

The following list refers to the page number in the workbook where the piece of equipment is described. This list is meant to accompany the one on pages 2 and 3.

EQUIPMENT	PAGE
Architect's scale	69, 70–73, 101
Metric scale	69, 77–79, 101
Engineer's scale	69, 81–83, 101
Compass	102, 148
Triangles (set squares)	91, 101, 102, 139
Lead holders or technical pencils	2, 3, 101
Protractor	87–88
Drafting tape or Masking tape	101, 138
Circle template	164, 173
Erasing shield	142
French curve	134, 242
Sandpaper pad	102, 148, 159
Drafting paper	101, 138
Dividers	214
Ames lettering guide	5

LINE TYPES

Remember to use different line thicknesses to represent different types of lines.

Thin lines that are dark and drawn with a 0.3 mm technical pencil include

Center lines,
Extension lines,
Dimension lines,
Leader lines,
Section lines (cross-hatching),
Hidden lines.

Thick lines that are dark and drawn with a 0.7 mm technical pencil include

Outlines.

Very thick lines that are dark and drawn with a 0.7 mm technical pencil include

Cutting plane lines.

Light lines drawn with a 0.3 mm technical pencil include
Guidelines used for lettering,
Construction lines.

Dark features drawn with a 0.7 mm technical pencil include
Arrowheads,
Lettering.

Figure A–1 shows a drawing illustrating the principal line types. Each line type has been labeled with a letter corresponding to its name on the accompanying list. Beside each line type are the thickness of line used and page number(s) referring you to the place in the workbook where it is described in more detail. This text refers to 0.3 mm lead for thin lines and 0.7 mm lead for thick lines. You could use a hard lead (3H or 4H) for thin lines and a softer lead (H or 2H) for thicker lines.

NOTE: All lines are dark except construction lines.

a. *Construction lines* are very thin and light (pages 36, 42, 55, 102, 141, 163).

b. *Object lines* are thick (page 44).

c. *Leader lines* are thin (pages 176, 177, 184).

d. *Hidden lines* are thin (pages 41, 44).

e. *Extension lines* are thin (pages 176, 177, 179, 183).

f. *Dimension lines* are thin (pages 176, 177).

g. *Break lines* may be thin or thick (page 285).

h. *Cutting plane lines* are very thick (pages 258, 260–262).

i. *Center lines* are thin (pages 44, 178).

j. *Section lines* (Cross-hatching) are thin and light (pages 258–260, 262–264).

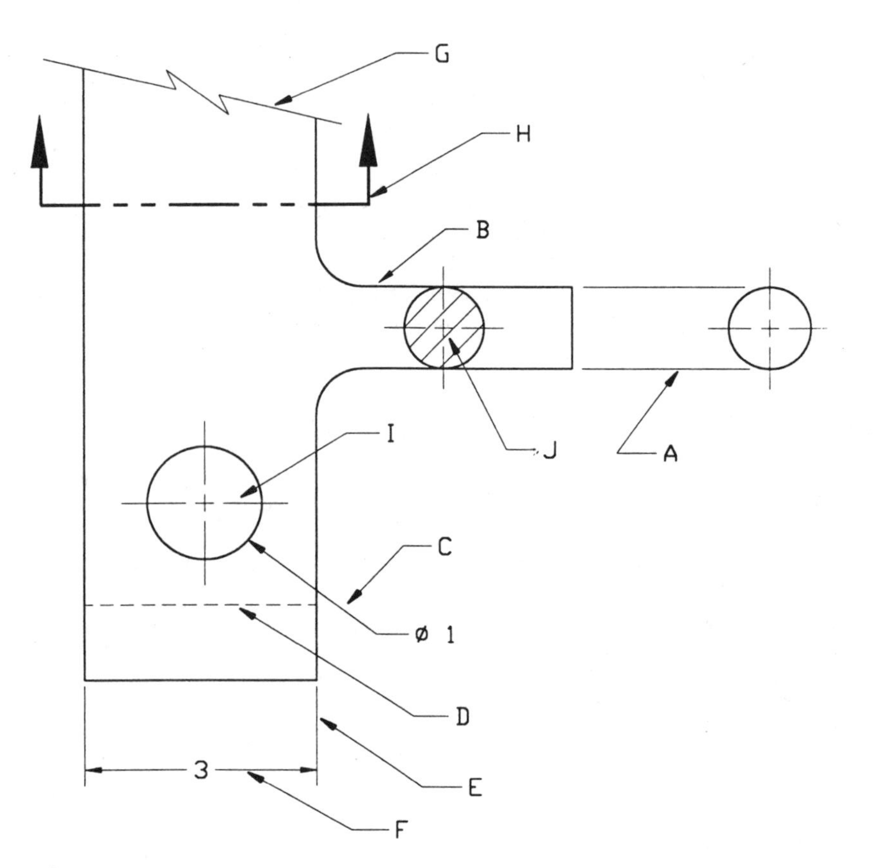

FIGURE A–1

TERMINOLOGY

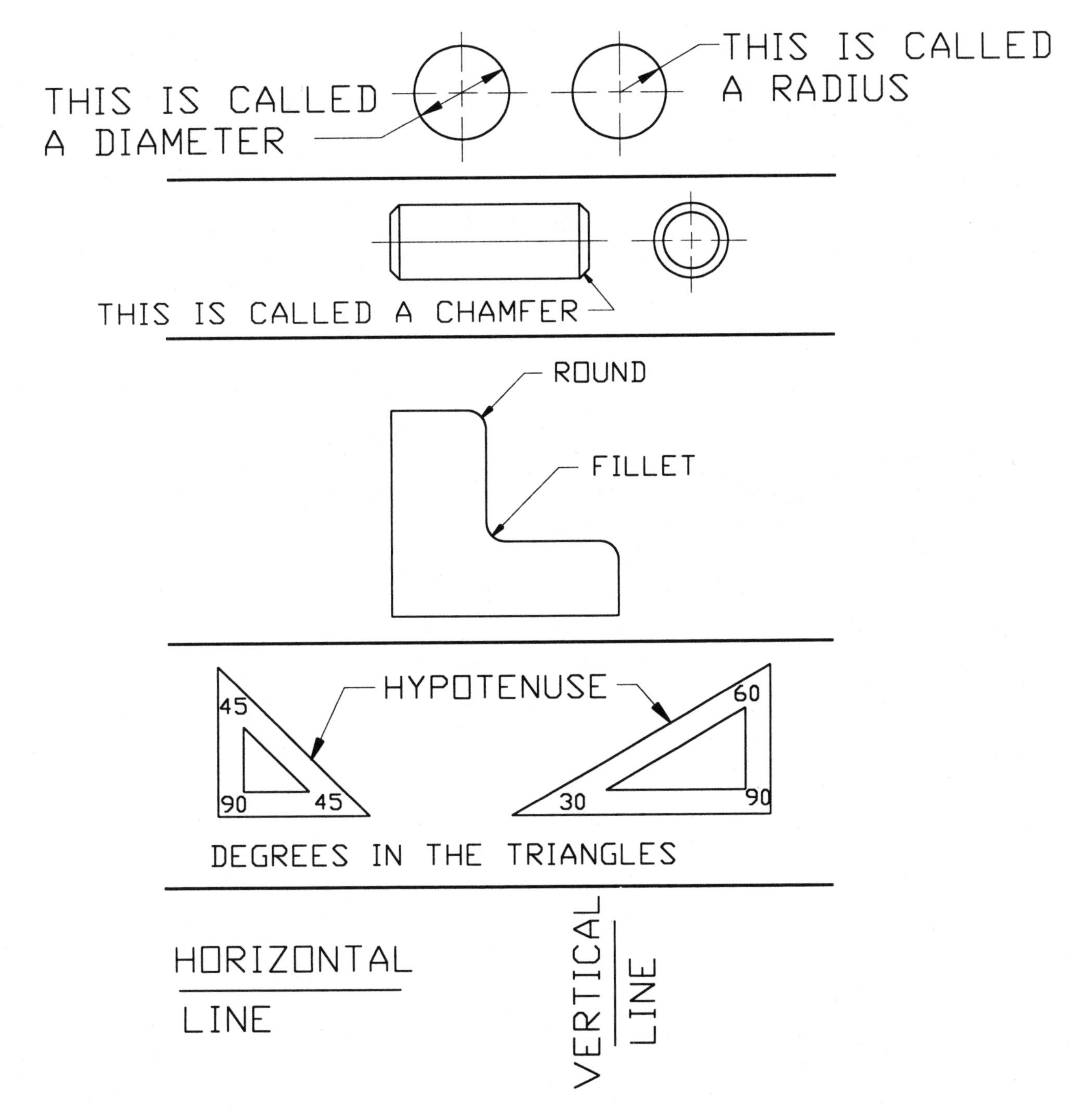

FIGURE A–2

HOW TO AVOID COMMON ERRORS IN DRAFTING

1. Guidelines should be used for *all* lettering, including numbers and letters.
2. Guidelines should be very *light.*
3. Object lines should be *dark* and *thick.*
4. Centerlines should be *thin*, as well as thinner than the outline.
5. Centerlines should extend beyond the object line by about 1/4".
6. An empty line should be left above and below a line of lettering.
7. The front view should be the most characteristic of the object.
8. Dimensions should not be placed too close to the object.
9. Dimensions should be placed on the view that shows the profile of the feature.
10. Units of measure, such as mm or ", should not appear after each individual dimension.
11. Leader lines are for holes and notes; they should not be used to dimension outside diameters.
12. The *diameter* of a hole should always be shown, and *not* the radius.
13. The drawing should not have a messy appearance. The drafting instruments and table must be clean. Use a sharp lead in your compass and pencil.
14. Object lines should be thick; centerlines, extension lines, and dimension lines should be thin.
15. The lines should be solid and not have a grey or fuzzy appearance.
16. All lines should be of consistent thickness. Use 0.3 mm and 0.7 mm leads; or use a thick lead holder, being sure to roll the pencil as you draw.
17. Corners should be sharp. Lines should touch but not overlap each other.

ANGLES

Figure A–3(a) through (f) illustrates some of the angles that you can make with your triangles.

An *acute* angle measures less than 90°. (See page 96.)

An *obtuse* angle measures greater than 90° and less than 180°. (See page 96.)

A *right* angle measures exactly 90°, as shown in Fig. A–3(a).

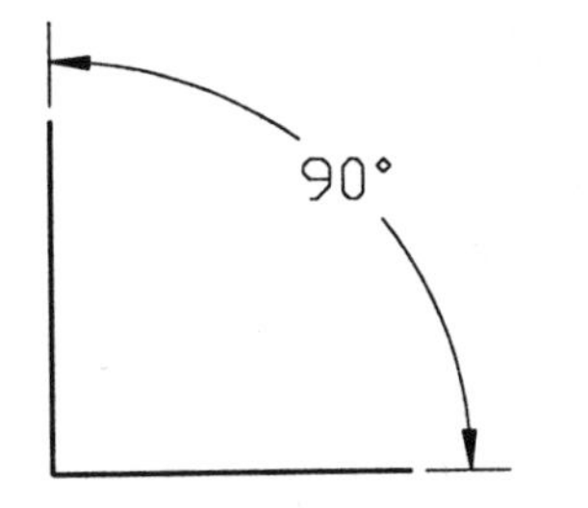

FIGURE A–3(A)

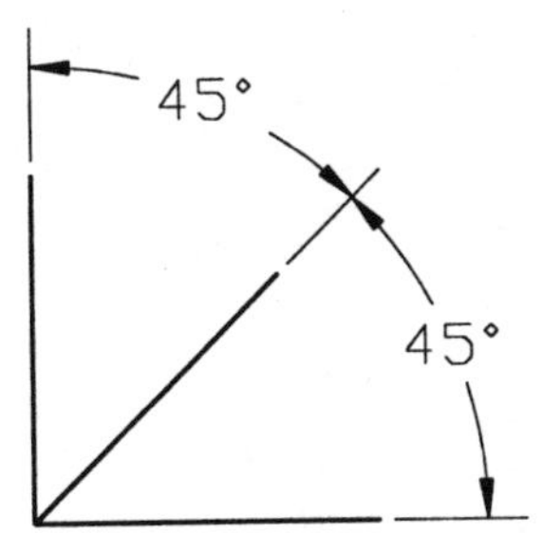

FIGURE A–3(B)

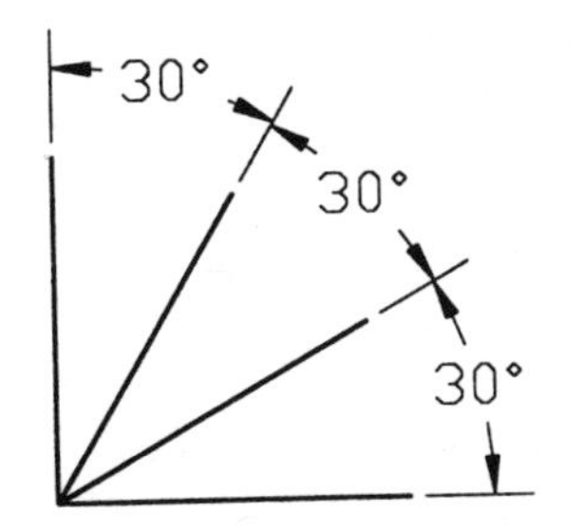

FIGURE A–3(C)

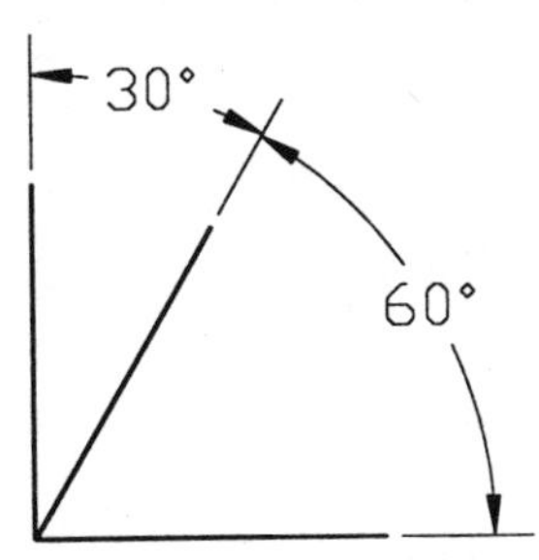

FIGURE A–3(D)

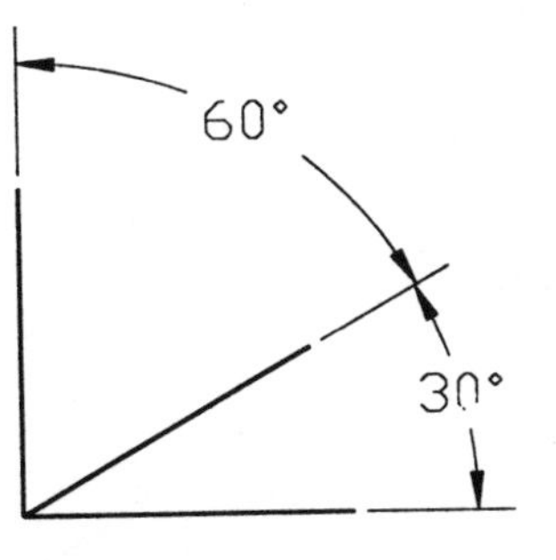

FIGURE A–3(E)

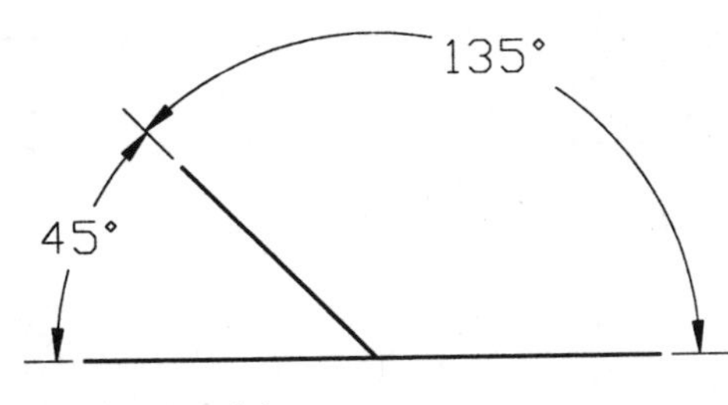

FIGURE A–3(F)

DECIMAL EQUIVALENTS TABLE

DECIMAL EQUIVALENTS — INCH-MILLIMETER CONVERSION TABLE

½	¼	⅛	1/16	1/32	1/64	Decimals	Millimeters	½	¼	⅛	1/16	1/32	1/64	Decimals	Millimeters
					1	.015625	.396875						33	.515625	13.096875
				1		.031250	.793750					17		.531250	13.493750
					3	.046875	1.190625						35	.546875	13.890625
			1			.062500	1.587500				9			.562500	14.287500
					5	.078125	1.984375						37	.578125	14.684375
				3		.093750	2.381250					19		.593750	15.081250
					7	.109375	2.778125						39	.609375	15.478125
		1				.125000	3.175000			5				.625000	15.875000
					9	.140625	3.571875						41	.640625	16.271875
				5		.156250	3.968750					21		.656250	16.668750
					11	.171875	4.365625						43	.671875	17.065625
			3			.187500	4.762500				11			.687500	17.462500
					13	.203125	5.159375						45	.703125	17.859375
				7		.218750	5.556250					23		.718750	18.256250
					15	.234375	5.953125						47	.734375	18.653125
	1					.250000	6.350000		3					.750000	19.050000
					17	.265625	6.746875						49	.765625	19.446875
				9		.281250	7.143750					25		.781250	19.843750
					19	.296875	7.540625						51	.796875	20.240625
			5			.312500	7.937500				13			.812500	20.637500
					21	.328125	8.334375						53	.828125	21.034375
				11		.343750	8.731250					27		.843750	21.431250
					23	.359375	9.128125						55	.859375	21.828125
		3				.375000	9.525000			7				.875000	22.225000
					25	.390625	9.921875						57	.890625	22.621875
				13		.406250	10.318750					29		.906250	23.018750
					27	.421875	10.715625						59	.921875	23.415625
			7			.437500	11.112500				15			.937500	23.812500
					29	.453125	11.509375						61	.953125	24.209375
				15		.468750	11.906250					31		.968750	24.606250
					31	.484375	12.303125						63	.984375	25.003125
1						.500000	12.700000	2	4	8	16	32	64	1.000000	25.400000

DRILLED HOLES

A drilled hole that does not go all the way through the part may be referred to as a "blind hole." The bottom of the hole is shaped like a triangle because of the shape made by the end of the drill. It is drawn at a 30° angle, as shown in Fig. A–4(a). We dimension a drilled hole as shown in Fig. A–4(b). "2 DEEP" specifies the depth of the drilled hole, which is from the top of the object to the horizontal line at the bottom of the hole (not to the point that the drill makes).

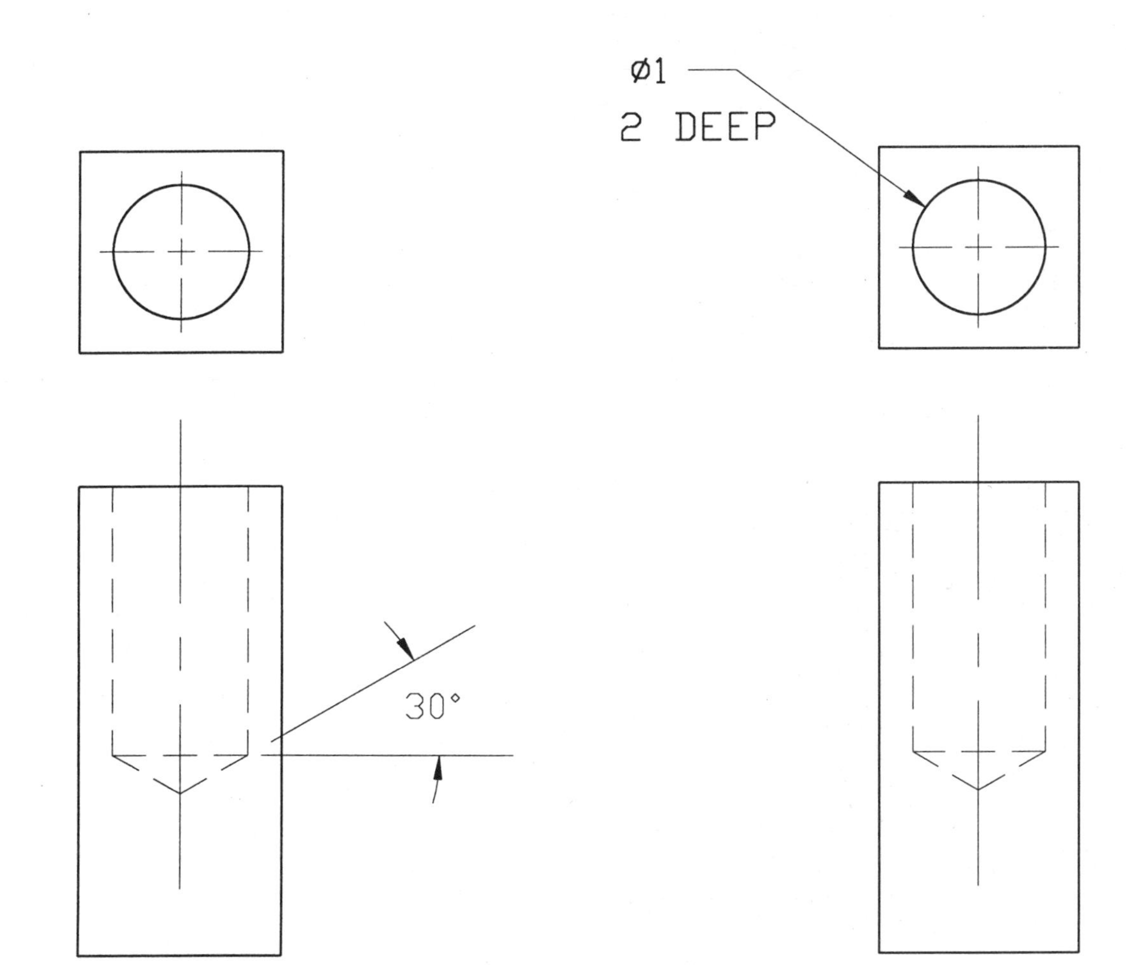

FIGURE A–4(A)

FIGURE A–4(B)

Drilled holes may have a countersink, a counterbore, or a spotface (Fig. A–5). The countersink and counterbore permit certain bolt heads to sit below the surface of the part. The spotface is used to make a smooth area around a hole, which provides a seat for the bottom of the bolt head.

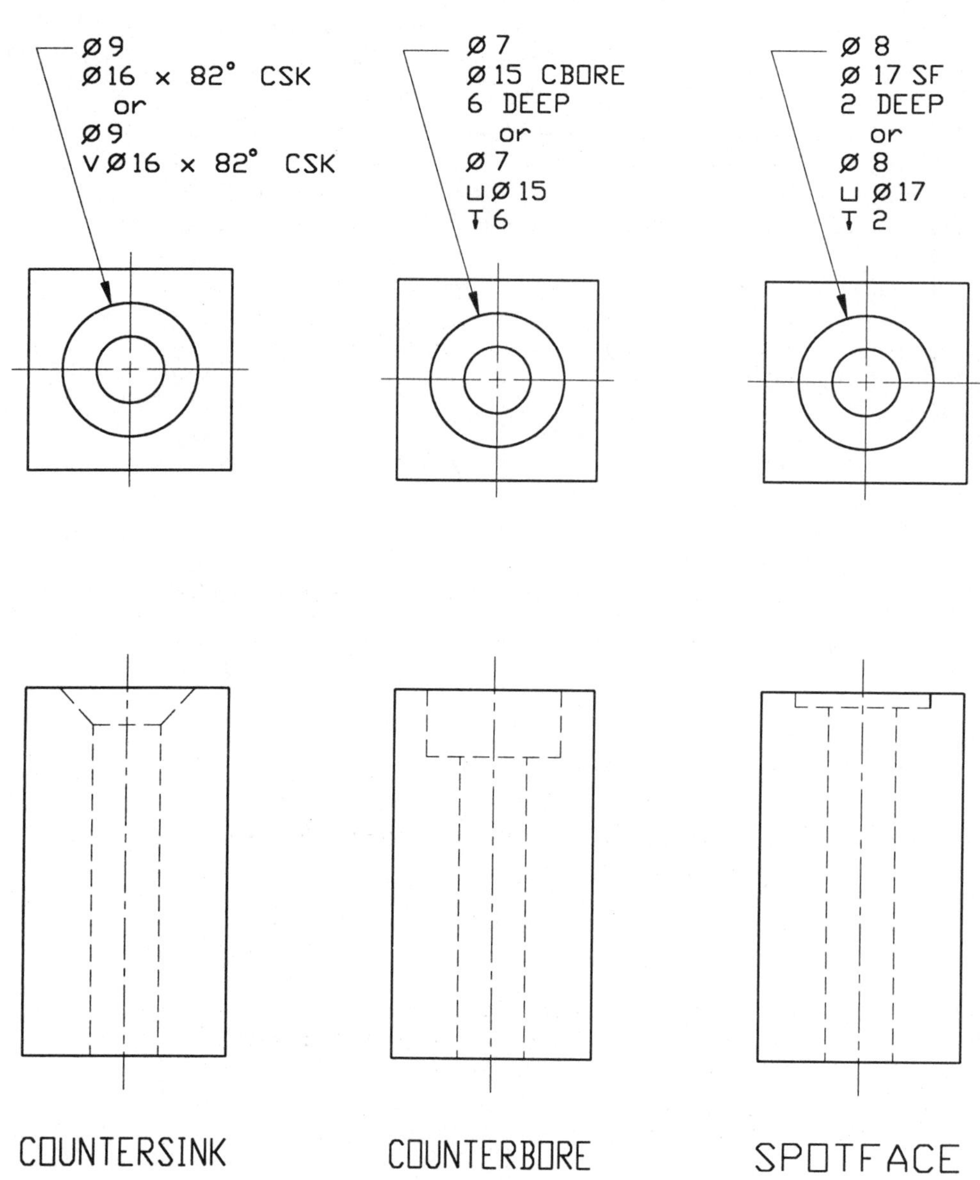

FIGURE A–5

STANDARDS

Most industries work to certain standards to ensure quality, safety, and interchangeability of products and information. Technical drawings are no exception. Drafting standards exist for certain countries and certain industrial areas. For example, the automotive and aerospace industries and the military have their own drafting standards. Companies in the United States generally work to the drafting standards of ANSI (American National Standards Institute). Canadian companies usually work to those of the CSA (Canadian Standards Association). A notable difference between these two standards is that ANSI addresses both the metric (SI) and inch systems of measurement (although the metric system is recommended), whereas CSA specifies the metric system as the standard. There also exists an international set of drafting standards published by ISO (International Organization for Standardization), which, like CSA, specifies the use of metric units only. CSA and ISO standards are almost identical for the purposes of this book. Where ANSI standards differ from the other two, the distinction is made on the relevant pages.

If you wish to purchase copies of the standards, you will find that they are expensive. Many large libraries and companies keep a set for use as reference, should you wish to review them. The standards are updated periodically, rendering the old ones obsolete.

This book addresses the directives of all three standards and notes any differences on the relevant pages. The important point to remember is to work to the standards of your company or school.

REFERENCE PUBLICATIONS

Specific standards that are relevant to technical drawing are listed below for your reference.

ANSI STANDARDS

ASME Y14.1M: *Metric Drawing Sheet Size and Format*
ASME Y14.2M: *Line Conventions and Lettering*
ANSI Y14.3: *Multi- and Sectional View Drawings*
ASME Y14.35M: *Revision of Engineering Drawings and Associated Documents*
ANSI Y14.5M: *Dimensioning and Tolerancing*

CSA STANDARDS

CAN3-B78.1-M83: *Technical Drawings—General Principles*
CAN3-B78.2-75: *Dimensioning and Tolerancing of Mechanical Engineering Drawings*
CAN3-B78.3-M77: *Building Drawings*
CAN/CSA-B78.2-M91: *Dimensioning and Tolerancing of Technical Drawings*

ISO STANDARDS

ISO 128: *Technical Drawings—General Principles of Presentation*
ISO 129: *Technical Drawings—Dimensioning—General Principles, Definitions, Methods of Execution and Special Indications*
ISO 2162: *Technical Drawings—Representation of Springs*
ISO 3098/1: *Technical Drawings—Lettering—Part 1: Currently Used Characters*
ISO 5455: *Technical Drawings—Scales*
ISO 6410: *Technical Drawings—Conventional Representation of Threaded Parts*

ISO 6433: *Technical Drawings—Item References*
ISO 7573: *Technical Drawings—Item List*
ISO 8048: *Technical Drawings—Construction Drawings—Representation of Views, Sections, and Cuts*

Addresses for the standards associations are as follows.

ANSI STANDARDS:

American Society of Mechanical Engineers (publishers)
United Engineering Center
345 East 47th Street
New York, NY 10017

OR

American National Standards Institute
1430 Broadway
New York, NY 10018

CANADIAN STANDARDS:

Canadian Standards Association (CSA)
178 Rexdale Boulevard
Rexdale (Toronto), Ontario, Canada
M9W 1R3
Regional Offices are in Vancouver, Edmonton, Winnipeg, Montreal, and Moncton.

ISO STANDARDS:

International Organization for Standardization
1, rue de Varembe
Postal Box 56
CH-1211
Geneva 20, Switzerland

Canadian suppliers of ISO standards are:
Standards Council of Canada
Suite 1200
45 O'Connor Street,
Ottawa, Ontario, Canada
K1P 6N7
Telephone: (613) 238-3222
(800) 267-8220

FORMULAS

Area of a circle: $A = \pi r^2$
Circumference of a circle: $C = 2\pi r$
Area of a rectangle: $A = bh$
Area of a triangle: $A = bh/2$
Area of a trapezoid: $1/2(b + B)h$
Ø means *diameter*
R means *radius*
DIA means *diameter*
BORE refers to *diameter*
DRILL refers to *diameter*
REAM refers to *diameter*
COUNTERBORE refers to *diameter*
COUNTERSINK refers to *diameter*
SPOTFACE refers to *diameter*

STANDARD SIZES OF DRAFTING PAPER

SIZE	INCH (MECHANICAL)	INCH (ARCHITECTURAL/ CIVIL)
A	8 1/2 x 11	9 x 12
B	11 x 17	12 x 18
C	17 x 22	18 x 24
D	22 x 34	24 x 36
E	34 x 44	36 x 48
F	28 x 40	
J (roll)	34 x 55 to 176	

SIZE	METRIC
A4	210 x 297
A3	297 x 420
A2	420 x 594
A1	594 x 841
A0	841 x 1189

ABBREVIATIONS

APPENDIX 1 • Abbreviations (ANSI Z 32.13)

Word	Abbreviation	Word	Abbreviation	Word	Abbreviation
Allowance	ALLOW	Elevation	ELEV	Not to scale	NTS
Alloy	ALY	Equal	EQ	Number	NO.
Aluminum	AL	Estimate	EST	Octagon	OCT
Amount	AMT	Exterior	EXT	On center	OC
Anneal	ANL	Fahrenheit	F	Ounce	OZ
Approximate	APPROX	Feet	(′) FT	Outside diameter	OD
Area	A	Feet per minute	FPM	Parallel	PAR.
Assembly	ASSY	Feet per second	FPS	Perpendicular	PERP
Auxiliary	AUX	Fillet	FIL	Piece	PC
Average	AVG	Fillister	FIL	Plastic	PLSTC
Babbitt	BAB	Finish	FIN.	Plate	PL
Between	BET.	Finish all over	FAO	Point	PT
Between centers	BC	Flat head	FH	Polish	POL
Bevel	BEV	Foot	(′) FT	Pound	LB
Bill of material	B/M	Front	FR	Pounds per square inch	PSI
Both sides	BS	Gage	GA	Pressure	PRESS.
Bottom	BOT	Gallon	GAL	Production	PROD
Brass	BRS	Galvanize	GALV	Quarter	QTR
Brazing	BRZG	Galvanized iron	GI	Radius	R
Broach	BRO	General	GEN	Ream	RM
Bronze	BRZ	Gram	G	Rectangle	RECT
Cadmium plate	CD PL	Grind	GRD	Reference	REF
Cap screw	CAP SCR	Groove	GRV	Required	REQD
Case harden	CH	Hardware	HDW	Revise	REV
Cast iron	CI	Head	HD	Revolution	REV
Cast steel	CS	Heat treat	HT TR	Revolutions per minute	RPM
Casting	CSTG	Hexagon	HEX	Right	R
Center	CTR	Horizontal	HOR	Right hand	RH
Centerline	CL	Horsepower	HP	Rough	RGH
Center to center	C to C	Hot rolled steel	HRS	Screw	SCR
Centigrade	C	Hour	HR	Section	SECT
Centigram	CG	Hundredweight	CWT	Set screw	SS
Centimeter	cm	Inch	(″) IN.	Shaft	SFT
Chamfer	CHAM	Inches per second	IPS	Slotted	SLOT.
Circle	CIR	Inside diameter	ID	Socket	SOC
Clockwise	CW	Interior	INT	Spherical	SPHER
Cold rolled steel	CRS	Iron	I	Spot faced	SF
Cotter	COT	Key	K	Spring	SPG
Counterclockwise	CCW	Kip (1000 lb)	K	Square	SQ
Counterbore	CBORE	Left	L	Station	STA
Counterdrill	CDRILL	Left hand	LH	Steel	STL
Counterpunch	CPUNCH	Length	LG	Symmetrical	SYM
Countersink	CSK	Light	LT	Taper	TPR
Cubic centimeter	cc	Machine	MACH	Temperature	TEMP
Cubic feet per minute	CFM	Malleable	MALL	Tension	TENS.
Cubic foot	CU FT	Manhole	MH	Thick	THK
Cubic inch	CU IN.	Manufacture	MFR	Thousand	M
Cylinder	CYL	Material	MATL	Thousand pound	KIP
Diagonal	DIAG	Maximum	MAX	Thread	THD
Diameter	DIA	Metal	MET.	Tolerance	TOL
Distance	DIST	Meter (Instrument or measure of length)	M	Typical	TYP
Ditto	DO	Miles	MI	Vertical	VERT
Down	DN	Miles per gallon	MPG	Volume	VOL
Dozen	DOZ	Miles per hour	MPH	Washer	WASH.
Drafting	DFTG	Millimeter	MM	Weight	WT
Drawing	DWG	Minimum	MIN	Width	W
Drill	DR	Normal	NOR	Wrought iron	WI
Each	EA			Yard	YD

ISOMETRIC GRID

Do not draw on this page. Use Fig. A–6 as a guide for drawing isometric views by placing it under a blank sheet of paper.

FIGURE A–6

CONVERSION TABLES

Conversion tables: Length conversions

Feet	$\times$ 12	= inches
	$\times$ 0.3048	= meters
Inches	$\times$ 2.54 $\times 10^{8}$	= Angstroms
	$\times$ 25.4	= millimeters
	$\times$ 8.333 33 $\times 10^{2}$	= feet
Kilometers	$\times$ 3.280 839 $\times 10^{3}$	= feet
	$\times$ 0.62	= miles
	$\times$ 0.539 956	= nautical miles
	$\times$ 0.621 371	= statute miles
	$\times$ 1.093 613 $\times 10^{3}$	= yards
Meters	$\times$ 1 $\times 10^{10}$	= Angstroms
	$\times$ 3.280 839 9	= feet
	$\times$ 39.370 079	= inches
	$\times$ 1.093 61	= yards
Statute miles	$\times$ 5.280 $\times 10^{3}$	= feet
	$\times$ 8	= furlongs
	$\times$ 6.336 0 $\times 10^{4}$	= inches
	$\times$ 1.609 34	= kilometers
	$\times$ 8.689 7 $\times 10^{1}$	= nautical miles
Miles	$\times 10^{3}$	= inches
	$\times$ 2.54 $\times 10^{2}$	= millimeters
	$\times$ 25.4	= micrometers
	$\times$ 0.61	= kilometers
Yards	$\times$ 3	= feet
	$\times$ 9.144 $\times 10^{1}$	= meters
Feet/hour	$\times$ 3.048 $\times 10^{4}$	= kilometers/hour
	$\times$ 1.645 788 $\times 10^{4}$	= knots
Feet/minute	$\times$ 0.3048	= meters/minute
	$\times$ 5.08 $\times 10^{3}$	= meters/second
Feet/second	$\times$ 1.097 28	= kilometers/hour
	$\times$ 18.288	= meters/minute
Kilometers/hour	$\times$ 3.280 839 $\times 10^{3}$	= feet/hour
	$\times$ 54.680 66	= feet/minute
	$\times$ 0.277 777	= meters/second
	$\times$ 0.621 371	= miles/hour
Kilometers/minute	$\times$ 3.280 839 $\times 10^{3}$	= feet/minute
	$\times$ 37.282 27	= miles/hour
Meters/hour	$\times$ 3.280 839	= feet/hour
	$\times$ 88	= feet/minute
	$\times$ 1.466	= feet/second
	$\times$ 1 $\times 10^{3}$	= kilometers/hour
	$\times$ 1.667 $\times 10^{2}$	= meters/minute
Feet/second^{2}	$\times$ 1.097 28	= kilometers/hour/second
	$\times$ 0.304 8	= meters/second^{2}

WEIGHTS AND MEASURES

Weights and measures

UNITED STATES SYSTEM

LINEAR MEASURE

Inches	Feet	Yards	Rods	Furlongs	Miles
1.0 =	.08333 =	.02778 =	.0050505 =	.00012626 =	.00001578
12.0 =	1.0 =	.33333 =	.0606061 =	.00151515 =	.00018939
36.0 =	3.0 =	1.0 =	.1818182 =	.00454545 =	.00056818
198.0 =	16.5 =	5.5 =	1.0 =	.025 =	.003125
7920.0 =	660.0 =	220.0 =	40.0 =	1.0 =	.125
63360.0 =	5280.0 =	1760.0 =	320.0 =	8.0 =	1.0

SQUARE AND LAND MEASURE

Sq. Inches	Square Feet	Square Yards	Sq. Rods	Acres	Sq. Miles
1.0 =	.006944 =	.000772			
144.0 =	1.0 =	.111111			
1296.0 =	9.0 =	1.0 =	.03306 =	.000207	
39204.0 =	272.25 =	30.25 =	1.0 =	.00625 =	.0000098
	43560.0 =	4840.0 =	160.0 =	1.0 =	.0015625
		3097600.0 =	102400.0 =	640.0 =	1.0

AVOIRDUPOIS WEIGHTS

Grains	Drams	Ounces	Pounds	Tons
1.0 =	.03657 =	.002286 =	.000143 =	.0000000714
27.34375 =	1.0 =	.0625 =	.003906 =	.00000195
437.5 =	16.0 =	1.0 =	.0625 =	.00003125
7000.0 =	256.0 =	16.0 =	1.0 =	.0005
14000000.0 =	512000.0 =	32000.0 =	2000.0 =	1.0

DRY MEASURE

Pints	Quarts	Pecks	Cubic Feet	Bushels
1.0 =	.5 =	.0625 =	.01945 =	.01563
2.0 =	1.0 =	.125 =	.03891 =	.03125
16.0 =	8.0 =	1.0 =	.31112 =	.25
51.42627 =	25.71314 =	3.21414 =	1.0 =	.80354
64.0 =	32.0 =	4.0 =	1.2445 =	1.0

LIQUID MEASURE

Gills	Pints	Quarts	U. S. Gallons	Cubic Feet
1.0 =	.25 =	.125 =	.03125 =	.00418
4.0 =	1.0 =	.5 =	.125 =	.01671
8.0 =	2.0 =	1.0 =	.250 =	.03342
32.0 =	8.0 =	4.0 =	1.0 =	.1337
			7.48052 =	1.0

METRIC SYSTEM

UNITS

Length—Meter : Mass—Gram : Capacity—Liter

for pure water at 4°C. (39.2°F.)

1 cubic decimeter or 1 liter = 1 kilogram

1000 Milli{*meters* (mm), *grams* (mg), *liters* (ml)} = 100 Centi{*meters* (cm), *grams* (cg), *liters* (cl)} = 10 Deci{*meters* (dm), *grams* (dg), *liters* (dl)} = 1{meter, gram, liter}

1000{meters, grams, liters} = 100 Deka{*meters* (dkm), *grams* (dkg), *liters* (dkl)} = 10 Hecto{*meters* (hm), *grams* (hg), *liters* (hl)} = 1 Kilo{*meter* (km), *gram* (kg), *liter* (kl)}

1 Metric Ton	= 1000 Kilograms
100 Square Meters	= 1 Are
100 Ares	= 1 Hectare
100 Hectares	= 1 Square Kilometer

HEXAGON HEAD BOLTS AND NUTS

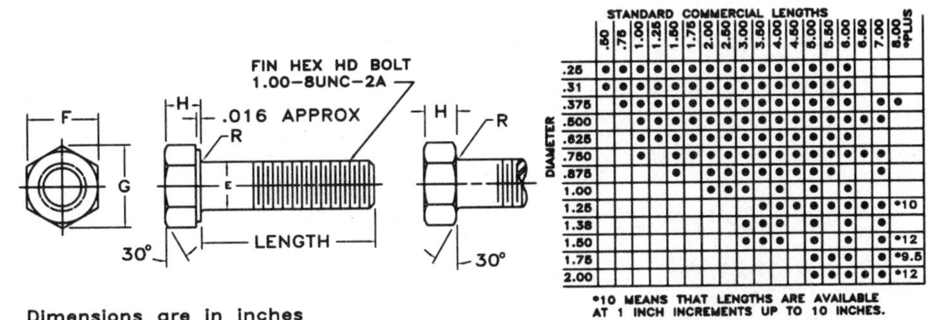

Dimensions are in inches

DIA	E Max.	F Max.	G Avg.	H Max.	R Max.
1/4	.250	.438	.505	.163	.025
5/16	.313	.500	.577	.211	.025
3/8	.375	.563	.650	.243	.025
7/16	.438	.625	.722	.291	.025
1/2	.500	.750	.866	.323	.025
9/16	.563	.812	.938	.371	.045
5/8	.625	.938	1.083	.403	.045
3/4	.750	1.125	1.299	.483	.045
7/8	.875	1.313	1.516	.563	.065
1	1.000	1.500	1.732	.627	.095

DIA	E Max.	F Max.	G Avg.	H Max.	R Max.
1-1/8	1.125	1.688	1.949	.718	.095
1-1/4	1.250	1.875	2.165	.813	.095
1-3/8	1.375	2.063	2.382	.878	.095
1-1/2	1.500	2.250	2.598	.974	.095
1-3/4	1.750	2.625	3.031	1.134	.095
2	2.000	3.000	3.464	1.263	.095
2-1/4	2.250	3.375	3.897	1.423	.095
2-1/2	2.500	3.750	4.330	1.583	.095
2-3/4	2.750	4.125	4.763	1.744	.095
3	3.000	4.500	5.196	1.935	.095

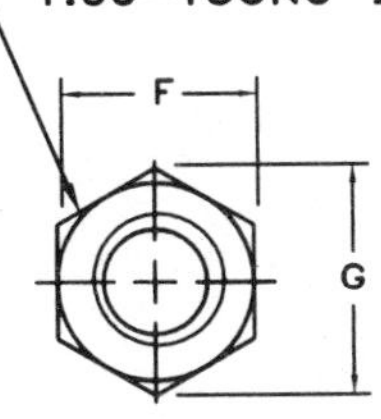

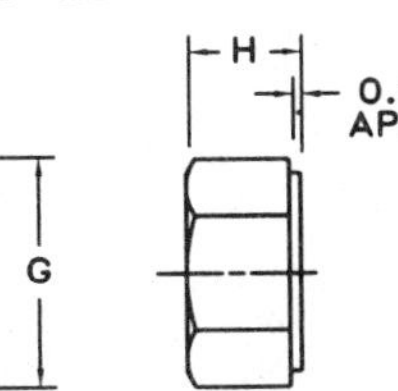

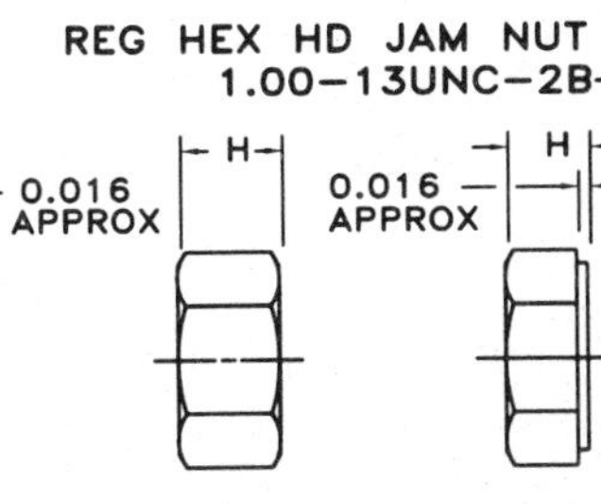

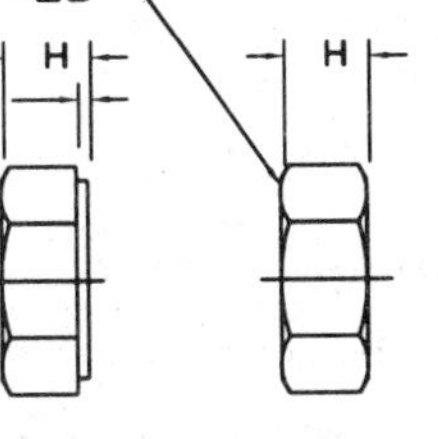

HEAVY HEX NUTS AND HEX JAM NUTS

REGULAR HEX NUT

HEX JAM NUT

MAJOR DIA		F Max.	G Avg.	H Max.	H Max.
1/4	.250	.438	.505	.226	.163
5/16	.313	.500	.577	.273	.195
3/8	.375	.563	.650	.337	.227
7/16	.438	.688	.794	.385	.260
1/2	.500	.750	.866	.448	.323
9/16	.563	.875	1.010	.496	.324
5/8	.625	.938	1.083	.559	.387

MAJOR DIA		F Max.	G Avg.	H Max.	H Max.
3/4	.750	1.125	1.299	.665	.446
7/8	.875	1.313	1.516	.776	.510
1	1.000	1.500	1.732	.887	.575
1-1/8	1.125	1.688	1.949	.899	.639
1-1/4	1.250	1.875	2.165	1.094	.751
1-3/8	1.375	2.063	2.382	1.206	.815
1-1/2	1.500	2.250	2.598	1.317	.880

PLAIN WASHERS

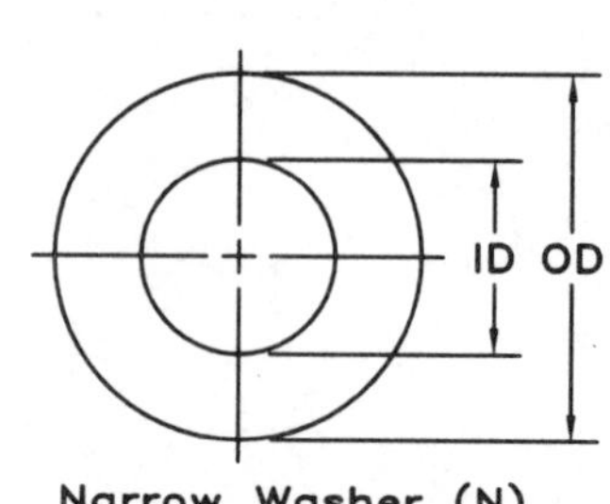

Narrow Washer (N)
TYPE A PLAIN WASHERS

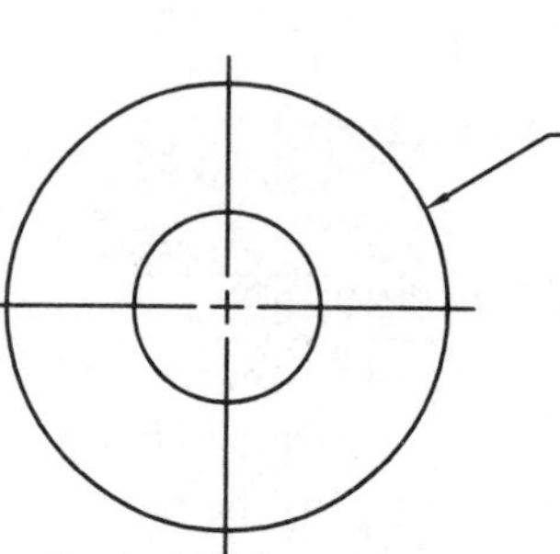

WIDE WASHER (W)

.938 X 2.25 X .165
TYPE A PLAIN WASHER

Dimensioned
Washer

In Screw Size Column
N= Narrow washer
W= Wide washer

SCREW SIZE	ID SIZE	OD SIZE	THICK-NESS
0.138	0.156	0.375	0.049
0.164	0.188	0.438	0.049
0.190	0.219	0.500	0.049
0.188	0.250	0.562	0.049
0.216	0.250	0.562	0.065
0.250 N	0.281	0.625	0.065
0.250 W	0.312	0.734	0.065
0.312 N	0.344	0.688	0.065
0.312 W	0.375	0.875	0.083
0.375 N	0.406	0.812	0.065
0.375 W	0.438	1.000	0.083
0.438 N	0.469	0.922	0.065
0.438 W	0.500	1.250	0.083
0.500 N	0.531	1.062	0.095
0.500 W	0.562	1.375	0.109
0.562 N	0.594	1.156	0.095
0.562 W	0.594	1.469	0.109
0.625 N	0.625	1.312	0.095
0.625 N	0.625	1.750	0.134
0.750 W	0.812	1.469	0.134
0.750 W	0.812	2.000	0.148

SCREW SIZE	ID SIZE	OD SIZE	THICK-NESS
0.875 N	0.938	1.750	0.134
0.875 W	0.938	2.250	0.165
1.000 N	1.062	2.000	0.134
1.000 W	1.062	2.500	0.165
1.125 N	1.250	2.250	0.134
1.125 W	1.250	2.750	0.165
1.250 N	1.375	2.500	0.165
1.250 W	1.375	3.000	0.165
1.375 N	1.500	2.750	0.165
1.375 W	1.500	3.250	0.180
1.500 N	1.625	3.000	0.165
1.500 W	1.625	3.500	0.180
1.625	1.750	3.750	0.180
1.750	1.875	4.000	0.180
1.875	2.000	4.250	0.180
2.000	2.125	4.500	0.180
2.250	2.375	4.750	0.220
2.500	2.625	5.000	0.238
2.750	2.875	5.250	0.259
3.000	3.125	5.500	0.284